The Palgrave Handbook of Research Methods and Ethics in Neurodiversity Studies

"This is a must-read book for anyone interested in participatory, and neuro-affirmative and neuro-inclusive research with potential to shift the narrative and improve outcomes with, and for, neurodivergent people. The authors describe in detail a range of methods used across a variety of disciplines in Neurodiversity Studies in a passionate and rigorous way."

—Georgia Pavlopoulou, *Associate Professor, UCL, and Director of Group for Research in Relationships and NeuroDiversity (GRRAND)*

"This scholarly, comprehensive and highly readable edited volume is the first of its kind dedicated to research methodology within Neurodiversity Studies. Informed by insider perspective, the work moves beyond the obvious and gets underneath thorny questions around topics such as ethics and co-production. It is a must-read for researchers, students and others and certainly merits priority inclusion on a wide range of reading lists."

—Nicola Martin, *Professor and Head of Research, Higher Degrees and Student Experience, London South Bank University, UK*

"Placing itself at the nexus of critical disability studies, participatory methodologies, and community care-work, this book traces the history of, and envisions a future for, neurodiversity studies as an essential program of research. The authors offer a wealth of community-informed expertise on research ethics and methodological innovation that should be required reading for current and future generations of neurodiversity scholars."

—Kristen Bottema-Beutel, *Professor of Special Education, Boston College, USA*

"There are many books relating to neurodivergence, and there are even more publications relating to research methods. This collection is different in that it connects the two. Rather than merely talking the talk, the editors and contributors are walking the walk of "nothing about us without us". The collection offers insights into successful co-construction, co-creation, participatory research and autoethnographic work that is so needed in the space. Recommended reading for anyone and everyone who wants to learn about ethics and practicalities of research within the context of neurodivergence."

—Nicole Brown, *Associate Professor and Head of Research Ethics and Integrity, Institute of Education, University College London, UK*

Hanna Bertilsdotter Rosqvist ·
David Jackson-Perry
Editors

The Palgrave Handbook of Research Methods and Ethics in Neurodiversity Studies

palgrave
macmillan

Editors
Hanna Bertilsdotter Rosqvist
School of Social Sciences
Södertörn University
Huddinge, Sweden

David Jackson-Perry
Queen's University
Belfast, UK

ISBN 978-3-031-66126-6 ISBN 978-3-031-66127-3 (eBook)
https://doi.org/10.1007/978-3-031-66127-3

Cover credit: Buena Vista Images/Getty Images

This Palgrave Macmillan imprint is published by the registered company Springer Nature Switzerland AG
The registered company address is: Gewerbestrasse 11, 6330 Cham, Switzerland

Contents

Notes on Contributors

Noah Adams lives in Toronto and the traditional territories of the Mississaugas of the Credit, Anishnabeg, Chippewa, Haudenosaunne, and Wendat peoples. He is a researcher, activist, and student at the University of Toronto, where he is completing a PhD in Adult Education and Community Development. His PhD work expands on his book (*Trans and Autistic: Stories from Life at the Intersection*) by exploring the development of transgender/autistic communities. He volunteers extensively and is currently working with the Transgender Professional Association for Transgender Health. Check out his website for more info (www.noahjadams.com).

Georgia Aitkenhead is a researcher at the Alan Turing Institute specialising in participatory research, citizen science, and neurodiversity, with a particular focus on autism. Her current research focus is co-creating an open, online platform, AutSPACEs, along with a community of autistic people, their supporters, and open-source developers. She is a current Software Sustainability Institute Fellow and a graduate of the Open Life Sciences Leadership Programme (2022). A core feature of her work is the inclusion of diverse participants in roles of power in research: direction-setting and decision-making roles at every stage of the research pipeline. She supports the AutSPACEs community via regular, open meet-up sessions, focus groups, workshops, close co-working, newsletters, co-presenting, co-design, and co-authorship. She has presented her work along with members of the AutSPACEs community at a variety of conferences and seminars, including at CogX (2020), Autistica Research Festival (2022), C*Sci (2023), and AI UK (2024). Prior to her role at the Turing, she was a civil servant on the UK's Digital, Data and Technology fast stream working across multiple departments including the Behavioural, Insights and Research Team at HMRC, the FCDO (then FCO), and Government Digital Services.

Mikael Åkerlund is a community researcher. He is a former employee at the post office who had a stroke in 2005. During the "Ihop-project", he has been employed at Karolinska Institute as a research assistant. He thinks that issues related to participation in society are important and affect all of us, but particularly people with disabilities.

Hanna Bertilsdotter Rosqvist is a neurodivergent (AuDHD) sociologist and professor in Social Work at Södertörn University. Her research is around neurodivergence, identity politics, and sexual, gendered and age normativity. She has published several papers on the themes of research methods and collaborative research methods. She is the editor of several books among them *Neurodiversity Studies. A New Critical Paradigm*, together with Nick Chown and Anna Stenning (London: Routledge, 2020), and *Exploring Autistic Sexualities, Relationality, and Genders: Living Under a Double Rainbow*, together with Anna Day and Meaghan Krazinski (London: Routledge 2024).

Monique Botha is an autistic and ADHD Leverhulme research fellow at the University of Stirling. Their research follows two veins: firstly, they focus on meta-research into the dehumanisation of autistic people by autism researchers and how that shapes the scientific process and standards that we allow for autism research. Secondly, their PhD research and other research have focused on minority stress in the autistic community and how it shapes autistic people's mental health, quality of life, and experiences in society. They are also a co-investigator on the Striving to Transform Autism Research Together Scotland project—a participatory project which brings together autistic people living in Scotland to create meaningful research which is of importance to autistic people in this context. Lastly, they also co-founded Community Against Prejudice Towards Autistic People which aims to use research networks and science to help tackle discrimination and victimisation of autistic people.

Amy Brown is Professor of Child Public Health at Swansea University, UK. Her research is on everything to do with becoming and being a new parent, with a particular focus on infant feeding, mental health, and normal baby behaviour. The ultimate aim of her research is to change society and culture, in order to see a future where women and families are respected, valued, and supported in growing, birthing, and caring for their babies.

Andrew Ivan Brown is a postdoctoral research associate in the Department of Philosophy at the University of Washington, Seattle. His interdisciplinary research examines ethical issues that arise from conceptual problems in biomedicine, especially surrounding mental disabilities, through both theoretical and ethnographic methodologies. He has recently done work on the sociophilosophical and affirmative dimensions of ADHD, the sociomaterial links between chronic traumatic encephalopathy and hypermasculinity in Canadian hockey leagues, and the discursive history of the Canadian "gay blood ban". At UW, he conducts qualitative research on participants in neural device studies (similar to Neuralink) as part of the Caring for BRAIN Pioneers NIH-funded project with the neuroethics group.

Karen Leneh Buckle is a postgraduate researcher at the University of Manchester. With a background in psychology, neuroscience, and bioethics, she is now undertaking groundbreaking community-led research on initiation impairments (inertia) in autistic people. She is autistic, a parent of four neurodivergent young people, and has been involved in autistic community for over 25 years. She has experience of involving autistic people in the neurotypical-dominated domain of autism research and often serves as a "translator" between autistic experts by experience and neurotypical researchers as she is fluent in the languages of both communities. She now finds herself juggling multiple roles as a research consultant on projects with organisations including: the National Autistic Society, the National Autistic Taskforce, Autistica, University of Cambridge, Autism@Manchester and research teams at several other universities, and most recently as co-lead of the public consultation on the controversial Spectrum 10K project.

Claire M. Chapman is a postgraduate researcher in Criminology within the School of Social and Political Sciences at the University of Glasgow. As an emerging scholar in the fields of criminology and disability studies, her doctoral research focuses on the experiences of autistic women in Scotland who have been victimised, exploring the nuanced aspects of victimology, intersectionality, disability, and harm within the context of support services. In addition to her research, she is a member of the Scottish Centre for Crime and Justice Research and Glasgow's Centre for Disability Research networks based at the university. Prior to undertaking her PhD, she obtained her MRes in Criminology (2022) at the University of Glasgow and her MA (Hons) Geography in 2019 from the University of Dundee. Outside of academia, she worked several years in social and economic research as a consultant in the private sector.

Isabelle Courcy is a professor at the Université de Montréal (Québec, Canada). She is a sociologist and an allistic researcher. She teaches qualitative research and mixed methods. Her research programme focuses on the social experience of autism, life course and services trajectories, quality of life, and social support networks of autistic adults. She values collaborative research with an emancipatory, feminist, and inclusive approach. She is a member of the research collective participation.

Jacquiline den Houting (they/them) is a research psychologist and autistic activist working in pursuit of social justice. An emerging autism research leader, they are committed to creating real-world change that benefits autistic people, in the areas that matter most to the autistic community. They believe that autism research can be most meaningful and impactful when it is conducted by and with those it is intended to serve—autistic people. They trained and worked as a psychologist in Brisbane, Australia, before completing their PhD and venturing into an academic career. Currently, they are a senior research fellow at La Trobe University in Melbourne, Australia. In 2019, they made their TEDx debut with a viral talk that has attracted more than 1.5 million views to date. They are proudly neurodivergent and queer, and share their life with their equally neurodivergent wife, an enthusiastic Labrador, and a disgruntled bearded dragon.

Marjorie Désormeaux-Moreau is an autistic researcher and occupational therapist who is currently an associate professor of Occupational Therapy at Sherbrooke University (Québec, Canada). They root their teaching, research, and community and activist involvement in their lived experiences. They are particularly interested in experiential knowledge and in the lived experience of Autistic adults, as well as in both the experiences and the intertwined experiential and professional knowledge of Disabled, Neurodivergent, and Mad occupational therapists. Their work tends to be rooted in a critical autism perspective (critical autism studies—and more broadly, critical disability studies). Outside of research, they are one of the founding members of Autisme Soutien, an autistically led non-profit organisation rooted in a neuroaffirmative, neuroinclusive, and anti-suicidal approach that aims to be an alternative to services that have been thought of and are offered by allistics (i.e. non-autistic people). They live and work on the unceded traditional territory of the Abenaki and Wabanaki Confederation.

Teddy G. Goetz (they/them) is a psychiatry resident at the University of Pennsylvania. Prior to earning their M.D. at Columbia, they studied biochemistry and gender studies at Yale, conducting research on a wide spectrum of biologically and socially determined aspects of gender-based health disparities, including earning their M.S. developing the first animal model of gender-affirming hormone therapy. Their current projects include implementation science research on LGBTQIA+ mental health. Their lived experience as a non-binary/trans, queer, neurodivergent (ADHD), chronically ill, Jewish person informs their writing, research, and clinical work. They have published the textbook *Gender-Affirming Psychiatric Care* (American Psychiatric Association Publishing, 2023), the graphic novel *Gender is Really Strange* (Jessica Kingsley Publishers, 2023), and the poetry chapbook *TRANS-abdominal Retrieval: a (gender)queer Jewish quest to make an embryo* (Beyond the Veil Press, 2023). Learn more about their work at teddygoetz.com.

Aimee Grant (she/they) is a senior lecturer in Public Health and a Wellcome Trust career development fellow at Swansea University. Her postdoctoral research has focused on experiences of pregnancy and early parenthood for marginalised groups, including white working-class and Disabled people. In 2019, she was diagnosed as Autistic and began researching Autistic reproductive health. She is interested in qualitative methodological development and has written two books to date: Doing Excellent Research With Documents

(Routledge, 2019) and Doing Your Research Project With Documents (Policy Press, 2022).

Jonathan Green is Professor of Child and Adolescent Psychiatry at the University of Manchester and an honorary consultant child and adolescent psychiatrist at the Royal Manchester Children's Hospital. He trained in medicine, paediatrics, and psychiatry and has led a specialist clinical service for children and adolescents with autism and other developmental conditions. He has researched in early autistic development and developed and tested early parent-mediated support for families of pre-school children with a diagnosis (PACT) and younger children before diagnosis (iBASIS): advocating an early developmental care approach that is being applied in UK and internationally. In recent years, he has undertaken increasing participatory and co-design work with the autistic community, including discussion on neurodiversity and the ethics of intervention (Green 2023), participatory approaches to autistic phenomenology (Murray et al. 2023; Green and Shaughnessy 2023), and co-design of a new anxiety intervention for young autistic children. He sat on the most recent UK NICE development group for autism. He is a senior investigator in NIHR, fellow of the UK Academy of Medical Sciences, and UK Global Senior Leader for autism for the International Society of Autism Research (INSAR).

Karen Henry is an autistic midwife and lecturer in Midwifery at the University of Suffolk. During midwifery training, she recognised that autistic women were often unprepared for a hospital admission with its unfamiliarity and unpredictability. She also recognised that maternity staff did not receive sufficient autism training to be able to support autistic parents. She created a series of support plans to enable accessible care through women sharing their support needs with maternity staff. The plans were developed further to incorporate maternity information sharing to further support preparation for pregnancy, birth, and the postnatal period. Whilst undertaking an MA in Autism, her dissertation focused on the notion that midwives with personal autism experience (through either being autistic themselves or being a mother of an autistic child) have skills, knowledge, and experience that are not taught in traditional midwifery training. From the findings, she developed a training framework to be co-produced and delivered by autistic women and autistic midwives to ensure that maternity staff training is meaningful, relatable, and memorable and equips them with ideas to support care adjustments for autistic women.

Kristien Hens is Associate Professor in the Department of Philosophy at the University of Antwerp. She has a PhD in Biomedical Sciences (Bioethics, 2010) and in Philosophy (2018). She was the PI of the ERC-StG-funded project NeuroEpigenEthics, which investigated the ethics of neurodevelopmental differences in the light of postgenomics. She now leads the team COMPOSTCollective, a group of researchers focusing on environmental and disability justice, concepts of biology and the importance of experience stories. She has written two books on autism: *Experiences of Adults Following an Autism Diagnosis* (with Raymond Langenberg, Palgrave Macmillan, 2018), *Towards an Ethics of Autism* (Open Book Publishers, 2021), and one in which she advocates for a posthuman approach to bioethics: *Chance Encounters* (Open Book Publishers, 2022). She teaches courses in media ethics and bioethics.

Willow Holloway is a late-diagnosed autistic/disabled woman, the parent and grandparent of a neurodivergent family. She has previously worked with autistic people in residential care and supported living. She has used her lived experience following her own diagnosis to increase understanding of autistic people's lives. She founded the Autistic Women's Empowerment Project in 2014 as an awareness-raising project. Since then, her personal roles and the AWE Project have evolved beyond recognition and she has gone on to campaign for recognition, acceptance, and equality for autistic people across Wales. She has undertaken peer support and mentoring training to support her roles. She is the Executive Chair of Autistic UK and is a director of several equality organisations in Wales including being Chair of Disability Wales and a trustee of Fair Treatment for the Women of Wales. She has delivered consultation events and training and has extensive knowledge of autism and disability from both a professional and personal perspective, facilitating interactive sessions which offer an informed insider perspective on autism. She also works at a strategic and policy level and is a member of the Ministerial Advisory Group on Neurodivergence and the Disability Rights Taskforce, as well as co-chairing the North Wales Integrated Autism Service Strategy Board.

Lill Hultman is a social worker and a senior lecturer in Social Work at Södertörn University. Her research interests concern critical disability studies and developing methods for participatory research processes with community researchers. During the "Ihop-project", she was employed as a post-doctor and affiliated researcher at Karolinska Institute. Her research focuses on disability research with a particular emphasis on critical disability studies and participatory action research.

Maya Hultman is an independent scholar. Her background is in art studies. She is a playwriter and photographer. She is also involved in a youth council that is part of an NGO addressing issues of childhood bullying.

Romy Hume is a neurodivergent researcher and independent scholar. Her research spans a wide array of education-related projects, including an in-depth exploration of neurodivergent children's experiences within Aotearoa New Zealand's education system. Her primary research interest lies in therapeutic relationship-building with autistic adults, a topic on which she is finalising a book. This upcoming publication weaves together creative writing and academic research, offering vivid insights into her findings. She also works with autistic adults in private practice, where she supports them to become their own best selves.

Cecilia Ingard is a neurodivergent (autism and ADHD) PhD student in Social Work at the University of Gävle. Her thesis is about how people with dementia could participate in decisions regarding their own life. She is a board member of a Swedish organisation that works for people with autism and ADHD and their societal rights. She is also a member of the co-creation group of R2D2, an international biomedical project.

David Jackson-Perry is a neurodivergent (ADHD) sociologist and visiting scholar at Queen's University, Belfast. His doctoral research explored autistic experiences of intimacy, sexuality, and authenticity using a participatory approach. He has presented widely and published several articles in these areas, as well as chapters in *Neurodiversity Studies: A New Critical Paradigm* (ed. Hanna Bertilsdotter Rosqvist, Nick Chown & Anna Stenning. London: Routledge, 2020) and *Working with Autistic Transgender and Non-Binary People* (ed. by Marianthi Kourti, Jessica Kingsley Publishers, 2021). He is also a specialist in Sexual Health and HIV project manager at the University Hospital of Lausanne, Switzerland.

Emily Jones is a professor at the Centre for Brain and Cognitive Development, Birkbeck, University of London. Her research interests centre on understanding the cognitive and neural mechanisms that drive variability in developmental trajectories. In this context, she runs a number of prospective longitudinal studies of typical and atypical neurodevelopment in infants and directs electrophysiological and eye-tracking acquisition across several large-scale European and Global Health studies of children and adults with neurodevelopmental conditions. She has published over 170 empirical papers in journals such as Cell, Science Translational Medicine, eLife, and Nature Communications, has editorial roles for Autism Research and the Journal of Child Psychology and Psychiatry Advances, and sits on several grant awarding bodies.

Steven K. Kapp is a senior lecturer in Psychology at the University of Portsmouth, where he introduced a module on Neurodiversity Studies for the Master's in Psychology and Learning Disabilities. His studies broadly consider the nature of autism and support for autistic people, conceptually, through participatory research, and considering the perspectives of autistic people, family members, practitioners, researchers, and others. As an autistic activist, he has supported systems change work for inclusive employment and influenced the DSM-5 autism diagnosis. A significant portion of his research activities is related to self-advocacy and the neurodiversity movement.

Helen Kara is a fellow of the Academy of Social Sciences and has been an independent researcher since 1999 and an independent scholar since 2011. She writes about research methods and research ethics and teaches doctoral students and staff at higher education institutions worldwide. Her books include Creative Research Methods: A Practical Guide and Research Ethics in the Real World: Euro-Western and Indigenous Perspectives for Policy Press, and Qualitative Research for Quantitative Researchers for SAGE. She also writes comics and fiction. In 2021, at the age of 56, she was diagnosed as autistic. Her neurodivergence explains her lifelong special interest in, and ability to focus on, words, language, and writing.

Jennifer Lowe is an early career public health researcher whose work focuses on palliative care, end-of-life services, grief, bereavement, and the everyday experiences of autistic adults across a range of healthcare and service contexts. She was formally diagnosed autistic and ADHD during her PhD candidature. Her interdisciplinary PhD research was conducted in collaboration with the Public Health Palliative Care Unit and Olga Tennison Autism Research Centre at La Trobe University. She is currently working as a research fellow in Palliative Care at the Perron Institute for Neurological and Translational Science. With a background in services marketing, she is passionate about using her industry experience and academic knowledge to improve real-world outcomes across disciplines. She has worked on research and evaluation projects commissioned by not-for-profit peak bodies and State Government Departments including Health and Education. Her multidisciplinary work has been published in peer-reviewed journals spanning the disciplines of marketing and public health. She has co-authored book chapters and presented her research at national and international conferences.

Damian Milton is a part-time lecturer at the University of Kent, having joined the Tizard Centre in 2017. He also works for the National Autistic Society as an autism knowledge and expertise consultant. He has been teaching in the further and higher education sectors since 2000 in a range of subject areas relating to autism and learning disabilities. His research is based predominantly on increasing the meaningful participation of autistic people and people with learning disabilities in research processes. He chairs the Participatory Autism Research Collective (PARC). He is involved in a number of projects for the Autism Education Trust and has been a consultant for the Transform Autism Education (TAE) project.

Katie Munday is a community researcher, whose current research explores the accessibility of healthcare for Neurodivergent people. Their main research interests are trans-Autistic narratives, Autistic culture, and Disabled people's experiences of healthcare services. They are Autistic, ADHD, Disabled, and queer.

Jeanette Nelson is a community researcher. In 2014, she suffered an aneurysm in her brain. She attended Alma Folk High school, in a study group for adults with aphasia. She is active in a Facebook community for stroke survivors. During the "Ihop-project", she has been employed at Karolinska

Institute as a research assistant. Previously, she worked as a flight attendant, health coach, and physiotherapist.

Christina Nicolaidis is a professor in the School of Social Work at Portland State University (PSU), Adjunct Professor in the Department of Medicine at Oregon Health & Science University (OHSU), the co-founder and co-director of the Academic Autism Spectrum Partnership in Research and Education (AASPIRE), and the founding Editor-in-Chief of the peer-reviewed journal, Autism in Adulthood. As a neurodivergent (ADHDer) general internal medicine physician and health services researcher, she uses community-based participatory research to improve the health and healthcare of marginalised populations. The majority of her current research projects focus on autism and neurodiversity. She also leads the Social Determinants of Health Initiative, teaches graduate students and residents, practices internal medicine, and mentors dozens of students, trainees, and faculty members locally and around the world.

Lindsay O'Dell is Professor of Critical Developmental Psychology at the Open University. She has a long-standing interest in researching and writing across neurodivergent issues and with diverse researchers and writers.

Amy Pearson is an assistant professor in Psychology in the Department of Psychology at Durham University. She is an autistic developmental psychologist with expertise in the field of autism and neurodiversity. Her current research focuses on understanding factors which impact on well-being among autistic people across the lifespan with a particular focus on interpersonal relationships and victimisation, social identity and stigma, and increasing accessibility for neurodivergent people in higher education.

Elizabeth Pellicano is committed to transforming autism science so that it more accurately reflects everyday autistic life. She is Professor of Autism Research at University College London (UCL), having recently been Professor at Macquarie University, Sydney, Australia. Previously, she was Director of the Centre for Research in Autism and Education (CRAE) at UCL Institute of Education. Her most recent research identifies ways to bridge the gap between lab and life and open up research to greater involvement of Autistic people themselves, with the aim of generating scientific discoveries that bring real benefits to Autistic people and their families.

Dora M. Raymaker is a research associate professor at Portland State University's Regional Research Institute for Human Services in the School of Social Work, co-director of the Academic Autism Spectrum Partnership in Research and Education (AASPIRE), and editor of Autism in Adulthood. Their research interests broadly include systems thinking, action research, measurement, ethics, and the dynamics at the intersection of society, disability, and technology. In application, they conduct intervention and services research in collaboration with disability communities to improve employment outcomes and reduce discrimination and stigma. In their remaining three minutes of free time, they enjoy writing fiction and making multimedia art, including science fiction novels Hoshi and the Red City Circuit and Resonance.

Ned Redmore is a postdoctoral researcher, at the Open University, interested in how social care and academia can better include autistic people with profound learning disabilities. Regarding social care, he focuses on how support services can better respond to the views and values of this group to improve cultures and practices. He researches with autistic people with profound learning disabilities in support services, exploring their experiences, views, and priorities through inclusive research approaches. Regarding academia, he is interested in how this group's marginalisation can be addressed through methodological development, intersectional collaboration, and critical awareness.

Fredrik Sandman is a community researcher. In 2013, he had a stroke and aphasia. He is a board member of the Aphasia Association in Stockholm County. During the "Ihop-project", he has been employed at Karolinska Institute as a research assistant. He is passionate about participating in this project and he wants to improve everyday life for people who have had a stroke and aphasia.

Abby Sesterka is a scholar of Indonesian languages, an accomplished translator, and a languages curriculum specialist. She teaches postgraduate courses for the Centre for Innovation in Learning and Teaching at Flinders University in South Australia and is an international leader in assessment of language acquisition. As an Autistic scholar and teacher, she is a regular facilitator for the Autism CRC's Sylvia Rodger Academy, and she has written broadly for grey literature publications on autistic well-being with her co-author, Dr. Erin Bulluss. The many and diverse elements of Abby's professional life are united by a fierce commitment to social justice, an unwavering belief in the value of collaboration, and a genuine enthusiasm for teaching and learning.

Nicola Shaughnessy is Professor of Performance at the University of Kent, specialising in participatory arts, creative health, and neurodiversity. Her work is interdisciplinary, informed by cognitive neuroscience and phenomenology as well as her lived experience of neurodivergence. She has been the lead researcher for four grants investigating autistic identities through creative practices and is also a co-investigator for three projects using arts-based methods for researching adolescent mental health. She is a neurodiversity consultant working with organisations such as the National Theatre and Audible. Her publications are wide ranging and cross disciplinary. They include Gertrude Stein (2007), Applying Performance: Live Art, Socially Engaged Theatre and Affective Practice (2012) and the edited collections Affective Performance and Cognitive Science: Body, Brain and Being (2013) and Performing Psychologies: Imagination, Creativity and Dramas of the Mind (co-edited with Philip

Barnard, 2019). She is editor and co-founder of Methuen's Performance and Science series.

Jonathan A. Smith is a professor of Psychology at Birkbeck University of London. He leads the Interpretative Phenomenological Analysis (IPA) research group comprised of staff, researchers and PhD students employing IPA to address a wide range of topics concerned with lived experience. He has himself employed IPA to examine a number of areas particularly in health, for example, the experience of chronic pain, decision making in clinical genetics, understanding depression. At Birkbeck, he teaches qualitative methods to psychology students at all levels, in addition to supervision a large group of PhD students. He is the lead author of the primary text on IPA (JA Smith, P Flowers, M Larkin [2022] Interpretative Phenomenological Analysis: Theory, Method, Research, London Sage [2nd ed]). He is a fellow of the Academy of Social Sciences and an honorary fellow of the British Psychological Society.

Paul Stenner is a professor of Social Psychology at the Faculty of Arts & Social Sciences, the Open University, and co-director of the Open Psychology Research Centre. He works at the School of Psychology and Counselling, the Open University. He does research in social psychology, theoretical psychology, and emotion and health psychology.

Lydia Stetson is a bisexual/queer adult-discovered/identified AuDHD ciswoman. She has a Master of Arts in Counseling Psychology and is currently pursuing her PhD in Clinical Psychology. Her interest in sexual and gender health related to Autistic experiences has grown steadily from her undergraduate experience as a sexual assault peer educator through her university's health centre, to her work with neurodivergent adults in a community-based programme, to her experience as a master's level mental health clinician privileged to build therapeutic relationships with GSRD Autistic youth and adults, and finally (for now) to looking to explore the lived experiences of queer Autistic individuals qualitatively in her doctoral dissertation.

Joke Struyf is a lesbian, mother, activist, and non-diagnosed non-neurotypical PhD candidate in Philosophy at the University of Antwerp. Since 2019, she has been researching normative motherhood, mothering, and sense of maternal agency. She is inspired by Black feminism, crip and queer theory, activism, and the mundane acts of everyday mothering. Next to that, she has worked on several European projects (B2-InF/R2D2-MH) under the supervision of Kristien Hens, through whom she became part of the NeuroEpigenEthics (now Compost Collective) team at University of Antwerp. The team engages in thinking about a more inclusive, neurodiversity-affirming way of doing research and organising academia. In 2023, she was involved in facilitating the adult co-creation group for R2D2-MH, a Horizon Europe project on resilience, risk, and mental health in developmental diversities.

Shruti Taneja-Johansson is an associate professor in Education and coordinator for the Platform for Research in Inclusive Education and School Development in the Department of Education and Special Education, University of Gothenburg. Her core areas of research interest are disability, education, and inclusion. In her research, she addresses issues of educational inequity among people with disabilities, across a range of educational settings and country contexts, with a specific interest in autism, ADHD, and intellectual disability. Understanding how mechanisms of inclusion and exclusion play out in different settings, and how the intersection of disability with other forms of disadvantage shape educational and life trajectories, is some of her areas of interest. Other important dimensions of her work are bringing to the fore individual narratives, collaborative research, and culturally and ethically sensitive research.

Sandra Thom-Jones is an autistic author, artisan, academic, and advocate. She is the author of *Growing in to Autism* (MUP, 2022) and *Autistics in Academia* (Cambridge University Press, 2024), and a passionate champion for the inclusion of autistic people in all aspects of society. She worked for more than two decades in the university sector as a researcher and senior leader, most recently as Pro Vice-Chancellor, Research Impact. She now works as a consultant providing a range of services for autistic people—as well as research and consultancy services for education, employment, and healthcare providers—through her website Autistic Professor. She is also an honorary professor at the University of Wollongong. She has undertaken numerous research projects in partnership with the autistic community on the experiences of autistic people, and the knowledge and attitudes towards autism in the broader community, including research that won the Autism CRC's 2019 and 2021 awards for research translation and 2022 award for inclusive research.

Bastian Greshake Tzovaras is a senior researcher in the Tools, Practices & Systems programme at the Alan Turing Institute, specialising in participatory, community-driven citizen science and open science practices. His research is particularly centred around how open science/source technologies and approaches can be used to empower communities and individuals to engage in their own knowledge production. Trained as a bioinformatician, his work has become highly interdisciplinary, bringing together participatory research methods with applied and theoretical approaches from data science, human-computer interaction, as well as critical science and technology studies. In his research, he has worked with a diverse set of (patient) communities, including around topics such as type-I-diabetes, cluster headaches, transgender gynaecological care, and sensory processing in autism. He also volunteers as a board member of the Open Humans Foundation—a non-profit that provides socio-technical infrastructures for personal data-driven citizen science—and the Open Bioinformatics Foundation, promoting best open science practices in data-heavy biological research.

Catherine Watson is a psychotherapist/counsellor, EMDR practitioner, and associate lecturer in Psychology and Counselling at the Open University. She holds an MSc in Experimental Psychology from the University of Sussex, where her research focused on discourse analysis of transcribed interviews on lived experience of women diagnosed with bipolar disorder. Her late diagnosis of Developmental Coordination Disorder (dyspraxia) led to her joining with the other authors of this chapter who share her passion for raising awareness and understandings of neurodivergence. She runs a monthly peer support group for those who have a diagnosis or self-diagnosis of DCD (dyspraxia) and plans to continue facilitating this, engaging in activism and increasing awareness locally, whilst reaching a wider audience with web-based information sharing and research contributions.

Sofia Österborg Wiklund is an international postdoctoral fellow in the Department of Historical, Philosophical and Religious Studies, Umeå University. Her project "Educating for a Better World: Historical Perspectives on Swedish Development Aid Education 1965–1995" is funded by the Swedish Research Council and hosted the Department of Archaeology, Conservation and History (IAKH), Faculty of Humanities, as well as the Department of Education, Faculty of Educational Sciences (IPED), at the University of Oslo, Norway. It specifically concerns the historical relationships between popular education, social movements, and development aid and cooperation. She has a Master's degree in Social Sciences and Cultural Analysis, and a PhD in Education from Linköping University. Her recent work has been about neurodiversity, policy analysis, and Swedish norm critical education. She has contributed to the initiation of Critical ADHD Studies by participating in an autoethnographic project aiming at rethinking ADHD from insider perspectives. She is now working on critical readings of historical portrayal of neurodivergence in children's literature, as well as folk high school education for neurodivergent participants.

Emma Williams trained as a developmental psychologist in the UK, where she completed a PhD on the development of early play in very young autistic children. She is currently a lecturer in Developmental Psychology at the University of Surrey where her teaching and research are primarily centred around the use of qualitative methods of data collection, analysis, and metasynthesis. Much of her research has focused on understanding the lived experience of autistic children and adolescents, particularly how they make sense of themselves in the context of school and the environmental, social, and intrapersonal factors that intersect to shape this. She is currently involved in several interdisciplinary collaborations that use participatory arts-based methods, in conjunction with quantitative and qualitative approaches, to gain new insights and understanding into adolescent lived experience and mental well-being, bringing together arts, science, and youth perspectives. These include an exploration of the identities and experiences of autistic girls and an investigation of adolescents' experiences around loneliness.

Gemma L. Williams is a research officer in the Faculty of Medicine, Health and Life Science at Swansea University, working on the Wellcome Trust-funded, longitudinal project "Autism: from menstruation to menopause". She is an autistic Early Career Researcher, whose linguistics PhD investigated the breakdowns in mutual understanding that can occur between autistic and non-autistic people, based on the idea of the "double empathy problem". Her doctoral research has also included a focus on loneliness in autism and the use of creative and participatory methodologies. Prior to joining Swansea University, she held an ESRC postdoctoral fellowship at the University of Brighton in social policy. She is a member of the Westminster Commission on Autism and an associate with the National Development Team for Inclusion where she's contributed to a number of commissioned reports, projects, and inquiries aimed at improving service provision for autistic and neurodivergent people within the UK, NHS England, and Local Authorities.

Kathryn Williams is an autistic and ADHD researcher currently undertaking an Economic and Social Research Council-funded PhD at Cardiff University's School of Social Science examining the role of communication in access to healthcare for autistic adults. She is also a voluntary non-executive director at Autistic UK CIC, an organisation run by autistic people for autistic people, where she leads their research participation. Her research position at Autistic UK first sparked her interest in inclusive participatory methodology, and she pursued an MSc in social science research methods for which she was awarded a distinction. Her other research interests include epistemic injustices faced by neurodivergent people, neoliberal-ableism, and neuronormativity within healthcare and social policies.

Alyssa Hillary Zisk (they/them) is a neurodivergent researcher with too many research interests and no intention of actually narrowing them down. Right now, most but not all of their research has to do with augmentative and alternative communication (AAC) as part of their work with AssistiveWare. Some of their AAC work is relevant to their own AAC use and needs, such as studying AAC use and relevant speech experiences for autistic people who use both AAC and speech. Other aspects of their AAC work, such as their dissertation on brain computer interfaces for people with amyotrophic lateral sclerosis or their work using large anonymous language use data collected from AAC apps, are essentially unrelated to their own use or needs. Outside research, they are a math teacher with the Art of Problem Solving, occasionally a neuroscience teacher at the University of Rhode Island, and an improviser at the Contemporary Theater Company. They are also on the admin team of the Facebook group, "Ask Me, I'm an AAC user! (24 Hour Rule!)".

List of Figures

List of Tables

PART I

Setting the Scene

CHAPTER 1

An Introduction to Research Methods and Ethics in Neurodiversity Studies

David Jackson-Perry and Hanna Bertilsdotter Rosqvist

Introduction–Some Background

The term neurodiversity—the "brute fact" of "the diversity of human minds, the infinite variation in neurocognitive functioning within our species" (Walker, 2014)—was long attributed to Judy Singer in her Honor's thesis in 1998, and in a book chapter soon after (Singer, 1999). Recent scholarship disputes this, suggesting that journalist Harvey Blume was the first to use the term in print (Botha et al., 2024). Whoever coined the *term* in print, however, there is no doubt that the co*ncept* of neurodiversity was circulating in autistic activist communities well before 1998 (Chapman, 2023; Botha et al., 2024; for discussion of how some autistic activists position the movement in a wider context see Tisoncik, 2020, and other contributions in Kapp, 2020). The earliest available (published) words closest to both the concept and term appeared in an autistic email list, Independent Living, in 1996:

> ...what allows the human race to progress socially and technologically is the *neurological diversity* of people. I.e., the atypical among a society provide the different perspectives needed to generate new ideas and advances, whether they

D. Jackson-Perry (✉)
Queen's University, Belfast, UK
e-mail: djacksonperry01@qub.ac.uk

H. Bertilsdotter Rosqvist
School of Social Sciences, Södertörn University, Huddinge, Sweden

H. Bertilsdotter Rosqvist and D. Jackson-Perry (eds.), *The Palgrave Handbook of Research Methods and Ethics in Neurodiversity Studies*,
https://doi.org/10.1007/978-3-031-66127-3_1

> be technological, cultural, artistic or otherwise (Tony Langon, in Dekker, 2023, emphasis added).

By definition, then, neurological diversity—what we now call neurodiversity—applies to everyone: just as all of nature is part of biodiversity, so all people belong to a neurodiverse group of humans, with each contributing to social functioning and evolution. None are inherently superior or disordered, and all are enabled or disabled according to the specific context in which they are situated. Singer (1999, p. 64) noted that:

> the key significance of the 'Autistic Spectrum' lies in its call for and anticipation of a politics of Neurological Diversity, or 'Neurodiversity'. The 'Neurologically Different' represent a new addition to the familiar political categories of class/gender/race and will augment the insights of the social model of disability.

Neurodiversity approaches today are indeed often aligned with the social model of disability, and "positioned in contrast to traditional medical models in which diagnostic criteria are based on perceived deficits and presumption of pathology" (Taylor, 2021, p. 101). Further, neurodiversity was a political as much as an ontological proposition which has since "inspired the creation of a rapidly growing civil rights movement based on the simple idea that the most astute interpreters of autistic behaviour are autistic people themselves" (Silberman, 2015, p. 16).

Those whose "neurocognitive functioning falls within the majority or dominant societal standards of 'normal'" are referred to as "neurotypical" (Walker, 2014), and Kassiane Asasumasu (no date) coined "neurodivergent" to describe those who are "neurologically divergent from typical". This too was foreshadowed decades earlier: at a 1993 conference, Jim Sinclair (2005) recalls Kathy Grant meeting a person with hydrocephalus, and exclaiming "cousins!", a word which was from then on used to describe people who were not autistic but "autistic-like" (Sinclair, 2005). Whoever may have initially used certain words or even concepts in writing, it is clear, to borrow from the title of Botha and colleagues' (2024) article, that "The neurodiversity concept was developed collectively".

Why This Book?

The basis for a social justice movement, a paradigm, and, as here, the emergent and flourishing field of Neurodiversity Studies (NDS), neurodiversity has become increasingly influential and visible. Academically and professionally, it has gained considerable traction in a wide range of fields, including medicine, human resource management, and education (see, for example, Clouder et al., 2020; Gobbo et al., 2019; Johnston-Tyler & Analla, 2015; Nyhan, 2018; Robinson, 2022).

With this popularity comes risks, one of which is what Neumeier (2018, no page) calls "neurodiversity Lite", whereby:

> neurotypical people, who either use the language of the movement in good faith or intentionally co-opt it, undermine its work by overlooking or outright contradicting its core concepts.

The danger here is of leaving "the dominant paradigm and political order unchallenged despite the shift in presentation" (Chapman, 2023, p. 8). Further, those who—knowingly or not—co-opt the language of neurodiversity without adhering to its core concepts are often those who already have high levels of visibility and credibility (Chapman, 2023; see also Bottema-Beutel, 2024). This poses important questions and challenges. As is discussed throughout this volume, academia is recognized as a site of much ethical and epistemic injustice towards neurodivergent people (Botha, 2021; Chapman & Carel, 2022; Orsini, 2022). We feel it important to showcase and reflect upon existing examples that resist, navigate, and reflexively seek solutions to this as they work towards a more just and creative research environment in the field of neurodiversity. In parallel to the rise of neurodiversity are increasing calls for researchers to centre neurodivergent experience, and to shift from "normal science" to "neurodiversity in autism science" (Pellicano & den Houting, 2022, p. 381). This is pertinent—and happening—beyond autism research, for example in ADHD (see Chapter 3, this volume; see also Sonuga-Barke, 2023; Sonuga-Barke et al., 2024; Goetz & Adams, 2024), dyslexia (Schaik, 2021), dementia (Örulv, 2023), acquired brain injury (see Chapter 8, this volume), and people with profound learning disabilities (see Chapter 15, this volume) amongst many other examples.

However, there is a "paucity" of resources concerning, for example, both the nuts and bolts and the "philosophical underpinnings" of even participatory research (Pickard et al., 2021, p. 160), a widely called for and long-standing methodology in the field of, for example, autism studies. How are those new to the field to carry out research within NDS in the face of this lacuna? What is the current state of ethical and methodological reflections and processes in this field?

With this volume, we seek to respond—albeit necessarily partially—to these questions. The chapters in this collection give wide-ranging and concrete examples of ways in which students and researchers—whether newcomers or experienced—may engage with NDS or carry out 'neurodivergent methodologies', and consider key methodological and ethical issues in the field. This volume introduces a variety of neurodivergent-affirmative research methods, gives illustrative examples of various levels and forms of participatory research, and presents some of the possibilities and challenges when researching and writing in 'cross-neuro-status' collectives.

What This Book Is...

Reviewing a previous collection on NDS (Bertilsdotter Rosqvist et al., 2020b), Dind (2021, no page) noted that its:

> focus on crafting neurodivergent methodologies is one of the aspects I was most excited about, as it opens up new possibilities for the work of autistic and neurodivergent researchers.

This volume goes further in this direction, intended specifically as a handbook of research methods and ethics within NDS. Contributors have therefore gone into considerable detail as to their research processes. This is not to suggest that the studies described here represent 'the models to be followed', but rather to be as transparent and reflexive as possible in sharing the challenges and successes of our various research journeys. In doing so, we hope to bring food for thought to students and those new and not so new to the field, as well as offer specific examples that may be borrowed from, added to, or rejected for something better suited to specific contexts.

While the title of this volume separates methods from ethics, this is something of an artificial distinction. The editors and contributors in this collection share their passion for, and the mechanics of, their participatory, neurodivergent, and neuro-affirmative work and address ethics alongside methods: methods are at the heart of ethical research in NDS. Participatory research, for example, necessarily involves an acknowledgement of, and an effort to mitigate, possible power dynamics between researchers and participants. Neurodivergent-affirmative research methods require rethinking what we think we know (see Chapter 22, this volume), resisting traditional scientific assumptions of the possibility of complete objectivity, or 'the view from nowhere' (Nagel, 1986), and acknowledging the complexity and potential of experience. This is particularly pertinent within NDS for two main reasons. Firstly, one contention of NDS as we see it is the importance of acknowledging knowledge as being situated. Drawing on Haraway (1988), we consider that all knowledge is partial and situated in both a particular social and historical context, and a particular bodymind implicated in what could be called epistemic power struggles. Following feminist standpoint theory, this acknowledges that "there is a relationship between knowledge and power" that legitimizes some (dominant) perspectives and de-legitimizes "marginally situated knowers" (Hall, 2017, p. 162). While both editors and most contributors to this volume are neurodivergent, we have encouraged all to reflect on how their social and/or neurological identities in a system in which neurotypicality is dominant might impact the phenomena under consideration (Milton & Green, 2024). We feel that this standpoint reflexivity is important for research in the field, whatever the neurotype of the researcher. Neurodivergent folk do not have a monopoly of 'truth' about neurodivergence: however, centring and elevating neurodivergent experience and expertise allows a multiplication of epistemic possibilities. Secondly, in a volume concerned largely with questions of access and differing and sometimes contradictory needs, how we operate cognitively and neurologically is substantively pertinent. To take autism as an example, while acknowledging that no core traits "have been found to be shared by all or most autistic people", Robert Chapman (2020, p. 424) also notes that "there may be very general statistical tendencies towards certain cognitive and neurological characteristics". We feel it important for researchers, whatever their neurotype, to explicitly reflect on how their own characteristics and those of their collaborators and/or participants might differ, how this differing might impact the research process, and how to navigate it to mutual advantage.

Challenging the view from nowhere, the notion of a non-situated, objective researcher, is therefore both an ethical and a methodological proposition. Throughout this volume, contributors explore various possibilities of carrying out NDS, offering alternatives to hegemonies of "the views of... 'neurotypical' professionals" (Wood et al, 2022, p. 3; see also Parsons et al, 2020), unreflective "neurotypical frames of analysis" (Schneid & Raz, 2020, p. 6), or the "neurotypical gaze" in research (McDermott, 2022, p. 51).

...and What This Book Is Not

This book is not an introduction to neurodiversity (or indeed NDS) more broadly. We have no ambitions to lay out a detailed history or overview of the development of neurodiversity approaches that make NDS possible: this has been admirably done elsewhere (see, for example, Milton et al., 2020; Kapp, 2020; Dwyer, 2022; Chapman, 2023). While we necessarily draw on and discuss concepts issue from neurodiversity approaches, this is a book about NDS, and more specifically about research methods and ethics within NDS.

Neither do we seek to create a definitive, exclusive definition of what constitutes NDS. Given that this is an emergent and developing field, any attempt to do so could anyway only result in failure. Therefore, while we suggest some features that we feel are relevant and necessary to the development of NDS at the time of writing, we actively encourage readers to reject, adapt, or otherwise navigate those as they see fit.

With these provisos in place, we now consider some aspects of NDS as it stands today, noting some elements we feel are central to the field, before concluding this chapter with a summary of the rest of the contents of this volume.

Neurodiversity Studies—An Emergent Field

The editors of *Neurodiversity Studies: A New Critical Paradigm* concluded that volume with some elements which they felt to be particularly pertinent to the emergent field at that time (Bertilsdotter Rosqvist et al., 2020b). We have reformulated these as follows:

- A commitment to developing interdisciplinary theoretical and methodological underpinnings that bring to the fore neurodivergent voices in research processes, decentring and taking a critical approach to neurotypical epistemic privilege.
- A recognition of the material reality of impairment and disability, while shifting focus from pathology and deficit to neurodivergent well-being and flourishing.
- An exploration of cross-neurotype interaction from a neurodivergent perspective. This implies challenging sometimes well-meaning 'business as usual' in research, and 'talking back to power' through questioning epistemic norms and practices and developing new emancipatory methodologies.
- The necessity of countering the imaginary ideal of a cognitive normal subject. This includes challenging dominant notions of what it is to be human while exploring what it is to be disabled.
- An acknowledgement of the importance of inclusivity: seeking to provide a theoretical and methodological home that can contain a wide range of neurodivergent experiences.

While that volume initially named and sketched out Neurodiversity Studies, it was founded upon previous neurodivergent theorizing and activism, but also other critical perspectives, foremost feminist, postcolonial, queer, and disability theorizing (for full discussion, see Bertilsdotter Rosqvist et al., 2020b). That collection was in part a response to a mainly United Kingdom Critical Autism Studies context in 2018–2020.[1] The group of editors, both autistic and AuDHD, sought to formulate inclusivity beyond autism and neurodiversity

[1] The *Neurodiversity Studies: A New Critical Paradigm* project arose during the summer 2018 as a result of discussions in the only Facebook group at that time for autistic autism researchers (Autistic Autism Researchers). At that time British autistic activists and academics were most active in that group. The final manuscript was submitted to the publisher Routledge in December 2019 and was then published in the spring 2020. Similarly, this book is the result of a broader neurodivergent researcher collaborative project "Autism and Social model". The broader project has been led by Bertilsdotter Rosqvist and was initiated by Richard Woods 8 July 2022 with a posting in the Facebook group Autistic Researchers Researching Autism (ARRA). In his posting Wood sought collaborators to "to do something to mark the 40th anniversary of the social model" (Mike Oliver's book *Social Work with Disabled People* (1983)), but also of the 30th anniversary of the neurodiversity movement (Jim Sinclair's presentation "Don't mourn for us" at the 1993 International Conference on Autism in Toronto (Sinclair, 1999). Several members of the group shown interest to participate in the project and the call "to do something" also

movements principally centred on autistic experiences. However, while that volume included a chapter on the Hearing Voices Movement (Hart, 2020) and a chapter reflecting upon neurodivergence and neuronormativity more broadly (Dieuwertje Huijg, 2020), it was heavily weighted towards autism. This is still the case for work around neurodiversity in general, and the question of wider neurodivergent representation remains a challenge to which we return below.

'Inclusion'? or 'Nothing About Us Without Us'?

Since the publication of *Neurodiversity Studies: A New Critical Paradigm*, NDS has been referred to in various ways. At its simplest, NDS can be seen as a cluster of research informed by neurodiversity approaches. This implies a 'difference approach' to neurodivergence (see, for example, Mitran, 2023; Wentz, 2021), drawing on social models of disability, and an acknowledgement of the importance of, for example, participatory research (Fletcher-Watson et al., 2021) and theorizing developed by neurodivergent scholars (Chown, 2020). Just as queer theorists have suggested we treat "homosexuality...as a position from which one can know...as a legitimate condition of knowledge" (Halperin, 1995, p. 60, original emphasis), so we propose we approach neurodivergence.

Hens and Van Goidsenhoven (2023, p. 10, original emphasis) note:

> Finally, we also want to mention *neurodiversity studies* here, a new field of inquiry that aims to find new ways to support including neurodivergent perspectives in knowledge production. It questions the theoretical assumptions surrounding idea of the neurotypical...It analyses the role of neuronormativity in theory and science and aims to contribute to redefining what it means to be human...We believe that any autism or developmental disability project should engage with fields such as neurodiversity studies or disability studies.

While we agree with much of this, we would like to question the notion of 'including neurodivergent perspectives in knowledge production'. Some autistic Critical Autism Studies (CAS) scholars have pointed out the problematic nature of autistic involvement in CAS as being too often limited to 'voices to be listened to' in line with inclusive research (c.f. Milton & Bracher, 2013): we must be vigilant to avoid "participation or engagement for the sake of crafting a thin veneer of legitimacy" (Orsini, 2022, p. 5).

The risk associated with such 'inclusion' is that there are echoes of business as usual: research is carried out primarily by assumed to be neurotypical researchers who choose, or not, or to varying extents, to 'include neurodivergent perspectives'. Business as usual considerably limits what is seen to be possible (Bertilsdotter Rosqvist et al., 2019; see also Chapter 5, this volume).

has been spread outside of the Facebook group. As the project idea started to unfold, Bertilsdotter Rosqvist took on the role as PI of the project.

Woods and colleagues (2018, p. 978), for example, cite Waltz' earlier (2014) arguing for "autistic persons to be allowed to take a more active paid role in the production of autism knowledge".

We do not feel this goes far enough. We would like to move away from 'being allowed' and turn this notion on its head. We therefore propose NDS as a research space in which *neurotypical* researchers willing to recognize and reflect upon their knowledge as being situated are those *being* included, not those *doing* the including. We therefore seek not just to question the role of neurodivergent folk in research processes, but also to explore the possibilities of evolving roles for *all* researchers regardless of neurotype. Pellicano (2020, pp. 233–234) has called for diverse collaborations, where complementary forms of knowledge, views and values, approaches, and levels of analysis are combined, suggesting that "(T)hese different forms of expertise are not at odds with each other; they are vital components of the same endeavor". However, while the 'endeavor' *may* be the same, research, however well-meaning, may also often be a site of tokenism or discord (Milton, 2014). When research funding is dependent on 'inclusion' or 'participation', for example, it is difficult to know to what extent those ambitions are genuinely shared by all parties (see, for example, Chapters 5, 8, 22, this volume).

We consider that NDS—and indeed this volume—needs to be informed by the 'nothing about us without us' slogan, largely constituting "neurodivergent research about neurodivergent people" (McDermott, 2023)—with neurotypical people 'included' and welcomed, as allies and as co-researchers. NDS explores and develops "neurodivergent modes of knowledge" (Marshall, 2022, p. 11), primarily from what Dieuwertje Huijg (2020, p. 216), refers to as a "neurodivergent standpoint". In this, we tentatively and conditionally join Elmadagli (2023, p. 75), who suggests considering "limiting the influence of non-autistic scholars" within CAS: conditionally and tentatively, because we see it as desirable to aim for *elevating* neurodivergent voices rather than *limiting* others.

None of this is to suggest that neurodivergent lay-people and researchers hold unique authority to the creation of knowledge in the field of neurodiversity. Rather, we feel that calls for 'participation' and 'inclusion' (always implicitly one-way, that is, inclusion *by* neurotypical researchers *of* neurodivergent folk) carry the risk of tokenism, and that there may be strategic and epistemological advantages to turning this notion on its head. NDS is therefore inclusive: but reframes the question of inclusion. Nor should NDS become a silo rejecting research coming from elsewhere than critical frameworks such as CAS or NDS. On the contrary, 'outside' frameworks and scholars are to be engaged with, to enrich our debates and for NDS scholars for example to critique neurodiversity research more broadly, as Guest (2020, p. 159) suggests we do with autism research, "by re-analysing the data from the autistic perspective to come up with alternative explanations".

A Critical Epistemological Approach and Neurodivergent-Affirmative Methodologies

Epistemic injustice (Fricker, 2007; Polhaus, 2012), whereby a person or group is wronged or undermined in their capacity as a knower, is a central theme in NDS, and is recurrent throughout this volume (see, for example, Chapters 8, 9, 12, 13, 19, 22, this volume). Epistemic injustice necessarily leads to, and draws upon, discussions of power. The "capacity for reason", as Fricker (2007, p. 44) points out, is such a site of contention, so central to our humanity, that it is unsurprising that:

> in contexts of oppression the powerful will be sure to undermine the powerless in just that capacity, for it provides a direct route to undermining them in their very humanity.

Following in the footsteps of Critical Autism Studies, NDS keeps a close eye on relations of power, between groups and between individual members of groups. NDS further proposes a critical epistemological approach to science, specifically the formulation of a critical approach to neurotypical epistemic privilege (eds Bertilsdotter Rosqvist et al., 2020), including the privilege inherent in 'neurotypical science' (see Chapter 25, this volume). By this we refer to theories and research methods informed by and filtered through neurotypical frames of analysis (Schneid & Raz, 2020): business as usual. Following this, we see NDS scholars as trouble shooters/makers—what Ahmed (2023) refers to as 'killjoys'—in research, challenging "our most basic assumptions about the nature of autism, broadening and altering our shared theoretical beliefs, values and techniques" (Pellicano & den Houting, 2022, p 389; see also Chapter 5, this volume). Following critical theorizing elsewhere (for discussion, see Bertilsdotter Rosqvist et al., 2020a), NDS stresses the need for situated scholarship, combatting research-based violence, and speaking back to epistemic injustice (Botha, 2021). Fricker (2007, p. 6) reminds us to "detect and correct" ways in which prejudice and stereotype hinder or limit scientific enquiry: we see these acts of correction and detection as central to NDS.

Also central to NDS is to re-imagine *how* research is carried out, how the shift from "normal science" to "neurodiversity in autism science" (Pellicano & den Houting, 2022, p. 381) may be operationalized. This may take a variety of forms, such as:

- Using neurodivergent-affirmative research methods or 'neurodivergent methodologies'.
- Implementing various levels and forms of participatory research.
- Researching and writing in 'cross-neuro-status' collectives.
- Unlearning and reimagining deficit-based narratives in existing neurodiversity research.

We borrow the concept of neurodivergent-affirmative from contexts of interventions and support (c.f. Chown et al, 2023; Hartman et al, 2023; Sonuga-Barke, 2023). Neurodivergent-affirmative research methods firstly consider that our neurotypes impact us as knowers and knowledge producers. They are therefore epistemological standpoints, acknowledging the potential of 'neurodivergent modes of knowledge' (Marshall, 2022; see also Chapter 25, this volume). Secondly, neurodivergent-affirmative research methods, as Sonuga-Barke (2023, p. 1406) points out, "produce neurodivergent affirmative experiences and promote [epistemological] thriving". This is clearly in accord with—and extends—NDS ambitions to seek out and correct epistemic injustice.

Debating Neurodiversity Studies from Within: Some Challenges for NDS

NDS is an emergent field developing through internal debates, mirroring both some of the strengths and limitations of this book. As editors, we have differing neurodivergent experiences, and in developing the content of this book we have sought to include a variety of neurodivergent perspectives. However, while NDS needs to be pertinent to all forms of neurodivergent experience (see Meinen, 2023; Radulski, 2022), it is still true to say that NDS remains largely 'autistic-centred' (Basten, 2023). Acknowledging current relative autistic privilege amongst neurodivergent people both politically and academically, and stressing the importance of different aspects of neurodivergence need to be integral to the forming of NDS (Botha & Gillespie-Lynch, 2022; Ne'eman & Pellicano, 2022).

Autistic-centredness within neurodiversity approaches has notably been challenged from ADHDers (c.f. Basten, 2023; Vreeling, 2021). However, we hope to see the development of a broader NDS, for example concerning dyslexic people (van Schaik, 2021), intersectional approaches including multiple neurodivergence (van Schaik, 2021; Toft, 2023; Hultman & Hultman, 2023; Radulski, 2022), possibilities of cross-pollination between the fields of NDS and critical dementia studies (c.f. Örulv, 2023; Sandberg et al., 2021), or Mad/Trauma studies (Perry, 2023). The current volume has made significant steps in this direction, with chapters from and/or about ADHDers (Hultman and Hultman, Goetz and Adams; Jackson-Perry), AuDHDers (Munday, Stetson), people with acquired brain injury (Hultman et al.), multiple neurodivergence (Hume; Struyf et al.), and autistic people with profound learning disabilities (Redmore). However, we have found it challenging to find contributors representing other forms of neurodivergence (innate and acquired alike), and this is an area in which progress needs to be made.

Similarly, this book reflects the field more broadly, in that "autistic people of colour and/or of non-western origins are few and far between" in both autistic community spaces and the neurodiversity movement more broadly (Kourti,

2022). We noted above that neurodiversity approaches were historically driven primarily by autistic advocates and researchers: these were also primarily White people from North America and the UK. In North America, where a high proportion of global autism research is undertaken, Black scientists, researchers, and clinicians (including in the field of autism) are considerably under-represented (Jones & Mandell, 2020). 'Whiteness' is one of the major stereotypes concerning autism (Botha et al., 2022), leading to disparities in diagnosis and care, with autistic Black children in the United States, for example, receiving later diagnoses and having reduced access to specialist health and educational resources compared to their White peers (Fombonne & Zuckerman, 2022).

We feel, as Giwa Onaiwu (2020, p. 271) assumes we might, "Pretty crappy" that, despite concerted efforts we have not managed to do better concerning the lack of representation of minoritized or racialized groups in this volume. We are also aware of a certain irony, considering that neurodiversity and decolonization approaches have much in common, notably, as Cheng et al., (2023, p. 573) suggest:

(1) Reconstructing ingrained structures of power and privilege.
(2) Pushing back against biomedical model beliefs and practices.
(3) Raising the knowledges, truths, and priorities of lived experiences to an equal status with insights generated from 'objective' scientific research.
(4) Providing space for communities and individuals to identify their own concerns and find their own solutions.

However, neurodiversity approaches developed from predominantly White, North American/UK autistic advocates. The Global South—or Global Majority—has a wide range of beliefs, socio-cultural norms, scholarship, and activist communities (see Cheng et al., 2023 for discussion), and it would be misguided to assume that 'Global North solutions' are translatable to 'Global South situations'. Indeed, Manase (2024) notes the colonialist history of importing Global North language and concepts, and warns against disregarding indigenous knowledge, while keeping an eye on how neurodiversity approaches could be usefully integrated. Imposing existing neurodiversity approaches as a global norm, a one-size-fits-all response to vastly differing contexts and worldviews, is arrogant at best. When calling for contributors to this volume, we specifically sought texts on NDS in Global South contexts. In hindsight, we should probably have sought to understand how neurodivergent researchers and advocates from the Global South see their *own* situations, how *they* are responding, and how we might learn from each other in a process of cross-pollination. Latterly, positive steps appear to have been made in this direction (see, for example, Chataika & Goodley, 2024; Manase, 2024), and we are confident that this bodes well for the future.

Chapter Contents

In terms of structure, this volume consists of three separate parts. **Part one** is ***setting the scene***, of which this introductory chapter is the first. In the two chapters making up the rest of this section, we zoom in on contextual elements that we consider important in Neurodiversity Studies. In Chapter 2, David Jackson-Perry maps out—albeit necessarily partially—Critical Autism Studies (CAS). This is perhaps the most important 'sub-discipline' contributing, in the wake of and alongside Critical Disability Studies, to the development of the principles underlying much of the practice and theory that will be discussed throughout this collection. Jackson-Perry gives an overview of the various 'branches' of CAS, restates the importance of CAS as a field in its own right, and looks to points of convergence between CAS and NDS. In Chapter 3, Andrew Ivan Brown, Hanna Bertilsdotter Rosqvist, and David Jackson-Perry turn their attention to ADHD, to ask what 'Critical ADHD Studies' might resemble. They summarize research largely concerned with pathology and diagnosis in the field, and then consider 'alternative narratives'. Without laying claim to what is or is not Critical ADHD Studies, the authors look for points of articulation between approaches holding potential for centring ADHDers as agents and tentatively propose some ambitions and guiding principles for the field.

We noted earlier that detailed descriptions of participatory methods and creative methods are relatively rare, and that this may constitute a barrier to researchers committing to them. **Part two of this volume, *knowing and doing: research ethics and methods***, responds to this lacuna. Many of the 19 chapters in this section describe in detail the methods used across a variety of disciplines in Neurodiversity Studies, and others reflect more broadly upon some of the themes we have evoked in the introduction, and to which we return in the conclusion.

In Chapter 4, Aimee Grant, Kat Williams, Karen Henry, Willow Holloway, Christina Nicolaidis, Helen Kara, and Amy Brown describe the development of the £2.4 million Wellcome Trust funded project, *Autism from menstruation to menopause,* which used a participatory model with autistic leadership at every stage of the research process. The authors describe how their participatory process shifted the focus of the study, shaped its design, and made it possible for a range of autistic participants with differing communication needs and preferences to take part in the study. The authors also give guidance to help autism researchers co-develop ethical and participatory research funding applications.

In Chapter 5, Joke Struyf, Cecilia Ingard, and Hanna Bertilsdotter Rosqvist, informed by a feminist disability approach, take the case of the early stages of another large-scale study, the ongoing international biomedical project on Risk and Resilience in Developmental Diversity and Mental Health (R2D2-MH) in which the authors have been involved. How did this biomedical project take neurodivergent knowledge production seriously and create a more emancipatory academic space? How can an international co-creation group integrate lived expertise into the research and how can this add to the debate around collaborative research? The authors consider the limits and possibilities of a neurodiverse research collaboration and reflect on how this way of doing research might produce a caring, epistemic healing—and learning—experience for all involved. In Chapter 6, Kristien Hens reflects upon her experiences as one of the work package leaders in R2D2, the study discussed in the previous chapter. Hens responds to Chapter 5, explaining the inner workings of the project as they relate to some of the frictions detailed in Chapter 5 and reflecting on lessons learned and her optimism for meaningful co-creation.

The neurodivergent and neurotypical authors of Chapter 7, Amy Pearson and co-authors, also report from their experiences of working together on a larger project exploring different ways of understanding the meaning of neurodiversity. Here, rather than discussing the larger project, the authors use an iterative, collaborative, and autoethnographic method of data collection and analysis to understand and critically analyse the experience of working together as a neuromixed group of authors, and the lessons learnt.

With Chapter 8, Lill Hultman, Jeanette Nelson, Fredrik Sandman, and Mikael Åkerlund continue with the theme of working together as a neuromixed group of authors, broadening the scope of this volume to bring adults with acquired brain injury into the discussion. The authors are an academic researcher (with ADHD) and three community researchers (who live with the consequences of stroke) on a three-year research project exploring experiences of participation and belonging for adult people with acquired brain injury. Through describing and analysing parts of the research process, the authors explore both the benefits of integrating experience-based knowledge and the challenges of doing research in a neuromixed space. Hultman and co-authors have a particular focus on modifications during the data collection and the analytical processes as a result of their collaboration. Here, the community researchers' own experience of stroke enabled an innovative data collection and analytical process in terms of interactive patterns between researchers and participants.

Marjorie Désormeaux-Moreau and Isabelle Courcy, in Chapter 9, describe the process related to the deployment of an online social support initiated by autistic citizens to which a study—led by an autistic researcher—was added. The research team consisted of five autistic citizens as co-knowledge producers, an autistic researcher and two non-autistic researchers. Désormeaux-Moreau and Courcy analyse the benefits and challenges associated with the 'dual positioning' of the research team and reflect on questions of epistemic injustice that emerged.

In the following chapter, Chapter 10, Sandra Thom-Jones and Jennifer Lowe, through an analysis of eight autism studies which adopted an 'autistic approach' at each stage of the research process, also take a critical look at the move towards adopting increasingly participatory practices. While accepting the good intentions behind this shift, the autistic authors note that concentration on 'partnership' alone may render invisible the structural, social, and organizational barriers faced by neurodivergent members of the academy. They detail a variety of examples of 'autistic approaches' to research design and set out the advantages of the meaningfully involvement of autistic people in all stages of the research process, from inception to dissemination.

With Chapter 11, we return to ADHD. Shruti Taneja-Johansson describes an in-depth, longitudinal, ethnographic study exploring young people with ADHD's transitions from schooling into adult education, higher education, employment, or other domains of adulthood. Johansson reflects on the possibilities and challenges of her methodology, paying particular attention to how she—as a neurotypical researcher—sought to develop mutually respectful and trusting relationships with participants, and reflexively developed the qualities she felt important in this regard.

Lill Hultman and Maya Hultman continue the theme of ADHD from yet another angle in Chapter 12, considering epistemic injustice as they reflect upon the limits and possibilities of ADHD-led participatory research processes. The authors consider the trade-off between a shared experience of living with ADHD bringing a sense of mutual understanding of experience and the need to overcome certain challenges related to ADHD. Hultman and Hultman also describe how an ADHD-led writing process can differ from a neurotypical one, and how they navigate these differences.

Navigating varying needs in the research process is also central to Romy Hume's work in Chapter 13, as is epistemic injustice. Hume considers the potential of Interpretive Interactionism—a qualitative methodology for insider researchers—to demonstrate how research can empower members of the social group to which the researcher and the participants belong, and level the playing field. Hume reports with concrete details on how the use of Interpretive Interactionism and creative methods in a study with multiply neurodivergent participants facilitated ongoing exchanges with her and considers the impact of power dynamics in the research process.

Nicola Shaughnessy and co-authors, in Chapter 14, combine leading expertise in, and experience across, neurodivergence and neurotypicality, clinical and research science, creative practices, medical humanities, sociology, education, and computers science. They reflect on the possibilities of an 'autistic phenomenology', describing how they sought depth of experience through Interpretative Phenomenological Analysis and breadth of experience through a citizen science digital platform.

In Chapter 15, Ned Redmore continues the theme of phenomenological inquiry, working in an area that has historically been under-researched from this angle in presenting his experience of researching (with) autistic people with profound learning disabilities. Redmore first gives an overview of the theoretical underpinnings of phenomenological enquiry. He goes on to describe his own research process in this context, concentrating on his fieldwork. The author thinks through both the advantages and the limitations of how he proceeded and questions his own positionality as a neurotypical former day service support worker.

Lydia Stetson, in Chapter 16, presents the interest—and the challenges—of using narrative research to understand the lived experiences and context of AuDHD individuals in a largely neurotypical world. Narrative researchers play integral roles in shaping the narrative and in co-constructing meaning from the stories with the narrators: how then is this impacted when there might be a 'mismatch' between researchers and narrators, particularly regarding communication, context, and experience of the world? What lessons does narrative inquiry have to teach us about the role of researcher reflexivity and co-construction of meaning?

With Chapter 17, Teddy G Goetz and Noah Adams consider how academic codes, cultural expectations, and values may be in tension for academics who are also neurodivergent community members and activists. The authors—autistic and ADHD—use 'duoethnography' to explore how their experience of academic structures and processes interface with neurodivergent individuals' experiences and expectations of such environments.

In Chapter 18, Katie Munday continues the theme of negotiating the neurotypical academy, taking a hard look at a little-explored but important question, that of navigating university research ethics applications as a neurodivergent academic. Munday draws on their own experience as an autistic and ADHD research master's student to explore how neuronormativity within ethics applications can create barriers to neurodivergent students. The author goes on to consider how making this process more accessible for neurodivergent students and scholars would positively impact the system for all.

Academic work drawing on materials such as blogs and podcasts produced by neurodivergent people outside academia is becoming more common. In Chapter 19, Alyssa Hillary Zisk takes a look at some of these forms of knowledge creation and ways in which they are being taken up within academia. The author discusses the importance—and the tensions—of engaging with and citing non-academic materials as a step towards epistemically just academic processes.

In Chapter 20, David Jackson-Perry details the participative methodology he used throughout his doctoral thesis exploring autistic accounts of sexuality, intimacy, and authenticity. While this is above all a detailed methodological chapter, the author reflects on his discomfort working as a neurodivergent but non-autistic scholar in the field. He reflects on what worked—and, just as importantly, what didn't—as he describes four participatory 'steps' in his research process: formal and informal community engagement; the creation of a research website; the recruitment and involvement of a Research Advisory Group; and an online consultation survey regarding methods and topics of interest.

Claire M. Chapman, in Chapter 21, also goes into considerable methodological detail, as she reports on her use of email interviewing in a study seeking to understand autistic/ADHD women's experiences of accessing support after being victims of crime. Chapman bases her discussion on this feminist disability study, underpinned by constructivist grounded theory, to describe the research design, the fieldwork she carried out, and to make recommendations for future research practice in neurodiversity research.

Concluding this section, in Chapter 22, David Jackson-Perry considers the intricate, long-standing, deeply entrenched, and powerful processes supporting and perpetuating epistemic injustice in autism research. He asks if, as things stand, the inclusion of autistic perspectives in autism research is sufficient to disrupt this. Using the relative invisibility of 'authenticity' in autism research as a case study in epistemic injustice, he proposes 'unknowing' as a methodological tool in NDS. The author concludes with concrete suggestions for an 'unknower's toolkit'.

The final section of this volume, **Part III/IV,** ***looking back to the future,*** consists of four chapters, the first three of which are commentary chapters by experts in the field.

Firstly, in Chapter 23, Andrew Ivan Brown takes a critical look at the notion of 'ADHD experts', considering who has been attributed this status and how this attribution has changed over time. Brown reflects on the tensions—and the possibilities—implicit in the evolution of the concept of expertise, and of partnerships between traditional 'ADHD experts' and critical ADHD collectives. In Chapter 24, Jacquiline den Houting, Abby Sesterka, and Elizabeth Pellicano consider how far we have come—and how far we still have to go—to achieve equitable co-production in autism research. After noting the enduring presence of epistemic injustice, and that a veneer of co-production may sometimes legitimize studies perpetuating it, the authors propose ways forward. They discuss capability building, and relationship building between communities, but never lose sight of the importance of grappling with power differentials that may still often render partnerships inequitable. Dora M. Raymaker and Christina Nicolaidis, the authors of Chapter 25, explore research informed by a neurodiversity approach as constituting a disruption to the 'normal science' of neurodivergence. They invoke a futures thinking approach to describe the nature of this disruption, its impact on the present state of research informed by neurodiversity approaches, and the imagined directions we could be pulled in as a 'new normal' in science takes shape. Finally, in Chapter 26, Hanna Bertilsdotter Rosqvist and David Jackson-Perry bring together the methodological and ethical themes that have emerged from individual chapters. The volume's editors first look back to contributions to identify concrete steps which they feel will be useful for students and academics working in the field. They provide an accessible toolbox for each theme to guide students and researchers as they set out on the voyage to creative and ethically and epistemologically robust research in the field. In conclusion, they return to the question of 'unknowing' discussed in Chapter 22 as a potential methodological tool and framework for carrying out research in Neurodiversity Studies.

References

Ahmed, S. (2023). *The feminist killjoy handbook.* Allen Lane.

Asasumasu, K. (no date). PSA from the actual coiner of "neurodivergent" [Tumblr]. *Tumblr*. Available at: https://sherlocksflataffect.tumblr.com/post/121295972384/psa-from-the-actual-coiner-of-neurodivergent

Basten, L. (2023). The codex is always on Crip time: ADHD (ness) and Reading. *Reception: Texts, Readers, Audiences, History, 15*, 89–96. Bertilsdotter Rosqvist et al, eds, 2020

Bertilsdotter Rosqvist, H., Kourti, M., Jackson-Perry, D., Brownlow, C., Fletcher, K., Bendelman, D., & O'Dell, L. (2019). Doing it differently: Emancipatory autism studies within a neurodiverse academic space. *Disability and Society, 34*(7–8), 1082–1101.

Bertilsdotter Rosqvist, H., Stenning, A., & Chown, N. (2020a). Introduction. In Neurodiversity Studies. A New Critical Paradigm, Bertilsdotter Rosqvist, H., Chown, N., Stenning, A. (eds.). London: Routledge. p 1–11

Bertilsdotter Rosqvist H., Stenning, A., Chown, N. (2020b). Neurodiversity studies: Proposing a new field of inquiry. I *Neurodiversity Studies. A New Critical Paradigm*, In H. Bertilsdotter Rosqvist, N. Chown, & A. Stenning (Eds.) (pp. 226–229). Routledge.

Botha, M. (2021). Academic, activist, or advocate? Angry, entangled, and emerging: A critical reflection on autism knowledge production. *Frontiers in Psychology*, 4196.

Botha, M., & Gillespie-Lynch, K. (2022). Come as you are: Examining autistic identity development and the neurodiversity movement through an intersectional lens. *Human Development, 66*(2), 93–112.

Botha, M., Chapman, R., Giwa Onaiwu, M., Kapp, S. K., Stannard Ashley, A., & Walker, N. (2024). The neurodiversity concept was developed collectively: An overdue correction on the origins of neurodiversity theory. Autism, 13623613241237871.

Botha, M., Dibb, B., & Frost, D. M. (2022). Autism is me": An investigation of how autistic individuals make sense of autism and stigma. *Disability and Society, 37*(3), 427–453.

Bottema-Beutel, K. (2024). Putting autism research in social contexts. Autism, 13623613241245642.

Chapman, R. (2020). The reality of autism: On the metaphysics of disorder and diversity. *Philosophical Psychology*, pp. 1–21.

Chapman, R., & Carel, H. (2022). Neurodiversity, epistemic injustice, and the good human life. *Journal of Social Philosophy.*

Chapman, R. (2023). *Empire of normality neurodiversity and capitalism.* Pluto Press.

Chataika, T., & Goodley, D. (2024). Introduction: Unpacking postcolonial disability studies. In *The Routledge handbook of postcolonial disability studies* (pp. 3–12). Routledge.

Cheng, Y., Tekola, B., Balasubramanian, A., Crane, L., & Leadbitter, K. (2023). Neurodiversity and community-led rights-based movements: Barriers and opportunities for global research partnerships. *Autism, 27*(3), 573–577.

Chown, N. (2020). Language games used to construct autism as pathology. In *Neurodiversity Studies. A New Critical Paradigm.* In H. Bertilsdotter Rosqvist, N. Chown, & A. Stenning (Eds.) (pp. 27–38). Routledge.

Chown, N., Murphy, S., & Suckle, E. (2023). The autism worldview dilemma: To normalise or not to normalise, that is the question. *Good Autism Practice (GAP), 24*(2), 5–12.

Clouder, L., Karakus, M., Cinotti, A., Ferreyra, M. V., Fierros, G. A., & Rojo, P. (2020). Neurodiversity in higher education: A narrative synthesis. *Higher Education, 80*(4), 757–778.

Dekker M. (2023). A correction on the origin of the term 'neurodiversity'. Independent living on the autistic spectrum: Where those who are different find that they're not alone. Martijn 'McDutchie' Dekker [Blog]. https://www.inlv.org/2023/07/13/neurodiversity-origin.html

Dieuwertje Huijg, D. (2020). Neuronormativity in theorising agency. Neurodiversity studies: A new critical paradigm, 213–217.

Dind, J. (2021). Review of Neurodiversity Studies: A New Critical Paradigm edited by Hanna Bertilsdotter Rosqvist et al. Disability Studies Quarterly, 41(1).

Dwyer, P. (2022). The neurodiversity approach (es): What are they and what do they mean for researchers? *Human Development, 66*(2), 73–92.

Elmadagli, C. (2023) The future of critical autism studies (CAS): Thinking through critical discourse studies and postcolonial feminism, ought. *The Journal of Autistic Culture, 5*(1), Article 11.

Fletcher-Watson, S., Brook, K., Hallett, S., Murray, F., & Crompton, C. J. (2021). Inclusive practices for neurodevelopmental research. *Current Developmental Disorders Reports, 8*, 88–97.

Fombonne, E., & Zuckerman, K. E. (2022). Clinical profiles of Black and White children referred for autism diagnosis. *Journal of Autism and Developmental Disorders, 52*(3), 1120–1130.

Fricker, M. (2007). *Epistemic injustice: Power and the ethics of knowing*. Oxford University Press.

Giwa Onaiwu, M. (2020). "They don't know, don't show, or don't care": Autism's white privilege problem. *Autism in Adulthood, 2*(4), 270–272.

Gobbo, K., Putney, V. T., & Shmulsky, S. (2019). Should neurodiversity culture influence how instructors teach? *Academic Exchange Quarterly, 23*(4), 12–17.

Goetz, T. G., & Adams, N. (2024). The transgender and gender diverse and attention deficit hyperactivity disorder nexus: A systematic review. *Journal of Gay and Lesbian Mental Health, 28*(1), 2–19.

Hall, K. Q. (2017). Queer epistemology and epistemic injustice. In The Routledge handbook of epistemic injustice (pp. 158–166). Routledge.

Halperin, D. (1995). *Saint foucault*. Oxford University Press.

Haraway, D. (1988). Situated knowledges: The science question in feminism and the privilege of partial perspective. *Feminist Studies, 14*(3), 575–599.

Hart, A. (2020). A new alliance?: The Hearing Voices Movement and neurodiversity. In *Neurodiversity Studies* (pp. 221–225). Routledge.

Hartman, D., O'Donnell-Killen, T., Doyle, J. K., Kavanagh, M., Day, A., & Azevedo, J. (2023). *The adult autism assessment handbook: A neurodiversity affirmative approach*. Jessica Kingsley Publishers.

Hens, K., & Van Goidsenhoven, L. (2023). Developmental diversity: Putting the development back into research about developmental conditions. *Frontiers in Psychiatry, 13*, 2994.

Hultman, L., & Hultman, M. (2023). Believe me, only I know how I feel. An autoethnographic account of experiences of epistemic injustice in mental health care. *Frontiers in Psychiatry, 14*, 1058422.

Johnston-Tyler, J., & Analla, Y. (2015) 'Addressing neurodiverse employment needs: Working with Autism Spectrum Disorder I, NVLD, ADHD, Learning and Mood Disordered Clients'., Career Planning & Adult Development Journal, 31(4).

Jones, D. R., & Mandell, D. S. (2020). To address racial disparities in autism research, we must think globally, act locally. *Autism, 24*(7), 1587–1589.

Kapp, S. K. (2020). Autistic community and the neurodiversity movement: Stories from the frontline. Springer Nature.

Kourti M. (2022). Autism, race, ethnicity and the whiteness of the neurodiversity movement. Presentation at the Autscape Conference, Derbyshire. http://www.autscape.org/2022/programme/presentations

Manase, N. (2024). Africanising neurodiversity. The Routledge Handbook of Postcolonial Disability Studies.

Marshall, B. A. (2022). Workplace neurodiversity: An exploratory study (Doctoral dissertation, Alliant International University).

McDermott, L. G. (2023). Neuroqueering physics: A new scientific literacy. arXiv preprint arXiv:2307.05531.

McDermott, C. (2022). Theorising the neurotypical gaze: Autistic love and relationships in The Bridge (Bron/Broen 2011–2018). *Medical Humanities, 48*, 51–62.

Meinen, L. E. (2023). Share the experience, don't take it: Toward attunement with neurodiversity in videogames. *Games and Culture, 18*(7), 919–939.

Milton, D., Ridout, S., Murray, D., Martin, N., & Mills, R. (2020). The neurodiversity reader: Exploring concepts, lived experiences and implications for practice. Pavilion.

Milton, D. E., & Green, J. (2024). Theorising autism. *Autism, 28*(4), 795–797. https://doi.org/10.1177/13623613241235786

Milton, D. (2014). Autistic expertise: A critical reflection on the production of knowledge in autism studies. *Autism, 18*(7), 794–802.

Milton, D., & Bracher, M. (2013). Autistics speak but are they heard? *A Journal of the BSA MedSoc Group, 7*(2013), 61–69.

Nagel, T. (1986). *The view from nowhere.* Oxford University Press.

Ne'eman, A., & Pellicano, E. (2022). Neurodiversity as politics. *Human Development, 66*(2), 149–157.

Neumeier, S. M. (2018).'To Siri With Love' and the problem with neurodiversity lite. Rewire News Group. https://rewirenewsgroup.com/2018/02/09/siri-love-problem-neurodiversity-lite/

Orsini, M. (2022). Who needs to (un) know? On the generative possibilities of ignorance for autistic futures. *International Journal of Qualitative Studies in Education*, 1–18.

Örulv, L. (2023). 18 Neurodiversity and dementia. Critical Dementia Studies, 247.

Parsons, S., Yuill, N., Good, J., & Brosnan, M. (2020). 'Whose agenda? Who knows best? Whose voice?' Co-creating a technology research roadmap with autism stakeholders. *Disability and Society, 35*(2), 201–234.

Pellicano, E. (2020). Commentary: Broadening the research remit of participatory methods in autism science—a commentary on Happé and Frith (2020). *Journal of Child Psychology and Psychiatry and Allied Disciplines, 61*(3), 233–235.

Pellicano, E., & den Houting, J. (2022). Annual research review: Shifting from 'normal science' to neurodiversity in autism science. *J Child Psychol Psychiatr, 63*, 381–396.

Perry, M. (2023). Voices in my head: Representations of mental illness in contemporary American musical theater.

Pickard, H., Pellicano, E., den Houting, J., & Crane, L. (2021). Participatory autism research: Early career and established researchers' views and experiences. *Autism*, 13623613211019594. https://doi.org/10.1177/13623613211019594

Pohlhaus, G. (2012). Relational knowing and epistemic injustice: Toward a theory of willful hermeneutical ignorance. *Hypatia, 27*(4), 715–735.

Radulski, E. M. (2022). Conceptualising autistic masking, camouflaging, and neurotypical privilege: Towards a minority group model of neurodiversity. *Human Development, 66*(2), 113–127.

Robinson, D. (2022). Neurodiversity in medical education: How can we improve postgraduate learning for neurodiverse doctors?. *Medical Teacher*, pp. 1–3.

Sandberg, L. J., Rosqvist, H. B., & Grigorovich, A. (2021). Regulating, fostering and preserving: The production of sexual normates through cognitive ableism and cognitive othering. *Culture, Health and Sexuality, 23*(10), 1421–1434.

van Schaik, M. (2021). Accept the idea that neurodiverse kids exist: Dyslexic narratives and neurodiversity paradigm visions. Unpublished theses, Wilfrid Laurier University. Theses and Dissertations (Comprehensive). 2403. https://scholars.wlu.ca/etd/2403

Schneid, I., Raz, A. E. (2020). The mask of autism: Social camouflaging and impression management as coping/normalization from the perspectives of autistic adults. *Social Science and Medicine, 248*, art. no. 112826.

Silberman, S. (2015). *Neurotribes*. Penguin.

Sinclair, J. (1999). Don't mourn for us. Autistic Rights Movement UK.

Sinclair, J. (2005). History of ANI. Autism Network International. https://www.autismnetworkinternational.org/History_of_ANI.html

Singer, J. (1999). Why can't you be normal for once in your life? From a 'problem with no name' to a new category of disability. In M. Corker & S. French (Eds.), *Disability discourse* (pp. 59–67). Open University Press.

Sonuga-Barke, E. J. (2023). Paradigm 'flipping' to reinvigorate translational science: Outlining a neurodevelopmental science framework from a 'neurodiversity' perspective. *Journal of Child Psychology and Psychiatry, 64*(10), 1405–1408.).

Taylor, G. (2021). Embracing neurodiversity in medicine. *Australian Journal of General Practice, 50*(3), 101–101.

Tisoncik, L. A. (2020). Autistics. org and finding our voices as an activist movement. Autistic community and the neurodiversity movement: Stories from the frontline, 65–76.

Toft, A. (2023). Telling disabled and autistic sexuality stories: Reflecting upon the current research landscape and possible future developments. *Sexes, 4*(1), 102–117.

Vreeling, L. L. (2021). All over the Place? A contextualized and experiential perspective of ADHD (Doctoral dissertation).

Walker, N. (2014) Neurodiversity: Some Basic Terms & Definitions. Available at: https://neurocosmopolitanism.com/neurodiversity-some-basic-terms-definitions.

Waltz, M. (2014). Worlds of autism: Across the spectrum of neurological difference. *Disability and Society, 29*(8), 1337–1338.

Wood, R., Crane, L., Happé, F., & Moyse, R. (2022). Learning from autistic teachers: Lessons about change in an era of COVID-19. *Educational Review.*
Woods, R., Milton, D., Arnold, L., & Graby, S. (2018). Redefining critical autism studies: A more inclusive interpretation. *Disability and Society, 33*(6), 974–979.

CHAPTER 2

Critical Autism Studies: Roots and Branches

David Jackson-Perry

Introduction[1]

In the previous chapter, we noted that after many years attributing the word and the concept of neurodiversity to a single person, it is now acknowledged that they emerged in a more complex and organic way through activist autistic communities, years before being named and claimed by or for a single source (Botha et al., 2024). We also pointed out the debt that Neurodiversity Studies (NDS) owes to Critical Autism Studies (CAS), which in turn also traces its roots to autistic activist communities. Ryan and Milton (2023) specifically credit the bases of CAS to the work of Autism Network International from 1992 on (for a fascinating and more detailed history of Autism Network International, see Sinclair, 2005), which laid the foundations for a critical approach to autism from the experiences of autistic people themselves (Woods et al., 2018; for an exploration of how certain autistic activists set their actions in a wider context see Tisoncik, 2020 and other contributions in Kapp, 2020). Indeed, although I refer to 'activists and academics' throughout this chapter, the distinction between the two may often be a distinction without a difference (for discussion, see Raymaker, 2020).

[1] This chapter extensively draws on, updates, and re-frames sections of the author's PhD thesis (Jackson-Perry, 2023).

D. Jackson-Perry (✉)
Queen's University, Belfast, UK
e-mail: djacksonperry01@qub.ac.uk

H. Bertilsdotter Rosqvist and D. Jackson-Perry (eds.), *The Palgrave Handbook of Research Methods and Ethics in Neurodiversity Studies*,
https://doi.org/10.1007/978-3-031-66127-3_2

The term Critical Autism Studies only started gaining *academic* traction following the *Interdisciplinary Workshop on Critical Autism Studies: Enabling Inclusion, Defending Difference*, which was held at Ottawa University in 2010, the first gathering to explicitly use the term. A paper published afterwards by the co-organizers (Davidson & Orsini, 2010) reviewed the workshop content and laid out some initial ambitions for CAS. While the participants at the conference were positioned by the authors as "critical autism scholars," they (2010, p. 131) noted that this "inclusive" term was "tentative," going on to recognize:

> the limitations of such labelling practices, not least the misleading suggestion of harmony in approach; workshop participants disagree, at times strongly, and on many and various fronts, not least because we did not always have a common disciplinary language on which to draw.

Beyond the lack, given the interdisciplinary ambitions of CAS, of a common disciplinary language, participants were also variously situated geographically and in their relationship to autism. Workshop attendees included autistic people, parents of autistic children, and researchers "with more ostensibly undemanding authorial identities and relations to the autism spectrum" (Davidson & Orsini, 2010, p. 132), foreshadowing some of the tensions and contestations that were to follow.

To explain this, as well as to give a sense of some of the principal positions and tensions in the field of CAS, this chapter considers the field as referring to three distinct (if not always discrete) branches: the Worlds of Autism branch, the autistic-led branch, and the Re-Thinking Autism branch. This order represents that in which CAS was formally proposed, described, (re)claimed, or (re)defined in publication by each branch. It is in no way to deny the theoretical or political underpinnings that made CAS possible, drawing as they do on critical theory more generally, criticism of disability studies, and early autistic theorizing that predate the use of the term CAS (Woods et al., 2018). The chapter concludes with a reflection on latter developments in the field, and points of articulation with NDS. The organization of this chapter responds in part to the question of whether "whether different subcultures are emerging within CAS" (Woods et al., 2018, p. 976).

To return to the caveat of Davidson and Orsini above, labelling of any kind must be tentative and comes with the risk of over-simplification, of presenting as stable something which has blurry and variable contours. Academics and activists in each of the branches I describe below had engaged with studies of autism that can be considered critical before CAS had been named or theorized, have collaborated in organizing and participating in events and in publications with those more solidly situated within different branches, and may move between branches at various times and in exploring various themes. To begin at (one of) the beginning(s), I start with the Ottawa workshop, what was to become what I call the Worlds of Autism branch.

The Worlds of Autism Branch

Although the 2010 Ottawa *Interdisciplinary Workshop on Critical Autism Studies: Enabling Inclusion, Defending Difference* was not "explicitly organized around themes of emotion and affect" (Davidson & Orsini, 2010, p. 131), the article by the workshop co-organizers took the place of emotions in CAS as its title. While I do not know if this was the authors' intention, this can be understood in (at least) two ways. Firstly, the two authors had previously "encountered numerous personal accounts of often profoundly emotional autistic experiences" in which they had observed, contrary to the way many clinical and research accounts 'narrated' autism, "that autistic capacity for, and expression of, emotion may be different, but it is certainly not entirely disabled" (p. 131). In parallel, though, the authors note that during the workshop itself "emotional undercurrents were clear and strong, often close to the surface" (p. 131): the authors go on to say:

> During the workshop, despite or sometimes because of the overt substantive focus, feelings were never submerged for long and discussion…was often emotionally charged. This was perhaps to be expected given the sheer variety of personal, as well as disciplinary, perspectives represented; participants differed markedly in the manner of their relation to the autism spectrum and included parents of autistic children as well as autistic people themselves, both often also academics. Here as always, such subjectivities intersect.

This is important for CAS more broadly, foreshadowing as it does the development of sometimes divergent priorities and approaches within the field. Indeed, this is something of a hallmark of CAS: as the authors of a much later article note, "autistic self-advocates, critical autism scholars and authors on this paper do not always agree about the 'nature or facticity' of autism" (Douglas et al., 2021, p. 612).

Various interrelated ambitions for the workshop and CAS were identified in the first published document discussing this field by name (Davidson & Orsini, 2010). Drawing on insights from Critical Disability Studies (CDS), a challenge to "deficit-focused and arguably disabling constructions of the autism spectrum" (p. 131) was a central theme of the workshop: indeed, its principal *raison d'être* had been "to address a strongly felt need among less mainstream autism scholars for an inclusive and interdisciplinary space where genuinely different and less disabling approaches to the spectrum might be explored" (p. 131). In doing so, participants were seeking "to expand narrowly drawn clinical models of autism." This requires, inter alia, that scholars "attend closely" to autistic voices, an approach conceptualized by one participant, Mark Osteen, as "empathetic scholarship" (p. 132) which should aim to identify and respond to the concerns of autistic people themselves.

In the edited volume *Worlds of Autism* (Davidson & Orsini, 2013), the editors and contributors built on the Ottawa workshop and the resulting article to more solidly propose the possible theoretical bases and ambitions of CAS, while making it clear that they saw the "contours" of CAS as being "emergent and in flux," and the aim of the collection as being "in part to interrogate rather than define or delimit the boundaries of this emerging field of study" (p. 11). The editors further note their "attempt to keep the space open and accessible for new and emerging scholars" (p. 11). I discuss below to what extent—and how—this ambition is echoed within the other branches.

The editors of *Worlds of Autism* identified three central elements of a critical approach to autism studies:

1. *Careful attention to how power relations shape the field of autism* (Orsini & Davidson, 2013, p. 12).

In both scientific literature and the popular imagination, claims about what autism 'is' proliferate. Yergeau (2018, p. 42), for example, noting that claims such as the "asociality" of autistic people are "not without power," goes on to ask "where…do such claims lie? How do they locate autism, and in what ways do they suspend the rhetorical capacities of autistic people?" These questions—where is epistemic authority situated and what are the effects of its distribution?—are (explicitly or not) present across the entire field of autism, whether concerned with diagnosis (Timimi & McCabe, 2016), cultural production (McGuire, 2016), or research theory and practice (Raymaker & Nicolaidis, 2013; Fletcher-Watson et al., 2019; Woods and Waltz, 2019). This awareness and interrogation of the situation and effects of power echoes CDS—and the queer theories and feminist scholarship to which CDS, in turn, owes much—as when Goodley (2017, p. 206) draws on Judith Butler to ask how "non-disabled bodies (are) made more viable than disabled bodies" and questions how "societal practices uphold the precarious higher status of non-disabled people through the abjection of disabled people."

2. *Concern to advance new, enabling narratives of autism that challenge the predominant (deficit-focused and degrading) constructions that influence public opinion, policy and popular culture* (Orsini & Davidson, 2013, p. 12).

Without fetishizing autistic experience or supporting a "celebratory approach to autism" that would risk denying "the idea that autism can be disabling," Orsini and Davidson (2013, p. 12) position *Worlds of Autism* within an "abilities framework." This framework, in "contrast to the prevailing deficit construction of autism, which is individualising and pathologising…not only respects the complex personhood of autistic individuals, but also reveals how the construction of autistic identities holds important insights for how to rethink, and extend, ideas associated with cognitive 'normalcy' (or 'ability') and difference" (O'Dell et al., 2016, p. 169). This second tenet makes both an ethical and an epistemological argument for rejecting deficit-based constructions of autism: on the one hand, a deficit framework reduces the recognizability of a person or a group as human by assuming an absence of qualities that are considered to be inherent to what it means to be human, such as empathy or theory of mind (Dinishak, 2016). On the other, and by the same sleight of hand, autistic people are considered within a deficit reading as being, at best, unreliable witnesses to their own experience. However, in a later publication, Orsini (2022, p. 5) notes that the choice of an abilities framework "could be justly critiqued for invoking ableist ideas about autistic ways of being, and for essentializing autistic traits as superior." In a neoliberal economy, in which supposedly typical autistic traits may be used to justify the 'usefulness' of some autistic people in certain jobs, there is clearly the risk of throwing many others 'under the bus' as it were, with those "who may not live up to the exacting standards of the neoliberal labour market economy (being) cast off as surplus, as excess, not quite up to speed" (Orsini, 2022, p. 5).

However, whatever fault lines may appear in an abilities framework, a deficit reading counters the finding that "traditional autism research is conducted on autistic subjects, not conducted with autistic people" (Raymaker & Nicolaidis, 2013, p. 183), excluding autistic people from being potential holders of knowledge. Further, and given the considerable interaction between many autistic people and research environments, it begs the question: "Within the context of a body of autism literature which is generally deficit-driven, is it then possible for autistic people to imagine themselves and their intimate experience other than through deficit?" (Bertilsdotter Rosqvist & Jackson-Perry, 2021, p. 340). This leads to the final tenet laid out by Davidson and Orsini.

3. *Commitment to develop new analytical frameworks using inclusive and non-reductive methodological and theoretical approaches to study the nature and culture of autism. The interdisciplinary research required (particularly in the social sciences and humanities) demands sensitivity to the kaleidoscopic complexity of this highly individualized, relational (dis)order* (Orsini & Davidson, 2013, p. 12).

This final element of CAS as laid out in *Worlds of Autism* requires a double reflection. Firstly, it asks scholars to reconsider their frameworks, not to limit themselves to analysis that draws on an either/or deficit reading of autism and autistic people. The attention paid to interdisciplinarity, to the political, to the relational, to the nature and culture of autism can again be seen in the extent to which CAS is informed by CDS (Goodley, 2017), although with the "focus of research…placed on issues specific to autism rather than on the broader construct of disability" (O'Dell et al., 2016, p. 168), an important point given that "autistic voices have been largely absent from discussions within disability studies" (Douglas et al., 2021, p. 51). This approach also requires rethinking how autism research is carried out, which, traditionally, has been "on autistic subjects, not with autistic people" (Raymaker & Nicolaidis, 2013, p. 183).

This has clearly been one of the driving factors behind the increasing move towards participatory methods in autism research.

The three points discussed above also go some way to answering the question "What's so critical about Critical Autism Studies?" (Orsini & Davidson, 2013, p. 9). The editors go on to note that while contributors to the volume "are cognizant of autism's complex history and take seriously the situated knowledge of autistic people themselves" (p. 9) and while a commitment to challenging deficit-based approaches is a point of convergence within the volume, an understanding of what that might mean and how it is best accomplished will vary. The question of what is critical about CAS has continued to be a point of contention amongst scholars in the field, starting with the autistic-led branch.

The Autistic-led Branch

In 2012, autistic academic and advocate Larry Arnold published the first edition of *Autonomy, the Critical Journal of Interdisciplinary Autism Studies* (Arnold, 2012). More recently, Arnold (2020) explained the choice of name:

> ...not only does it encapsulate what the thrust of my thinking had been, agency and autonomy for autistic people, it contains the same 'Aut' root of Autism, which has been used before for organizations and events such as Autreat, Autscape, and Autreach. Of its subtitle (the Critical Journal of Interdisciplinary Autism Studies), 'Critical' was to embody a questioning and examining of the prevailing paradigms of autism research. 'Interdisciplinary' indicated that it wished to include contributions from a variety of academic fields.

Autonomy aimed to publish peer-reviewed articles mainly written by autistic contributors and to highlight important autistic events and texts such as those written by advocate Jim Sinclair (1993, 2013), indicating—as had Orsini and Davidson in 2010—the contribution of social movements to the nascent field of CAS. Just as "Critical Disability Studies...emerged from the disability movement, driven primarily by disabled people and used as a means of critiquing the various ways in which systems and structures act to exclude them from work and society" (MacLeod, 2019, p. 51), something analogous is true of CAS and is most firmly asserted from within this branch. For example, in their article *Redefining Critical Autism Studies: A more inclusive interpretation*, the authors (autistic scholars Larry Arnold, Steve Graby, Damian Milton, and Richard Woods) note "the historic fact of autistic leadership of CAS" (Woods et al., 2018, p. 976). These authors put forward a definition of CAS proposed by Waltz (2014, p. 1337), acknowledging its proximity to that proposed in Orsini and Davidson's introduction to *Worlds of Autism*:

> The 'criticality' comes from investigating power dynamics that operate in Discourses around autism, questioning deficit-based definitions of autism, and

> being willing to consider the ways in which biology and culture intersect to produce 'disability.'

While the neurodiversity paradigm is a principal tributary of CAS generally, this is most put to the fore by the autistic-led branch is, as Woods and colleagues (2018, p. 974) clearly indicate in the first lines of their article:

> Critical Autism Studies has come a long way since Jim Sinclair's…seminal speech with autistic activist-academics critiquing the deficit-laden, pathologising autism discourses favoured by the medical community.

It is a small step from these points to the central tenets of CAS proposed in *Worlds of Autism* and cited above: the attention to power relations, the search for new, enabling narratives, and the injunction to study the nature and culture of autism all find an echo in Neurodiversity Studies.

In summary, the autistic-led branch shares similar definitions and values as the *Worlds of Autism* branch. However, Woods and colleagues add that a central feature of CAS is that it "is literally critical, as it is a turning point, the pivotal juncture when autistic persons reclaim autism narratives for ourselves" (2018, p. 977). This can be compared to the criticality laid out in *Worlds of Autism*, which Orsini and Davidson (2013, p. 15) state "derives also from an active engagement with autistic communities." However, while for the autistic-led branch, 'autistic persons' are squarely centred and indeed arguably stand as sole legitimate arbiters of CAS (see Guest, 2020 for discussion), within the *Worlds of Autism* branch the neurostatus of the scholars actively engaging with autistic communities is left undefined. In contrast to both these branches, the Re-Thinking Autism network branch seeks to completely redefine the interest and scope of Critical Autism Studies.

The Re-Thinking Autism Network Branch

The autistic authors of *Redefining Critical Autism Studies: A more inclusive interpretation* wrote that article partly in response to an edited collection, *Re-thinking Autism: Diagnosis, Identity, and Equality* (Runswick-Cole, Mallett & Timimi, 2016a, 2016b, 2016c). "We have rigorously debated," Woods and colleagues (2018, p. 976) considered their:

> appropriate response, as autistic authors, to the RAN [Re-Thinking Autism Network). Options considered included splitting from CAS to form another discipline, such as 'Transformative Autism Studies' or 'Emancipatory Autism Studies', or insisting that autistic authors should be counted as key authors in CAS. Either of these options could theoretically lead to inclusive academic practices as part of the Disability Studies literature, the latter in particular meaning that CAS articles would need to be inclusive of the autistic voice and to filter out non-emancipatory CAS scholarship. However, we discarded the former option because it is undesirable to surrender CAS to purely non-autistic perspectives.

What was it in this new volume that produced such a strong reaction from the autistic-led branch—to the extent that some of them considered abandoning a framework in which they consider themselves and their theoretical ancestors the initiators?

Far from *Worlds of Autism*, looking to "interrogate rather than define or delimit the boundaries of this emerging field of study" (Orsini & Davidson, 2013, p. 10), *Re-Thinking Autism* seemed to be redefining the contours of CAS. It is something of a jarring first note struck in the opening lines of *Re-Thinking Autism*, where the editors claim to have produced the first edited collection in the "previously inexistent" field of CAS (Runswick-Cole, Mallett & Timimi, 2016a, 2016b, 2016c, p. 7). This is of course demonstrably and blatantly not the case, as discussed above. However, this incoherence is managed by redefining the field, and the editors suggest a highly specific and constricted CAS definition constituting just two principal components: questioning whether a diagnosis of autism is scientifically valid, and whether it is meaningful to those labelled as such.

Given that, as discussed above, CAS was far from 'inexistent' in 2016, it is difficult to see this as other than provocative. While it was perhaps not intentionally divisive, the editors could not be unaware that:

> Autistic neurodiversity activists have defined Critical Autism Studies not in terms of being critical of autism's existence (unlike many non-autistic thinkers outside the movement), but of the power dynamics that marginalize autistic scholars, pathologize autism, and overlook social factors that contribute to disability in autistic people (Kapp, 2020, p. 9).

Many of the themes running through *Re-Thinking Autism* are similar to those found in *Worlds of Autism*: an understanding of autism as (to greater or lesser extents) a cultural production as much as a medical one; ways in which autism speaks to, challenges, and is challenged in turn by, what it means to be human. However, ultimately, all these discussions come full circle to the central tenet of the book in its conclusion (Runswick-Cole, Mallett & Timimi, 2016a, 2016b, 2016c, p. 318):

> Crucially, labelling and the diagnosis of autism has been consistently identified by authors within this text as acting in disabling, rather than enabling, ways in the lives of people labelled with autism.

Such a broad statement ignores research that consistently notes the number of ways in which (notably adult) diagnosis is identified as holding considerable benefit by and for those who seek it (Bertilsdotter Rosqvist, 2012; Hickey et al., 2018; Huang et al., 2020; Jackson-Perry, 2023). That *Re-Thinking Autism* inclines to the view of an autism diagnosis as necessarily and consistently disabling may to some extent be explained by the fact that a high proportion of contributors to *Re-Thinking Autism* are involved, clinically and/or research-wise, in child services, and that the question of childhood diagnosis is indeed fraught with tension (see Goodley, 2014 for discussion, for a similar discussion in the context of ADHD see Chapter 3, this volume). However, explicitly redefining and reducing CAS to a discussion of the (in)validity of diagnosis, redefines and reclaims an entire field, implicitly negating the work of many scholars who were defined (by themselves or others) as working within it, which appears somewhat disingenuous in such a sensitive field.

Back to the Future?

As a field, CAS has had something of a bumpy ride, and certainly for me has not always made for a comfortable theoretical home. Debates within CAS have sometimes been virulent and gone to the core of what CAS 'is' or 'should' include. Many CAS scholars, for example, are also proponents of participatory research (PR) methodologies: indeed, PR is a feature, a logical extension of the precepts laid out in *Worlds of Autism* (Davidson & Orsini, 2013; Raymaker & Nicolaidis, 2013) and is positioned as central to CAS in the introduction to a later volume (Milton & Ryan, 2023, discussed further below). However, while acknowledging that "the term 'Critical Autism Studies' may still resonate for those activists and researchers interested in foregrounding perspectives that situate autism outside of pathologizing frameworks," the initiator of the Ottawa Workshop and co-editor of *Worlds of Autism*, Michael Orsini recently (2022, p. 5) seemed to call into question participatory practices, cautioned that the "sudden interest in bringing autistic voices or perspectives to the table may reproduce what many have decried in the past: participation or engagement for the sake of crafting a thin veneer of legitimacy." Participatory research is core to much of CAS (Woods et al., 2018), and this challenges to PR therefore conceivably something of a body blow. However, it should perhaps be read more as an incitement to keep a critical eye on how and why PR is carried out rather than an outright rejection of its underlying principles or potential usefulness (this is certainly the case in Chapter 22, this volume).

The publication of *Re-Thinking Autism* (Runswick-Cole, Mallett & Timimi, 2016a, 2016b, 2016c) and the response by Woods and colleagues (2018) described above perhaps represented a more existential threat to CAS than Orsini's questioning. Around 2016–2018, the bad feeling between different branches of CAS appeared to have reached a point of no return. However, far from the implosion that might have been expected, there has since been a flurry of CAS activity, with, for example, the publication of *The Routledge International Handbook of Critical Autism Studies* (Milton & Ryan, 2022). In the introduction, the editors specify that a "key aim" of the volume was to "coalesce the split sub-cultures within Critical Autism Studies by including contributions from a wide range of perspectives" (Ryan & Milton, 2022, p. 5). While the lead editor, Damian Milton, had been one of the co-authors of the 'autistic-led' response to *Re-Thinking Autism*, the

Routledge handbook includes chapters from editors and contributors to the latter.

In their introduction to the handbook, Ryan and Milton (2023) also anchor CAS to its early autistic activist roots. Through contributions, the editors note that they sought to give "activism a legitimised voice, a level of respect that activism is often not granted" (p. 5). This recalls, as does Sonya Freeman Loftis in a recent special edition on CAS in *Ought: The Journal of Autistic Culture* (2023, p. 11), that CAS has always been "an ideological space in which traditional scholarship and contemporary activism meet, sometimes productively and sometimes uncomfortably." With CAS, academia and activism more or less explicitly interact with, feed off, and play with each other, similarly to queer theory and HIV activism (Halperin, 1995). In the same edition, Alyssia Hillary Zisk (2023, p. 18) also underlines these blurry lines, noting the importance—and legitimacy—of resources beyond academia for autism research and providing an annotated list (see also chapter 19, this volume). Zisk looks to Mitzi Waltz' earlier (2014) definition of criticality that invokes investigating the power dynamics evident in autism research, troubling deficit readings, and considering how biology and culture converge to "produce 'disability'?" to remind us that:

> Disability activists in general, and autistic activists associated with the neurodiversity movement and/or paradigm specifically, do these things whether or not we are also academics.

This insistence on recalling and reiterating the importance of activism within CAS is recurrent in the special edition of *Ought*. Cansu Elmadagli (2023, p. 75), noting the risks of tokenism and "colonization" of autistic epistemic spaces, suggests that "limiting the influence of non-autistic scholars within CAS might have to be considered." While Elmadagli explicitly refutes an "us and them" approach, this is perhaps hard to avoid with such a proposition. It is also, of course, difficult to know how this limiting could be carried out, or by whom, and as discussed elsewhere (see Chapters 1 And 26, this volume), we prefer to think of elevating and putting to the fore neurodivergent scholars rather than limiting others. Orsini (2022, pp. 6–7) is understandably extremely wary of "the accumulation of autistic knowledge…being mobilized in the service of non-autistic knowers," worrying that this is itself a form of "epistemic exploitation," leaving the burden of educating the latter on the shoulders of the former. He goes on to say that while "(R)eplacing knowledge about autism by non-autistic knowers with knowledge advanced by autistic knowers is not inherently harmful," it might "promote or, at the very least, reinforce, a liberal fantasy of authentic or meaningful participation for those who have won the inclusion lottery." This is certainly pertinent regarding participatory or inclusive methods in the current climate, where considerable care is needed (see Chapter 22, 24, this volume for discussion). However, neurodivergent academics and activists currently appear to a certain degree to be in step here, be it within ADHD studies, CAS, or NDS. We are perhaps at a moment in time when the line between activism and the academy, never completely clear, is increasingly becoming both blurred and legitimized, and indeed Raymaker (2020, p. 138), for example, describes her work as "scientific activism."

Concluding Remarks: Neurodiversity Studies and Critical Autism Studies: in Competition or Walking Together?

NDS and CAS share much, from their roots to their wide-ranging subjects of enquiry, covering art (Moraru, 2023), politics (Chapman, 2023), literary, film, and cultural criticism (Krazinski, 2023; Murray, 2012; Stenning, 2024; Waltz, 2008), knowledge production (Gillespie-Lynch et al., 2017; Kourti, 2021; Milton, 2014)...the list is probably endless. However, I conclude with the question of inclusivity/exclusivity, which today appears to be (re)surfacing as powerfully—and certainly as critically—as ever, in both NDS and CAS.

It might be expected that NDS, definitionally including neurodivergence of any type and sharing common activist and theoretical groundings as well as drawing on many of the ambitions of CAS, would render the latter unnecessary, or be in competition with it. Contrary, in all honesty, to my expectations, co-editing or writing this and other chapters in the current volume have buried this idea for me. Far from imploding, CAS seems to have come through its skirmishes reinvigorated, with recent major publications taking on the mantle. CAS and NDS appear synergetic rather than in competition, and CAS clearly has its own *raison d'être*. Loftis (2023, p. 12) notes that CAS, differently to NDS, is a:

> field specifically for rethinking autism, in particular, as a disability that has drawn public attention and controversy, as well as a field of study that has drawn a large number of autistic scholars and activists who want to share in the production of knowledge about autism and autistic experience.

That NDS and CAS can coexist, bleed, and feed into each other seems self-evident. Indeed, they are joined by far more than that which divides them, starting with the question of a wary inclusion of non-autistic and neurotypical contributors. In the previous chapter, we noted that we wished to turn the notion of inclusivity 'on its head': Hanna Bertilsdotter Rosqvist and I have written critically about the notion of inclusion (see Chapters 1 and 26, this volume) after reading and reviewing the contributions to the current volume, suggesting that NDS be conditionally inclusive of neurotypical researchers. CAS scholars are suggesting "limiting" non-autistic participation (Elmadagli, 2023, p. 75). Dyi Dieuwertje Huijg (2024) is organizing a Special Edition for the Canadian Journal of Disability Research, *Critical and Intersectional ADHD Thought: ADHDers Think Back* ("only for ADHDers this time").

These moves are in similar directions, in step: we apparently still—again?—feel the need for spaces 'by us and for us.' This is categorically not to say that work by neurotypical or non-autistic scholars is not valid, or that scholars of another neurotype cannot produce important or relevant research around autism or neurodivergence. As an ADHD academic, I have often felt considerable discomfort working in autism research (see Chapter 20, this volume). I now feel that this discomfort is an important part of the toolbox for researchers studying other neurotypes, encouraging in-depth community engagement, and self-questioning: we should stay with that discomfort, use it creatively, productively, and humbly (see Chapter 22, this volume). Nor is this to reject autism or neurodiversity research coming from outside critical frameworks such as CAS or NDS: on the contrary, they are to be engaged with, for example "for autistic researchers to critique work by non-autistic authors by re-analysing the data from the autistic perspective to come up with alternative explanations" (Guest, 2020, p. 159). Although I cannot know if this is what Elmadagli (2023, p. 75) meant for CAS, our call for conditional inclusivity of neurotypical researchers in the introductory and concluding chapters of the current volume is above all to resist the notion of neurodivergent people *being included*, maintaining the status quo whereby neurotypical researchers are generally the ones *doing* the including. It is, in a sense, a call for a temporary strategic step too far, necessary to turn the notion of inclusivity on its head as we aim for more ethically and epistemologically just research in the field of neurodiversity.

References

Arnold, L. (2012). Introduction. *Autonomy, the Critical Journal of Interdisciplinary Autism Studies*, 1(1).

Arnold, L. (2020). Autonomy, the critical journal of interdisciplinary autism studies. In S. K. Kapp (Ed.), *Autistic community and the neurodiversity movement: Stories from the frontline* (pp. 211–220). Springer.

Bertilsdotter Rosqvist, H. (2012). Normal for An Asperger: Notions of the meanings of diagnoses among adults with Asperger syndrome. *Intellectual and Developmental Disabilities, 50*(2), 120–128.

Bertilsdotter Rosqvist, H., & Jackson-Perry, D. (2021). Not doing it properly? (Re) producing and resisting knowledge through narratives of autistic sexualities. *Sexuality and Disability, 39*(2), 327–344.

Botha, M., Chapman, R., Giwa Onaiwu, M., Kapp, S. K., Stannard Ashley, A., & Walker, N. (2024). The neurodiversity concept was developed collectively: An overdue correction on the origins of neurodiversity theory. Autism, 13623613241237871.

Chapman, R. (2023). *Empire of normality neurodiversity and capitalism*. Pluto Press.

Davidson, J., & Orsini, M. (2010). The place of emotions in critical autism studies. *Emotion, Space and Society, 2*(3), 131–133.

Davidson, J., & Orsini, M. (Eds.). (2013). *Worlds of autism: Across the spectrum of neurological difference*. University of Minnesota Press.

Dieuwertje Huijg, D. (2024). Dr. Dyi Dieuwertje Huijg. https://ddhuijg.com/, https://ddhuijg.com/

Dinishak, J. (2016). The deficit view and its critics. *Disability Studies Quarterly*, *36*(4).

Douglas, P., Rice, C., Runswick-Cole, K., Easton, A., Gibson, M. F., Gruson-Wood, J., Klar, E., & Shields, R. (2021). Re-storying autism: A body becoming disability studies in education approach. *International Journal of Inclusive Education, 25*(5), 605–622.

Elmadagli, C. (2023). The future of critical autism studies (CAS): Thinking through critical discourse studies and postcolonial feminism. *Ought: The Journal of Autistic Culture*, *5*(1), Article 11.

Fletcher-Watson, S., Adams, J., Brook, K., Charman, T., Crane, L., Cusack, J., Leekam, S., Milton, D., Parr, J. R., & Pellicano, E. (2019). Making the future together: Shaping autism research through meaningful participation. *Autism, 23*(4), 943–953.

Freeman Loftis, S. (2023). Critical autism studies: The state of the field. *Ought: The Journal of Autistic Culture*, *5*(1), Article 5.

Gillespie-Lynch, K., Kapp, S. K., Brooks, P. J., Pickens, J., & Schwartzman, B. (2017). Whose expertise is it? Evidence for autistic adults as critical autism experts. *Frontiers in Psychology, 8*, 438.

Goodley, D. (2014). *Dis/ability studies: Theorising disablism and ableism*. Routledge.

Goodley, D. (2017). *Disability studies: An interdisciplinary introduction* (2nd ed.). SAGE Publications Ltd.

Guest, E. (2020). Autism from different points of view: Two sides of the same coin. *Disability and Society, 35*(1), 156–162.

Halperin, D. (1995). *Saint Foucault*. Oxford University Press.

Hickey, A., Crabtree, J., & Stott, J. (2018). 'Suddenly the first fifty years of my life made sense': Experiences of older people with autism. *Autism, 22*(3), 357–367.

Huang, Y., Arnold, S. R., Foley, K. R., & Trollor, J. N. (2020). Diagnosis of autism in adulthood: A scoping review. *Autism, 24*(6), 1311–1327.

Jackson-Perry, D. (2023). Exploring autistic accounts of sexuality, intimacy, and authenticity (Doctoral dissertation, Queen's University Belfast). Available at: https://pure.qub.ac.uk/en/studentTheses/exploring-autistic-accounts-of-sexuality-intimacy-and-authenticit

Kapp, S. K. (2020). Autistic community and the neurodiversity movement: Stories from the frontline (p. 330). Springer Nature.

Kourti, M. (2021). A critical realist approach on autism: Ontological and epistemological implications for knowledge production in autism research. *Frontiers in Psychology, 12*, 713423.

Krazinski, M. (2023). Sharp stick grasps at autistic women's liminal vulnerability. *Ought: The Journal of Autistic Culture*, *5*(1), Article 7.

MacLeod, A., Lewis, A., & Robertson, C. (2014). "CHARLIE: PLEASE RESPOND!" Using a participatory methodology with individuals on the autism spectrum. *International Journal of Research & Method in Education, 37*(4), 407–420.

McGuire, A. (2016). *War on autism: On the cultural logic of normative violence*. University of Michigan Press.

Milton, D. (2014). Autistic expertise: A critical reflection on the production of knowledge in autism studies. *Autism: The International Journal of Research and Practice*, *18*(7), 794–802.

Milton, D., & Ryan, S. (Eds.). (2022). The Routledge International Handbook of Critical Autism Studies. Taylor and Francis.

Moraru, C. (2023). The cultural category of neurodiversity: Art spaces for creating identities and understanding neurominorities. *Espace, 133*, 6–16.

Murray, S. (2012). *Autism*. Routledge.

O'Dell, L., Bertilsdotter Rosqvist, H., Ortega, F., Brownlow, C., & Orsini, M. (2016). Critical autism studies: Exploring epistemic dialogues and intersections, challenging dominant understandings of autism. *Disability and Society, 31*(2), 166–179.

Orsini, M., & Davidson, J. (2013). Introduction. In J. Davidson, & M. Orsini (Eds.) *Worlds of autism*. Minneapolis: University of Minnesota Press.

Orsini, M. (2022). Who needs to (un) know? On the generative possibilities of ignorance for autistic futures. *International Journal of Qualitative Studies in Education*, 1–18.

Pukki, H., Bettin, J., Outlaw, A. G., Hennessy, J., Brook, K., Dekker, M., Shaw, S. C., Bervoets, J., Rudolph, S., Corneloup, T., & Yoon, W. H. (2022). Autistic perspectives on the future of clinical autism research. *Autism in Adulthood, 4*(2), 93–101.

Raymaker, D., & Nicolaidis, C. (2013). Participatory research with autistic communities: Shifting the system. In *Worlds of autism: Across the spectrum of neurological difference* (pp. 169–188). University of Minnesota Press.

Raymaker, D. M. (2020). Shifting the system: AASPIRE and the loom of science and activism. Autistic community and the neurodiversity movement: Stories from the frontline, 133–145.

Runswick-Cole, K., Mallet, R., & Timimi, Sa. (2016). Introduction. In *Re-thinking autism: Diagnosis, identity and equality* (pp. 7–15). Jessica Kingsley Publishers.

Runswick-Cole, Katherine, Mallet, Rebecca, & Timimi, Sami. (2016). Conclusion. In *Re-thinking autism: Diagnosis, identity and equality* (pp. 317–319). Jessica Kingsley Publishers.

Runswick-Cole, K., Mallett, R., & Timimi, S. (Eds.). (2016a). *Re-Thinking Autism: Diagnosis*. Jessica Kingsley Publishers.

Ryan, S., & Milton, D. (2022). *Critical autism studies: An introduction*. Routledge.

Sinclair, J. (1993). Don't Mourn For Us, Autreat.com. Available at: https://www.autreat.com/dont_mourn.html.

Sinclair, N. (2005). History of ANI. Autism Network International. https://www.autismnetworkinternational.org/History_of_ANI.html

Sinclair, J. (2013). Why I dislike "person first" language. *Autonomy, the Critical Journal of Interdisciplinary Autism Studies*, 1(2).

Stenning, A. (2024). *Narrating the many autisms: Identity*. Taylor and Francis.

Timimi, S., & McCabe, B. (2016). Autism screening and diagnostic tools. In K. Runswick-Cole, R. Mallett, & S. Timimi (Eds.) Re-thinking autism: Diagnosis, identity and equality (pp. XX). Jessica Kingsley Publishers.

Tisoncik, L. A. (2020). Autistics. org and finding our voices as an activist movement. Autistic community and the neurodiversity movement: Stories from the frontline, 65–76.

Waltz, M. (2008). Autism = Death: The social and medical impact of a catastrophic medical model of autistic spectrum disorders. *Journal of Popular Narrative Media, 1*(1), 13–24.

Waltz, M. (2014). Worlds of autism: Across the spectrum of neurological difference. *Disability and Society, 29*(8), 1337–1338.

Woods, R., & Waltz, M. (2019). The strength of autistic expertise and its implications for autism knowledge production: A response to Damian Milton. *Autonomy, the Critical Journal of Interdisciplinary Autism Studies*, *1*(6).

Woods, R., Milton, D., Arnold, L., & Graby, S. (2018). Redefining critical autism studies: A more inclusive interpretation. *Disability and Society, 33*(6), 974–979.

Yergeau, M. (2018) *Authoring autism.* Duke University Press.

Zisk, A. H. (2023). Critical autism studies beyond academia: An annotated list. *Ought: The Journal of Autistic Culture*, *5*(1), Article 6.

CHAPTER 3

An Introduction to Critical ADHD Studies

Andrew Ivan Brown, Hanna Bertilsdotter Rosqvist, and David Jackson-Perry

Introduction

What can be seen as a nonlinear, intertangled mass of alternative approaches challenging traditional medical models on ADHD has both similarities and differences to the debates within Critical Autism Studies (CAS: see Chapter 2, this volume). In the following, we[1] set the scene for alternative approaches to ADHD by first describing two principal backstories. These consist of a story of the pathology paradigm of ADHD on the one hand, and its main opponent on the other, what could be referred to as the diagnosis critical paradigm. The latter bears a clear resemblance to the Re-Thinking Autism network branch, including its primary concern with childhood diagnosis and the implicit biases of this line of thought (see Chapter 2, this volume), and may be viewed as sharing medical sociological origins (c.f. Brown, 2023; Conrad, 2006).

[1] The 'we' in this chapter consists of people with a shared experience of ADHD. Some of us have a formal ADHD diagnosis (ADHDer). Some of us are autists but also acknowledge their ADHD traits (AuDHDer).

A. I. Brown (✉)
University of Washington, Seattle, Washington, United States
e-mail: aibrown@uw.edu

H. Bertilsdotter Rosqvist
School of Social Sciences, Södertörn University, Huddinge, Sweden

D. Jackson-Perry
Queen's University, Belfast, UK

H. Bertilsdotter Rosqvist and D. Jackson-Perry (eds.), *The Palgrave Handbook of Research Methods and Ethics in Neurodiversity Studies*,
https://doi.org/10.1007/978-3-031-66127-3_3

We then consider how we perceive what we for now refer to as the emerging field of Critical ADHD Studies (CADS), a term coined by Dieuwertje Huijg (2021, n.p.). We suggest three closely related streams of more or less emergent alternative approaches to ADHD, all of which have considerable overlap, including in their perceptions of the limitations of the pathology and diagnosis critical paradigms: strength-based approaches to ADHD; cognitive difference approaches to ADHD; and CADS phenomenological and community theorizing approaches. We see these streams not as being completely discrete, but as informing, bleeding into each other, and as holding a variety of possibilities to re-story ADHD discourse. Without seeking to narrowly define what is and what is not CADS, we think of these streams as sharing certain ambitions, as holding potential for allyship despite sometimes disparate visions of what ADHD 'is.' This is therefore not intended as a gatekeeping exercise. Rather, we propose a tentative definitional tool we find useful in the process of mapping out alternative approaches to ADHD. Whether or not they explicitly build upon neurodiversity approaches or the cognitive difference approach to neurodivergence, all seek to centre ADHDers as agents in varying ways.

The Pathology Paradigm of ADHD and Diagnosis Critical Paradigm of ADHD

We begin this section by briefly distinguishing at least three ways in which 'mental disorders' are taken up in the pathology paradigm: diagnostic criteria; psychopathological models; and research on neurobiological or genetic biomarkers or aetiologies. We also briefly touch on some philosophy of science that has been used to clarify how notions of 'truth' and 'existence' operate in the pathology paradigm (c.f. Brown, 2023).

Diagnostic criteria are grounded in statistical symptom clusters and are designed to assist with diagnostic practices (Lovett & Harrison, 2021). In the case of ADHD, these practices are both relatively homogeneous, insofar as they follow standardized criteria found in the DSM-5-TR or ICD-11 (depending on country), and heterogeneous, insofar as different geographical regions or institutions (schools, health facilities, specialized 'ADHD clinics,' etc.) employ varying techniques to assist with diagnosis (for detailed examples of these practices, see Brown, 2023).

In contrast to diagnostic practices, psychopathological models or theories of ADHD—such as executive dysfunction models—often have as their primary aim the development of better therapeutic interventions. Although these models usually do not intend to capture the 'truth' about what the condition is, they nevertheless become imbued with 'truth value' when said interventions are proven to be successful in scientific trials (Hacking, 1983). These statistical and psychological models also underpin neuroscientific and genetic research on biomarkers and/or aetiologies, which are then "evoked as the final proof of the scientific validity of the psychological theories themselves" (De Vos, 2016: 35)—i.e. biological truths (Thome et al., 2012).

We now provide a brief overview of some psychopathological models of ADHD. Most of these models are associated with ongoing neuroscientific or genetic research (with no reliable biomarker for ADHD yet found), but for the sake of brevity and clarity we avoid discussing these models in neuroscientific or genetic terms. The literature on this topic is abundant, so this should be viewed as a selection illustrating what these sorts of pathologizing theories/perspectives look like. ADHD cognition and symptomatology are commonly defined through models of cognitive deficits, focusing on "externally observed manifestations rather than its experienced phenomenology" (Murray et al, 2023, p. 221, discussing autism) of these underlying deficits. Within this paradigm, an assumed "reduced functional connectivity" (Alexander & Farrelly, 2018, p. 238) in ADHD brains leads to dysfunction in executive function, reward processing, attention networks, and default networks (ibid, p. 237). Different models have been developed throughout the years on the neuropsychology of ADHD, the most dominant being the model of executive function deficits (Barkley, 1997) whereby ADHD has been constructed as a disorder of impaired executive functions (Brown, 2006).

The ADHD subtypes, inattention and hyperactivity, serve as the phenotype for ADHD (Mueller & Tomblin, 2012). Core deficits suggested are a "divided, focused or sustained attention deficit" (Kuntsi & Stevenson, 2000, p 17) and an association "between a slow inhibitory process and hyperactivity [that] may reflect a generally slow, variable in speed and inaccurate pattern of responding." Like autism, ADHD is associated with "multisensory abnormalities" (Panagiotidi et al., 2017, p. 11). In contrast to autism, associated with bottom-up processing, centring a detail-oriented cognitive style (Valla & Belmonte, 2013), distractibility in ADHD, it is suggested, "may stem from a failure to efficiently engage top-down control" (Friedman-Hill et al., 2010, p. 93).

Closely connected to cognitive processing deficits are time-related deficits, including time processing/estimation which have consequences for time management and planning abilities (Tobia et al., 2021). Nielsen (2017, p. 260) has described this "impairment in time" as "an embodied experience of being 'out of sync'" between the individual and its surroundings. Attentional deficits in ADHD are understood in terms of deficits in abilities to sustain attention, or the problem of selective attention which "presupposes, concomitantly, orientation and concentration directed towards a stimulus, ignoring or decreasing the emphasis on other concurrent stimuli" (dos Santos Assef et al., 2007, p. 34). Gallagher and Blader (2001, p. 148) have stressed a combination, whereby people "with ADHD may have a particular profile of slowed performance in tasks of sustained attention and set-shifting, and their use of working memory may be particularly impaired."

Social and educational sciences and professions closely align with this cognitive deficit approach, what Prosser (2015, p. 597) has referred to as "psycho-medical discourses" on ADHD, and are commonly represented as auxiliary disciplines to neuro- and psychological scientific disciplines and health professions. The tasks of educational professionals in this case are to discover early signs of atypical cognitive development and, following diagnosis, to carry out social and educational intervention and support informed by cognitive deficit approaches to ADHD (Österborg Wiklund et al, 2023).

In opposition to the pathology paradigm, a stream within social sciences, particularly medical sociology, as well as some high-profile pushback from psychiatrists (Frances, 2013; Timimi et al., 2004), underlines the importance of critical perspectives on ADHD. Within the diagnosis critical paradigm of ADHD, the meaning of 'critical' is in line with the Re-Thinking Autism network branch of CAS, questioning whether a diagnosis of ADHD is scientifically valid and whether it is meaningful to those so labelled. The critical perspectives on ADHD in this context are informed by theoretical approaches from medical sociology emerging as far back as the mid-1970s (c.f. Conrad, 2006).

These perspectives constitute a "critical attack on the ADHD diagnosis" (Foreman, 2018, p. 251). ADHD is represented as a product of a general medicalization process (Bianchi et al, 2016) or pharmacologization process in society (Frances, 2013), partly as a response to changes in education systems, or "medicalization of education" (Signor et al, 2017), or, previously, a crisis of social control in education (c.f. Brown, 2023). Researchers in this field have argued that ADHD diagnoses are a clear case of the medicalization of childhood, leading to an overdiagnosis and overtreatment of ADHD (Ortega & Müller, 2020). Timimi (2005, p. ix), for example, considers that the increase of 'so-called neuro-developmental diagnoses' and the corresponding increase in prescriptions of psychotropes is culturally based, representing 'the institutionalization of Western cultures' hostility towards children.

From this critical approach to diagnoses, ADHDers are commonly approached as objects or victims of medicalization and diagnoses are viewed as stigmatizing labels (Hjörne & Evaldsson, 2015). For example, Nilsson Sjöberg (2021, p. 243) has argued that a diagnosis can be seen as "a form of onto-epistemological violence," through which "the unique subject is forced into an object and a second-class citizen who undergoes instrumental techniques of behaviour modification." While these critics often ignore the agency of ADHDers, and view ADHD from the same deficit approach as the pathology paradigm discussed above, they are nevertheless, at least in some cases, writing from social justice and anti-oppression paradigms, such as antipsychiatry (Harwood, 2006) or mad studies (Joseph, 2021). Although they have in mind the dissolution of ADHD as a diagnosis and psychological category, their end goal is not to leave individuals to deal with their attention, hyperactive, and impulsive difficulties on their own, without social, institutional, or therapeutic support. Rather, these critiques often have alternatives to pathologizing diagnoses in mind, such as calling on psychology and psychiatry to start perceiving 'mental distress' in all its forms as a normal part of being human, with the aim of helping individuals when they need help, but all the while not pathologizing them with discrete labels or categories (Rose, 2018; for a similar approach in paediatrics, see Gillberg, 2010).

However, as mentioned, the diagnosis critical paradigm also largely relies on a deficit approach to ADHD, stressing diagnosis of ADHD as a problem—either for society or for the individual and their surroundings, as within the pathology paradigm. They do not consider how ADHD-affirmative perspectives might also provide solutions to many of the same issues they identify. Having broadly laid out the paradigms most concerned with the question of deficit, we now take a closer look at alternative approaches critical to the above, which we consider to move towards CADS as we understand it.

Strength-Based Approaches to ADHD

As a way of opposing one-sided perspectives of the two backstories described previously, some researchers within social sciences have moved to strength-based approaches to ADHD. From these approaches, those who have been diagnosed with ADHD are considered to have some agency through an emphasis on their strengths as individuals, or by emphasizing the strengths their ADHD can bring to the table in certain contexts. ADHDers are also perceived as agentic in terms of their ability to reflexively engage with different contextual ideas of ADHD in their sense making of themselves and their ADHD (Fleischmann & Miller, 2013; Nielsen, 2018). A strengths-based approach to ADHD is also central in ADHD advocacy movements and considerations of the epistemic legitimacy of experience-based expertise in contrast to (or at least in addition to) scientific expertise (Rabeharisoa et al., 2014; Versteeg et al., 2018). Indeed, strength-based approaches have been widespread in ADHD self-advocacy and self-help literature, reaching back as far as the early 1990s (examples include Hallowell & Ratey, 1994/2011; and many more).

Illustrating one version of a strength-based approach, where individuals are reminded that they are 'more than just their diagnosis,' Vehmas has argued that while ADHD's "underpinning neuro-cognitive impairment affects one's cognitive, emotional and moral conduct," moral responsibility also depends "on social and relational factors" (Vehmas, 2011, p. 156). Illustrating another version of a strength-based approach, from an interactionist lens, Prosser (2015) considers ADHD identities from the perspective of "identity development as a dialectical process of interaction between individuals and social constructions." This, he argues, "can open up new opportunities to see the creation, recreation, appropriation and evolution of ADHD identities" (Prosser, 2015, p. 603). In this version, ADHD is seen as providing strengths to individuals in certain contexts. For example, Lasky et al., (2016, p. 160) describe ADHD symptomology as "context-dependent," noting that in some contexts, "participants report feeling better able to focus; in others, their symptoms -such as high energy levels-become strengths rather than liabilities" (c.f also Bertilsdotter Rosqvist et al., 2023a).

The broader sea change towards strength-based approaches has been associated with disability activism following the biosocial model of disability. This model draws on Rabinow's concept of 'biosocial' (1996), as a:

> process by which biology, particularly biological similarity, becomes a focus for community, association and collective action. In this context one can speak of biological or perhaps even, in some cases, genetic citizenship (Roseand Novas 2004). (Hughes, 2009, p. 678)

Aligned with the biological citizenship approach is research situating ADHD as divergent thinking and variable attention (Hallowell & Ratey, 2022). From this cognitive difference approach, the centrality of hyperfocus (or deep flow and interest-based attention and motivation) has been stressed. From this approach, selective attention (a neurotypical norm naturalized in psychological discourse as a cognitive capacity universal to all humans) is decentred so that alternative, and equally capable, neurodivergent forms of attention are recognized (see Brown, 2023, pp. 245–251, 254–255 for an example of such decentring).

Within this paradigm, researchers have explored an association between ADHD and creativity (Healey & Rucklidge, 2006) or innovative thinking (White & Shah, 2016). In line with this, Hupfeld et al., (2019, p. 191) note that "adults with ADHD often report episodes of long-lasting, highly focused attention." Summarizing previous research, Hupfeld et al. (2019), illustrating how these different approaches are far from discrete among themselves, stress the multi-dimensional aspects of hyperfocus (HF), that includes engagement in certain tasks in certain environments, raising similar questions over the role of environmental cues on ADHD behaviours, where "only certain situations induce HF" (Hupfeld et al., 2019, p. 192). Similar to autistic theorizing about flow state (McDonnell & Milton, 2014) or the theory of monotropism (Murray et al., 2005), both of which offer alternative, neurodivergent affirmative forms of attention, HF among people with ADHD is defined by Hupfeld et al., (2019, p. 192) as:

> a state of heightened, intense focus of any duration, which most likely occurs during activities related to one's school, hobbies, or 'screen time' (i.e. television, computer use, etc.); this state may include the following qualities: timelessness, failure to attend to the world, ignoring personal needs, difficulty stopping and switching tasks, feelings of total engrossment in the task, and feeling 'stuck' on small details.

What does all the above indicate for the potential development of Critical ADHD Studies?

Critical ADHD Studies (CADS)

A limitation of both the pathology paradigm of ADHD and the diagnosis critical paradigm of ADHD is what can be seen as a focus on ADHD rather than ADHDers, thereby ignoring their agency and potential as autonomous actors in—and credible witnesses of—their own lives. This is what autistic theorist Yergeau (2018, p. 10) has formulated as rhetorical involuntariness:

> It's important to note that whatever the placeholder—whether schizophrenia, autism, depression, cerebral palsy, ADHD, bipolar—mental disability signals a kind of rhetorical involuntariness.

In contrast to the rendering of rhetorical involuntariness of ADHDers, further following Yergeau, we stress the problem of a "medicalized storying of lack" (Yergeau, 2018, p. 7) and the possibilities for re-storying ADHD. Paraphrasing Yergeau (p. 25), this implies "methods for queering futures, for projecting [ADHD] desires and [ADHD] ideals."[2]

In 2021 (n.p.) Dieuwertje Huijg noted that:

> there is little development in 'the field of ADHD Studies' towards an engagement with the diversity and complexity of ADHD experiences and perspectives, and of epistemological, methodological and theoretical orientations.

Dieuwertje Huijg went on to call for "a field of Critical ADHD Studies, which is ADHD-affirmative, intersectional, and produced by ADHDers themselves." This is a call we take up and explore here.

One of the starting points of CADS, then, is to simultaneously reject the pathologization of ADHD while also taking seriously critiques of ADHD grounded in social justice and anti-oppression paradigms; in other words, to consider how affirmative ADHD perspectives can provide new solutions to these old problems. A strengths-based approach to ADHD (as outlined in the previous section) stresses the importance of ADHDers as knowers of their own lives and the importance of both listening to ADHD voices and acknowledging the contextuality of experiences of ADHD. However, even if the importance of the ADHD voice and experiential knowledge in ADHD research is stressed, strength-based approaches offer little concrete challenge to non-ADHD epistemic privilege. As such, they still often rely on outsider (e.g., educator, therapist, parent) or 'privileged ADHDers' (see below) approaches to ADHD, and often fail to consider valid social and ethical critiques raised by diagnosis critical perspectives.

CADS is particularly concerned—in line with the autistic-led and Worlds of Autism branches in CAS—with research where non-ADHD epistemic privilege

[2] Yergeau was referring to autism, and not ADHD.

is challenged, stressing ADHD as an epistemic standpoint holding experiential knowledge important to research (c.f, Sonuga-Barke, 2023, Sonuga-Barke et al., 2024).

Similar to the CAS autistic-led branch, the CADS stream moves the issue of agency and autonomy further; from the strength-based approaches to ADHD as described above, where mainly non-ADHD researchers seek to listen to ADHDers and situate them as subjects and agents in their own lives, into a new field of CADS. Research within CADS is critical towards neuronormativity (for example critiques of the psychological normalization and naturalization of selective attention) and is carried out with ADHD communities and led by ADHDers (Meadows, 2021; Dieuwertje Huijg, 2021). From this emerging ADHD epistemic standpoint, Meadows (2021, n.p.) proposes:

> more analysis of the power dynamics at work in the ADHD world, more criticism of the research that is presented to us as fact in the media, more answers that affirm and empower us without politically pacifying us.

Following the calls from Dieuwertje Huijg and Meadows, Bertilsdotter Rosqvist et al. (2023a, 2023b, 2023c), Basten (2023), and Brown (2023) are among scholars both pointing out limitations of current research on ADHD and setting out further contours of 'something else': CADS. From this approach, the importance of epistemic insider positions, in line with the disability studies approach of 'nothing about us, without us,' is stressed. This means a move in research and professional practice from talking *about* ADHD communities to talking *with* or *from* ADHD communities. Beyond the common ground of ADHD-led research and ADHD epistemic standpoints questioning epistemic violence and epistemic authority of non-ADHDers in ADHD research, researchers associated with CADS may develop their form of CADS from different conceptual toolboxes (see the conclusion to this chapter for some examples).

CADS of course overlaps with, and takes inspiration from, strength-based, cognitive difference approaches to ADHD and pays special attention to the questions of *why* ADHD-affirmative, intersectional, and ADHD epistemic standpoints are not only needed, but inseparable from one another. We provide brief answers to these questions here, not to put an end to the discussion, but to stress the need for future researchers to further consider these factors when doing CADS research.

The pathologization of ADHD goes beyond stigma relating to the label or diagnosis; it is a pathologization of ADHD-associates traits and behaviours, often generalized as inattention, hyperactivity, and impulsivity (see Brown, 2023 for a distinction between ADHD stigma and ADHD misrecognition). This latter pathologization becomes compounded in cases of racialized diagnoses (Reyes II, 2021), where, for example, Black or Indigenous youth who exhibit ADHD symptoms are frequently diagnosed not with ADHD but

Fetal-Alcohol Syndrome Disorder (FASD), thus effectively linking inattention, impulsivity, and hyperactivity to brain damage (Charlesworth, 2021; Ergun et al., 2021). Similarly, clinicians may have a tendency to diagnose Black children with conduct disorder but White children with ADHD (Harwood, 2006, p. 24). In such cases, certain marginalized groups are excluded from accessing certain accommodations afforded to ADHDers, but also from whatever benefits are conferred via strength-based approaches. These issues are compounded further, as Huijg (2021) points out, due to the overwhelming whiteness of ADHD research. It is important not to let racialized diagnostic practices seep into CADS research through processes of inclusion/exclusion, making intersectionality an important part of CADS. Black, Indigenous, and other People of Colour (BIPOC) voices must be part of CADS research, as their communities have a very real stake in the practical outcomes of ADHD research and diagnostic practices. This is an area in which both CAS and NDS have thus far struggled (see Chapters 1, 2, this volume).

Intersectional approaches also include consideration of socio-economic backgrounds, and who has access to what resources or strategies. As Meadows (2021) notes, strength-based approaches to ADHD, especially in self-help literature, are often written from privileged ADHDer perspectives. Criticizing one of Hallowell's recent books, she writes,

> [the book] says, you're special, creative, among the ranks of geniuses and CEOs! You have a 'race car brain' and you just need to learn how to harness it so you can be a successful entrepreneur, too. He proposes that all we really need is love, connection, and better career choices, as if everyone has the option to choose where they work, to shape their own life.

Intersectional approaches to ADHD stress the need to move from individualized strength-based approaches (which are often inaccessible to BIPOC and lower socio-economic status ADHDers) to broader, collective challenges to non-ADHD epistemic privilege.

So, while strength-based approaches sometimes emphasize and centralize ADHD voices, and while neurodiversity approaches emphasize depathologization, CADS emphasizes a constant awareness of how challenging non-ADHD epistemic privilege goes hand-in-hand with intersectional approaches, and how depathologization and ADHD-affirmative world-making will remain incomplete as long as intersectional voices are ignored.

CADS Phenomenological and Community Theorizing Approach

As Brown (2023) has previously argued, ADHD-affirmative approaches require several 'starting points' that CADS could draw on. One such approach, we suggest, is the phenomenological and community theorizing approach.

One of the authors in Bertilsdotter Rosqvist et al., (2023b, p. 2) noted that "everything I had heard about ADHD before was impossible to relate to," and expressed the need to be "separated from the stereotype" and form ADHD subjectivities from ADHD standpoints. Lack of nuanced descriptions of the lived experience of ADHD does not stop with the urgency to gain legitimacy for experience-based knowledge within ADHD research, but further stresses the need for developing ADHD epistemic standpoints and new theories and methods for ADHD research (see Goetz & Adams, 2022; Chapters 11, 12, 23, this volume).

Bertilsdotter Rosqvist et al. (2023a, 2023b, 2023c) frame their research method as their ADHD collective, autoethnographic re-storying process, aiming at re-storying meanings of ADHD and forming an ADHD epistemic standpoint from their own embodied standpoints as ADHDers and AuDHDers. They situate their research within the neurodiversity paradigm, aligned with a cognitive difference approach to ADHD, and building upon an emerging strand within Neurodiversity Studies; phenomenological inquiries into neurodivergence (Murray et al., 2023; also see Chapters 14,15, this volume), here with a particular focus on ADHD bodyminds.

In contrast to approaches exploring (and defining) neurodivergencies based on social and cognitive expressions or external manifestations (Murray et al., 2023), theorists within this field suggest an exploration of neurodivergence and developments of neurodivergent theorizing based on what Fergus Murray (2018, n.p.) has referred to as theorizing "drawn at least in part from internal observations," or "experienced phenomenology." Bertilsdotter Rosqvist et al. (2023a, 2023b, 2023c), firstly, stress the importance of a combination of ADHD community theorizing (theories developed by ADHDers within the ADHD community) exemplified by new theoretical models based upon the lived experience of ADHD, such as Dodson´s (2022) "interest-based nervous system" and Hallowell and Ratey's (2022) Variable Attention Stimulus Trait (VAST) and echoing theoretical concepts from autistic-led theorizing/ concepts within autism studies (for example monotropism, Murray et al., 2005). Secondly, informed by the collective storytelling method of Jackson-Perry et al. (2020), the importance of collectivity in ADHD knowledge production is stressed. In summary, a CADS phenomenological and community theorizing approach develops alternative understandings of ADHD in the context of collective ADHDer spaces, aimed at steering the research field of ADHD more generally in a neuroaffirmative direction, and producing knowledge of ADHD which is possible for ADHDers themselves to produce and relate to.

Bertilsdotter Rosqvist et al. (2023a, 2023b, 2023c) also combine neuroqueer theorizing with CADS, stressing the possibilities of a CADS lens in research exploring neurotypicality and neuronormativity (see also Dieuwertje Huijg, 2020). Following Yergeau (2018,p. 181) arguing for the need for "interbodily inventions" in autism studies, whereby "embodied communication is not a site for intervention, as many clinicians would have us believe, but is rather a site of invention," this approach proposes ADHDers' embodied experiences of ADHD as a site of invention rather than intervention.

In CAS, there are calls to 'unknow' what is assumed to be true about autism (see Orsini, 2022; see also Chapter 22, this volume), to move away from assumptions of deficit and towards more epistemically and ethically just approaches to conditions that have largely fallen under the remit of psychiatric or medical disciplines. This is equally important in the field of ADHD and requires both re-storying what is seen as "the problem," but also troubling notions of who is legitimate to narrate those stories (c.f. Bertilsdotter Rosqvist & Nygren, 2023). Similarly, Basten (2023, p. 90) notes that discussion of ADHD "outside of neurodivergent communities can be daunting," as the "name itself is misleading." Basten writes:

> Apart from the fact that ADHD is not a 'disorder' there is a lot to be said for the argument that 'it' is named after external presentations and not the experiences of ADHDers themselves. More importantly, the rhetorical insult of the misnomer can be taken as a pars pro toto for decades of research in favor of medical, individualistic models of disability, resulting in stigmatization and a heavily distorted public image of ADHD, as well as a complete disregard for the shared cultural experiences of ADHDness.

To this could be added the difficulty for ADHDers to project and position themselves as credible holders of knowledge, given the deficit-infused backstories described above. Concerning autism, Bertilsdotter Rosqvist and Jackson-Perry (2021, p. 341) wondered how—or if—it is possible for someone to separate their:

> 'lived experience' from that same person's 'epistemic infection' by a body of (collective) knowledge that defines their actions as being the product of deficit?

We suggest that this separation is essential for re-storying ADHD, and can be sought in a variety of ways: systematic unknowing of much existing literature (see Chapter 22, this volume); joining affirmative ADHD community spaces (c.f. Belek, 2023; see also Chapter 23, this volume); or accessing neurodivergent affirmative professionals and other allies (Jackson-Perry et al., 2020).

While Basten's earlier analysis focuses specifically on ADHD ways of reading, in her proposal of ADHDness based on the experiences of ADHDers themselves are echoes of many of the elements we have identified above as being integral elements of what we are calling CADS: a challenge to neuronormative notions, an allyship between CADS and neuroqueer or crip theorizing, and a questioning of who can legitimately contribute to ADHD storying.

Conclusion

Within CAS, much is made of what is critical about the different strands of Critical Autism Studies, leading to considerable tension within the field (see Chapter 2, this volume). We feel it is important to learn from this in conceptualizing what we are for now calling CADS, and to think about commonality, about how different approaches may to some extent pull together, all the while perhaps pulling in different directions and drawing on different starting points. We see more value in calling for a broad field that can include often disparate and sometimes contradictory positions, than in attempting to lay claims to a definitive 'owning' of CADS. We finish this chapter with some thoughts on what this pulling together might include or look like.

- Situated knowledge: We consider that the epistemic value of knowledge brought by ADHD bodyminds, the experience of ADHDness, must be central to deepen further understanding of ADHD, both to challenge 'knowledge' that is often narrowly confined within—and limited by—disciplinary silos and to produce alternative and affirmative discourses in the field.
- Intersectionality as a 'given': A paucity of intersectionality has been a feature of CAS (and NDS so far). CADS can learn from this and consider intersectionality as a necessary part of a framework seeking to challenge epistemic injustice and seek out depathologization and ADHD-affirmative world-making.
- Interdisciplinarity: We do not consider that the remit for 'who can talk about ADHD' should be limited to psychiatry or medical sociology, which can only stultify other lines of enquiry. Philosophy, queer and crip theory, postcolonial and feminist and materialist approaches to name but a few clearly have their place in discussion of any object considered to be pathological.
- Looking back at—and talking back to—the norm: Whatever else we consider ADHD to be, we consider it—similarly to autism—as holding the potential to reveal and question hegemonic behavioural norms and pressures, and the 'policing' of those norms, without *reducing* ADHD to a theoretical tool.
- Being comfortable with uncertainty: We seek to hold in tension that while social/normative structures may impact and even make possible a diagnosis of ADHD, a 'something' exists that we would recognize as ADHD ways of being, as "a different form of life" as Chapman (2019, p. 421) say about autism, that predates the existence of diagnosis and discourses—not an either/or, but a both/and proposition.
- A need to 'unknow' and 're-know': Without denying the materiality of manifestations of ADHD that might, to differing degrees and at different times and contexts, be disabling, to unknow and then to seek to re-know

what ADHD 'is,' how it has been and how it could be re-storied, and to understand how this re-storying impacts those living with ADHD.

- Allyship: Just as LGBTQIA + individuals may occupy different positionings within this alphabet soup with differing specificities and priorities, they may also share similar combats. What divides us, such as our individual needs and priorities, should not impede us from seeing beyond the differences to arrive at an understanding of the commonality of much of what needs resisting. Whichever approach we most identify with (or not), to be able to move forward we need to find points of articulation wherein there is room for an understanding of both our individual specificities and our shared challenges.

References

Alexander, L., & Farrelly, N. (2018). Attending to adult ADHD: A review of the neurobiology behind adult ADHD. *Irish Journal of Psychological Medicine, 35*(3), 237–244.

Barkley, R. A. (1997). *ADHD and the nature of self-control.* Guilford press.

Basten, L. (2023). The codex is always on crip time: ADHD (ness) and Reading. *Reception: Texts Readers, Audiences, History, 15*, 89–96.

Belek, B. (2023). 'A smaller mask': Freedom and authenticity in autistic space. *Culture, Medicine, and Psychiatry, 47*(3), 626–646.

Bertilsdotter Rosqvist, H., Hultman, L., Österborg Wiklund, S., Nygren, A., Storm, P., & Sandberg, G. (2023b). Naming ourselves, becoming neurodivergent scholars. *Disability and Society*, 1–20.

Bertilsdotter Rosqvist, H., Hultman, L., Österborg, S., Nygren, A., Storm, P., & Sandberg, G. (2023a). Intensity and variable attention: Counter narrating ADHD, from ADHD deficits to ADHD difference. *The British Journal of Social Work, 53*(8), 3647–3664.

Bertilsdotter Rosqvist, H., Hultman, L., Österborg, S., Nygren, A., Storm, P., & Sandberg, G. (2023c). ADHD in higher education and academia. *Canadian Journal on Disability Studies, 12*(3), 1–26.

Bertilsdotter Rosqvist, H., & Jackson-Perry, D. (2021). Not doing it properly? (Re) producing and resisting knowledge through narratives of autistic sexualities. *Sexuality and Disability, 39*(2), 327–344.

Bertilsdotter Rosqvist, H., & Nygren, A. (2023). I am that name? naming neurotypical imaginaries of the sole autist in autistic/autism fiction. *Canadian Journal of Disability Studies, 12*(1), 117–140.

Bianchi, E., Ortega, F., Faraone, S., Gonçalves, V. P., & Zorzanelli, R. T. (2016). Medicalization beyond physicians: Pharmaceutical marketing on attention deficit and hyperactivity disorder in Argentina and Brazil (1998–2014). *Saúde e Sociedade, 25*, 452–462.

Brown, A. I. (2023). The univocity of attention: ADHD and the case for a renewed self-advocacy (Doctoral dissertation, York University). Available at: https://yorkspace.library.yorku.ca/items/47c9e86c-c9b8-4951-adf1-c8b81499272a

Brown, T. E. (2006). Executive functions and attention deficit hyperactivity disorder: Implications of two conflicting views. *International Journal of Disability, Development and Education, 53*(1), 35–46.

Chapman, R. (2019). Autism as a form of life: Wittgenstein and the psychological coherence of autism. *Metaphilosophy, 50*(4), 421–440.

Charlesworth, J. (2021). *Excluded: Increasing understanding, support and inclusion for children with FASD and their families* (p. 148). Representative for Children and Youth. https://rcybc.ca/wp-content/uploads/2021/04/RCY-FASD_Report-Apr2021_FINAL_highres.pdf

Conrad, P. (2006). *Identifying hyperactive children: The medicalization of deviant behavior*. Ashgate Publishing, Ltd.

Dieuwertje Huijg, D. (2020). Neuronormativity in theorising agency. Neurodiversity Studies: A new critical paradigm, 213–217.

Dieuwertje Huijg, D. (2021). ADHD whiteness: An exploration of the (absent) role of race in adult ADHD research. https://ddhuijg.com/blog-post-adhd-whiteness/#more-19. Downloaded April 27, 2024.

Dodson, W. (2022). Secrets of your ADHD brain. ADDitude. https://www.additudemag.com/secrets-of-the-adhd-brain/ Downloaded April 27, 2024.

dos Santos Assef, E. C., Capovilla, A. G. S., & Capovilla, F. C. (2007). Computerized Stroop test to assess selective attention in children with attention deficit hyperactivity disorder. *The Spanish Journal of Psychology, 10*(1), 33–40.

Ergun, G., Schultz, M., & Rettig, E. (2021, June 1). Fetal alcohol spectrum disorder—Issues of misdiagnosis and missed diagnosis in black youth: A case report. *Innovations in Clinical Neuroscience*. https://innovationscns.com/fetal-alcohol-spectrum-in-black-youth/

Fleischmann, A., & Miller, E. C. (2013). Online narratives by adults with ADHD who were diagnosed in adulthood. *Learning Disability Quarterly, 36*(1), 47–60.

Foreman, D. M. (2018). Attention-deficit hyperactivity disorder (ADHD): Progress and controversy in diagnosis and treatment. *Irish Journal of Psychological Medicine, 35*(3), 251–257.

Frances, A. (2013). *Saving normal: An insider's revolt against out-of-control psychiatric diagnosis, DSM-5, big pharma, and the medicalization of ordinary life (9780062229274)*. Brandon Sanderson.

Friedman-Hill, S. R., Wagman, M. R., Gex, S. E., Pine, D. S., Leibenluft, E., & Ungerleider, L. G. (2010). What does distractibility in ADHD reveal about mechanisms for top-down attentional control? *Cognition, 115*(1), 93–103.

Gallagher, R., & Blader, J. (2001). The diagnosis and neuropsychological assessment of adult attention deficit/hyperactivity disorder: Scientific study and practical guidelines. *Annals of the New York Academy of Sciences, 931*, 148–171.

Gillberg, C. (2010). The ESSENCE in child psychiatry: Early symptomatic syndromes eliciting neurodevelopmental clinical examinations. *Research in Developmental Disabilities, 31*(6), 1543–1551.

Goetz, T.G., & Adams, N. (2022). The transgender and gender diverse and attention deficit hyperactivity disorder nexus: A systematic review. *Journal of Gay and Lesbian Mental Health.*

Hacking, I. (1983). *Representing and intervening: Introductory topics in the philosophy of natural science*. Cambridge University Press. https://doi.org/10.1017/CBO9780511814563

Hallowell, E. M., & Ratey, J. J. (1994/2011). *Driven to distraction (revised): Recognizing and coping with attention deficit disorder*. Anchor.
Hallowell, E. M., & Ratey, J. J. (2022). ADHD 2.0: New science and essential strategies for thriving with distraction--from childhood through adulthood. Ballantine Books.
Harwood, V. (2006). *Diagnosing "disorderly" children: A critique of behaviour disorder discourses*. Routledge.
Healey, D., & Rucklidge, J. J. (2006). An investigation into the relationship among ADHD symptomatology, creativity, and neuropsychological functioning in children. *Child Neuropsychology, 12*(6), 421–438.
Hjörne, E., & Evaldsson, A. C. (2015). Reconstituting the ADHD girl: Accomplishing exclusion and solidifying a biomedical identity in an ADHD class. *International Journal of Inclusive Education, 19*(6), 626–644.
Hughes, B. (2009). Disability activisms: Social model stalwarts and biological citizens. *Disability and Society, 24*(6), 677–688.
Hupfeld, K. E., Abagis, T. R., & Shah, P. (2019). Living "in the zone": Hyperfocus in adult ADHD. *ADHD Atten Def Hyp Disord, 11*, 191–208.
Jackson-Perry, D., Rosqvist, H. B., Annable, J. L., & Kourti, M. (2020). Sensory strangers: Travels in normate sensory worlds. In Neurodiversity Studies (pp. 125–140). Routledge.
Joseph, A. (2021). Psychiatric power and the ADHD experience. *Canada Watch Summer 2021: Critical Perspectives on Mental Health/Mad Studies*, 11–12.
Kuntsi, J., & Stevenson, J. (2000). Hyperactivity in children: A focus on genetic research and psychological theories. *Clinical Child and Family Psychology Review, 3*(1), 1–23.
Lasky, A. K., Weisner, T. S., Jensen, P. S., Hinshaw, S. P., Hechtman, L., Arnold, L. E., W. Murray, D., & Swanson, J. M. (2016). ADHD in context: Young adults' reports of the impact of occupational environment on the manifestation of ADHD. *Social Science and Medicine, 161*, 160–168.
Lovett, B. J., & Harrison, A. G. (2021). Assessing adult ADHD: New research and perspectives. *Journal of Clinical and Experimental Neuropsychology, 43*(4), 333–339.
McDonnell, A., & Milton, D. (2014). Going with the flow: reconsidering 'repetitive behaviour 'through the concept of 'flow states'.
Meadows, J. (2021). We Need Critical ADHD Studies Now. https://jessemeadows.medium.com/we-need-critical-adhd-studies-now-52d4267edd54. Jul 5, 2021. Downloaded July 28, 2022
Mueller, K. L., & Tomblin, J. B. (2012). Diagnosis of attention-deficit/hyperactivity disorder and its behavioral, neurological, and genetic roots. *Topics in Language Disorders, 32*(3), 207–227.
Murray, F. (2018). Me and monotropism: A unified theory of autism. Psychologist. https://www.bps.org.uk/psychologist/me-and-monotropism-unified-theory-autism. Downloaded 30th April 2024
Murray, D., Lesser, M., & Lawson, W. (2005). Attention, monotropism and the diagnostic criteria for autism. *Autism, 9*(2), 139–156.
Murray, D., Milton, D., Green, J., & Bervoets, J. (2023). The human spectrum: A phenomenological enquiry within neurodiversity. *Psychopathology, 56*(3), 220–230.
Nielsen, M. (2018). Experiences and explanations of ADHD: An ethnography of adults living with a diagnosis. Routledge.

Nielsen, M. (2017). ADHD and temporality: A desynchronized way of being in the world. *Medical Anthropology, 36*(3), 260–272.

Nilsson Sjöberg, M. (2021). Reconstructing truth, deconstructing ADHD: Badiou, onto-epistemological violence and the diagnosis of ADHD. *Critical Studies in Education, 62*(2), 243–257.

Orsini, M. (2022). Who needs to (un) know? On the generative possibilities of ignorance for autistic futures. *International Journal of Qualitative Studies in Education*, 1–18.

Ortega, F., & Müller, M. R. (2020). Global mental health and pharmacology: The case of attention deficit and hyperactivity disorders in Brazil. *Frontiers in Sociology, 5*, 535125. https://doi.org/10.3389/fsoc.2020.535125

Österborg, S., Bertilsdotter Rosqvist, H., Bagger, A., & Allan, J. (2023). Waiting for discovery and support?: Neurodivergent subjectivities in the Swedish educational landscape. *European Journal of Inclusive Education, 1*(1), 15–29.

Panagiotidi, M., Overton, P. G., & Stafford, T. (2017). Multisensory integration and ADHD-like traits: Evidence for an abnormal temporal integration window in ADHD. *Acta Psychologica, 181*, 10–17.

Prosser, B. (2015). Attention deficit hyperactivity disorder in Australia: Perspectives from the sociology of deviance. *Journal of Sociology, 51*(3), 596–612.

Rabeharisoa, V., Moreira, T., & Akrich, M. (2014). Evidence-based activism: Patients', users' and activists' groups in knowledge society. *BioSocieties, 9*(2), 111–128.

Rabinow, P. (1996). Artificiality and enlightenment: From sociobiology to biosociality. *Essays on the anthropology of reason* (pp. 91–112). Princeton University Press

Reyes II, R. (2021, July 9). *ADHD Latinxs as nepantleras* [Webinar]. Intersectional neurodiversity and disability reading groups. https://intersect-nd-dis-rg.wixsite.com/rg-site/reyes-adhd-latinxs-as-nepantleras

Rose, N. (2018). *Our psychiatric future*. Wiley.

Signor, R. D. C. F., Berberian, A. P., & Santana, A. P. (2017). The medicalization of education: Implications for the constitution of the subject/learner. *Educação e Pesquisa, 43*(3), 743–763.

Sonuga-Barke, E. J. (2023). Paradigm 'flipping' to reinvigorate translational science: Outlining a neurodevelopmental science framework from a 'neurodiversity' perspective. *Journal of Child Psychology and Psychiatry, 64*(10), 1405–1408.

Sonuga-Barke, E. J., Chandler, S., Lukito, S., Kakoulidou, M., Moore, G., Cooper, N., Matejko, M., Jackson, I., Balwani, B., Boyens, T., Poulton, D., Harvey-Nguyen, L., Baker, S., & Pavlopoulou, G. (2024). Participatory translational science of neurodivergence: Model for attention-deficit/hyperactivity disorder and autism research. *The British Journal of Psychiatry*, 1–5. https://doi.org/10.1192/bjp.2023.151

Thome, J., Ehlis, A.-C., Fallgatter, A. J., Krauel, K., Lange, K. W., Riederer, P., Romanos, M., Taurines, R., Tucha, O., Uzbekov, M., & Gerlach, M. (2012). Biomarkers for attention-deficit/hyperactivity disorder (ADHD). A consensus report of the WFSBP task force on biological markers and the World Federation of ADHD. *The World Journal of Biological Psychiatry, 13*(5), 379–400. https://doi.org/10.3109/15622975.2012.690535

Timimi, S., & 33 Coendorsers. (2004). A critique of the international consensus statement on ADHD. *Clinical Child and Family Psychology Review, 7*(1), 59. https://doi.org/10.1023/B:CCFP.0000020192.49298.7a

Timimi, S. (2005). *Naughty boys: Anti-social behaviour, ADHD and the role of culture*. Palgrave Macmillan.

Tobia, V., Landis, T., & Graziano, P. (2021). Examining temporal cognition in preschoolers with attention deficit hyperactivity disorder: Insights from parent-child interactions. *Journal of Child and Family Studies, 30*(9), 2315–2327.

Valla, J. M., & Belmonte, M. K. (2013). Detail-oriented cognitive style and social communicative deficits, within and beyond the autism spectrum: Independent traits that grow into developmental interdependence. *Developmental Review, 33*(4), 371–398.

Vehmas, S. (2011). Disability and moral responsibility. *Trames, 15*(2), 156–167.

Versteeg, W., te Molder, H., & Sneijder, P. (2018). "Listen to your body": Participants' alternative to science in online health discussions. *Health (united Kingdom), 22*(5), 432–450.

De Vos, J. (2016). *Metamorphoses of the brain: Neurologisation and its discontents*. Springer.

White, H. A., & Shah, P. (2006). Uninhibited imaginations: Creativity in adults with attention-deficit/hyperactivity disorder. *Personality and Individual Differences, 40*, 1121–1131.

Yergeau, R. M. (2018). *Authoring autism: On rhetoric and neurological queerness*. Duke University Press.

PART II

Knowing and Doing: Research Ethics and Methods

CHAPTER 4

Co-developing an Autism Research Funding Application to Facilitate Ethical and Participatory Research: The Autism from Menstruation to Menopause Project

Aimee Grant, Kathryn Williams, Karen Henry, Willow Holloway, Christina Nicolaidis, Helen Kara, and Amy Brown

Introduction: Autism, Health and Gender

Around 3% of people are Autistic (Centers for Disease Control & Prevention, 2023), with age and gender differences resulting from historical under-diagnosis, rather than innate differences (Halladay, 2015). Autism is a naturally occurring part of cognitive diversity, resulting in changes in communication style and sensory processing, which can be Disabling in a society that promotes neuronormativity (Milton, 2012). One area in which Autistic people are disadvantaged is in access to healthcare. Autistic people have increased morbidity and early mortality (Bishop-Fitzpatrick & Kind, 2017). There is not a simple explanation for this, with multifactorial causes responsible. For example, widespread societal stigmatisation is directed towards Autistic people (Cage et al., 2018)—including by health professionals (Corden et al., 2021) who may communicate awkwardly with Autistic patients (Mason et al., 2019).

A. Grant (✉) · A. Brown
Centre for Lactation Infant Feeding and Translational Research, Swansea University, Swansea, UK
e-mail: aimee.grant@swansea.ac.uk

K. Williams · W. Holloway
Autistic, CIC, UK

K. Williams
School of Social Sciences, Cardiff University, Cardiff, UK

H. Bertilsdotter Rosqvist and D. Jackson-Perry (eds.), *The Palgrave Handbook of Research Methods and Ethics in Neurodiversity Studies*,
https://doi.org/10.1007/978-3-031-66127-3_4

It has long been established that being marginalised by more than one characteristic, such as being both a woman and from a minoritised ethnicity, compounds the impact of that marginalisation; this is known as intersectionality (Crenshaw, 1989). However, much research on Autistic adults has focused on Autistic men's experiences, failing to consider the additional marginalisation of Autistic women. This is despite Autistic women and girls being more than four times as likely to commit suicide than Autistic men and boys (aIRR: 8.551 V 1.93; Kõlves et al., 2021). Autistic women are also more likely to have chronic pain conditions (Asztély et al., 2019), neurological conditions (Da Walt et al., 2021) and be at higher risk of dying in hospital than non-Autistic people and Autistic men (Akobirshoev et al., 2020). These health disparities could potentially be explained by Autistic women and girls being more likely to experience societal stigma for presenting as their authentic selves, resulting in "Autistic masking", which is energy intensive (Miller et al., 2021) and contributes to women and girls going unrecognised as Autistic or not receiving the support they need (Wood-Downie et al., 2021). Furthermore, when Autistic people become parents, particularly to Autistic children, they may be subjected to involuntary social work intervention based on their "perplexing presentations" and ableist notions of what constitutes "good motherhood" (Benson, 2023), although variation in healthcare systems can potentially reduce this stigma (Ellis et al., 2023).

We know the care given to UK-based women in a general population during the perinatal period and in relation to gynaecological healthcare results in unmet needs (Brown, 2021). Due to Autistic people's known differences in bodily experiences relating to interoception, pain and sensory perception (Mason et al., 2019), Autistic women and other people with wombs' (APWW) reproductive health experiences and needs are likely to vary compared to their non-Autistic peers. For example, we know that many APWWs experience overstimulation during menstruation (Gray & Durand, 2023) and that APWWs are less likely to attend cervical screening than non-Autistic peers (Nicolaidis et al., 2013). Even so, there is a dearth of research on the reproductive health needs of APWW. For example, recent systematic reviews of Autistic pregnancy (Samuel et al., 2021) and infant feeding (Grant et al., 2022) identified only six and eight peer reviewed studies, respectively, with many including limited

K. Henry
School of Nursing and Midwifery, University of Suffolk, Boston, UK

C. Nicolaidis
School of Social Work, Portland State University, Oregon, USA

Department of Medicine, Oregon Health & Science University, Oregon, USA

Academic Autism Spectrum Partnership in Research and Education, Oregon, USA

H. Kara
We Research It Ltd, London, UK

data. We were unable to find any published research on Autistic people's experiences of pregnancy loss to contextualise a recent autoethnography of an ectopic pregnancy (Grant, 2023). Accordingly, we conducted an online survey of this area, and maternity care more widely, identifying many inadequacies in existing healthcare (Grant et al., 2023a). However, we felt that there was a significant and urgent need for additional research in this area, with the aim of improving healthcare to reduce the current disparity in both quality of life and life expectancy of Autistic people.

Autistic Representation in Research

The importance of Autistic involvement in research is clearly articulated in Critical Autism Studies, which highlights the social nature of Autistic barriers to full involvement in society (O'Dell et al., 2016; see also Chapter 2, this volume). Accordingly, to reduce health inequalities faced by Autistic people, it is necessary for us to be front and centre in research about our health. However, Autistic adult experiences of health and healthcare have been severely under-explored in research (Nicolaidis et al., 2019; Pellicano et al., 2014); as a high proportion of Autism research funds are directed to research on Autistic children and their parents (Nicolaidis et al., 2019) or to biological differences rather than Autistic people's needs (Emerson, 2021). Furthermore, Autism research is often carried out without relevant infrastructure, authenticity and respect (Fletcher-Watson et al., 2019), including a lack of meaningful input from the Autistic community (Pellicano et al., 2014). This is despite the Autistic community being clear that they want to collaborate on the design of Autism research (Fletcher-Watson et al., 2019; Nicolaidis et al., 2019). Furthermore, it has been argued that participatory approaches to research are not incentivised by funding bodies and in academic promotions processes (Keating, 2021). This situation means that there is insufficient evidence to argue for Autistic needs, partially accounting for policy and government services that fail to meet Autistic needs (Grant et al., 2023b).

Researchers have used a variety of participatory research approaches with marginalised communities since the 1960s. For example, in Participatory Action Research (PAR), problems and potential solutions are proposed by the community with researchers acting as facilitators; this should occur from the very beginning of the project and throughout (Smith, 2013). Similarly, in Community-Based Participatory Research (CBPR), academics and community members strive to work together as equal partners in every phase of the research (Israel et al., 2008). Over the past two decades, significant pockets of good practice in participatory Autism research have grown organically, resulting in clusters, including in the UK, Australasia and the United States of America. One of the founding groups leading the way in participatory Autism research was AASPIRE, a long-standing collaboration, established in 2006, between Autistic and non-Autistic academics and members of the Autistic community. The AASPIRE team have completed a plethora of research projects, including a focus on burnout (Raymaker et al., 2020), employment (Raymaker et al., 2023) and access to healthcare (Nicolaidis et al., 2016) using a CBPR approach (Nicolaidis et al., 2011). They have also developed an "inclusion toolkit" (see: https://aaspire.org) and practice-based guidelines for the inclusion of Autistic people as team members or study participants in research (Nicolaidis et al., 2019). Alongside this, the Participatory Autism Research Collective (PARC) is a group of Autistic researchers and activists from across the UK who aim to link those who wish to see more significant Autistic involvement in Autism research (see: https://participatoryautismresearch.wordpress.com/). Furthermore, in Aotearoa New Zealand, researchers worked with Autistic people to co-produce national Autism research priorities, identifying that research should focus on quality of life issues and move away from biological research (Emerson et al., 2023). In relation to our grant application's originally planned focus on the maternity period, PAR approaches have been used in a limited way with a general population of participants to understand breastfeeding (Condon & Ingram, 2011).

In addition to participatory research methods, over the past decade "Citizen Science" (European Citizen Science Association, 2015) has been used to involve the public in data collection and analysis, often through created open data repositories. Much of this has *not* involved marginalised communities, or been based on their priorities (Curtis, 2018). That said, the Parenting Science Gang successfully completed a number of projects focused on the research priorities of parents in the "Gang" (Collins et al., 2020). To date, citizen science has been used in a limited way in Autism research (see also Chapter 14, this volume). For example, the "AutSPACEs" project is a collaboration between Autistica and the Alan Turing Institute. It aimed to co-create open source data on sensory processing for Autistic people (see: https://github.com/alan-turing-institute/AutSPACEs), but had not resulted in publications at the time of writing.

Using Creative Methods in Research with Marginalised Communities

In this section, we will describe the use of creative methods and the "ethics of choice". Both will be described, including what they are and how they have been used with Autistic and Disabled communities. "Creative methods" are a range of qualitative techniques that incorporate methods that move beyond traditional researcher-centric data collection and analysis (Kara, 2020). In doing so, they aim to engage participants more effectually in data co-production, analysis and dissemination. A sub-set of creative methods is "visual methods", which involve the use of some sort of artefact, such as photographs, everyday objects or participant-created documents (e.g., timelines, drawings, collages) which participants select or create based on guidance provided by researchers. These then either become data or are used as an elicitation tool during interviews to co-produce data (Mannay, 2016). This approach often has the stated aim of reducing the power imbalance between participants and researchers (Kara, 2020), although when they are tightly controlled by the researcher, they can reinforce existing power imbalances (Pell et al., 2020).

To date, the use of creative methods with Autistic people is in its infancy, although it has been used with success with young people in a range of studies. For example, Autistic teens using photo-elicitation to describe their experiences of sleeping (Pavlopoulou, 2021), life charts developed across multiple interviews to talk about school experiences and non-Attendance (Moyse, 2024) and walking interviews focused on educational transitions (Shepherd and Shepherd, 2015). In using these approaches to data production, Autistic participants shared content that was not expected by the researcher (Moyse, 2024). Furthermore, interviewers have used emojis as part of interviews with Autistic young people about their experiences of school, but also used them to understand how participants were perceiving the interview encounter itself (Peuravaara, 2021). These creative methods have also been used with young people who have learning disabilities (Shepherd & Shepherd, 2015) and who use Augmentative and Alternative Communication (AAC) (Teachman & Gibson, 2018), who have often been neglected in Autism research.

Alongside collaborative techniques of data production seen in creative methods, it has increasingly been argued that when working with marginalised groups it is essential from an ethical standpoint to offer a choice of ways to participate, known as "the ethics of choice". In Aimee's research with marginalised pregnant women, for example, she offered a range of visual methods approaches, including using researcher dyad interviews (Grant et al., 2019). In another study, Disabled participants were reminded multiple times that they could choose to complete a range of visual methods tasks or to bring along anything they preferred, which included medical notes and objects related to treatment being shared during interviews (Phillips et al., 2018). This approach of maximising participants' choice has been called a "palette of methods" in co-produced research with young people focused on alcohol use where participants could choose from interviews, peer interviews, diaries, mobile phone interviews, text messaging and participant observation (Wilkinson & Wilkinson, 2018). Wilkinson and Wilkinson (2018) offered participants the choice for their data production to be synchronous (live with the researcher present) or asynchronous (prepared when alone and then shared, sometimes occurring as a conversation over time), as well as offering methods that allowed for speaking, typing and writing. The need to offer choice in how Autistic participants take part in research is well established (Nicolaidis et al., 2019). This can include the communication mode and location for interviews (Moyse, 2024), including a choice of interactive and welcoming sensory environments (Brown et al., 2016). Moreover, it has been suggested that in online interview studies, Autistic participants should be reminded that they may turn off their camera, ask for interviewers to turn off their camera, exclusively use the chat function, and also that they may use any accommodations to improve the sensory environment (Kaplan-Kahn & Caplan, 2023). The use of repeat interviews without a defined structure imposed by the researcher has also been identified as a way of building trust when discussing sensitive topics with Autistic participants (Stone et al., 2023).

Arguably, the literature on Autism research is moving beyond these opportunities to provide choice and beginning to explore notions of individualising data collection tools and even research methods to meet the needs of individual participants. This has been seen across a range of research designs. First, usually associated with a positivist research paradigm, structured interview instruments used in an evaluation of care homes utilised scales modified per participants' usual communication preferences, such as using either a traffic light system or thumbs up or thumbs down (Long & Clarkson, 2017). Second, in interview studies with non-Autistic interviewers, lay Autistic "co-interviewers" have attended interviews and had the opportunity to ask additional questions that may not have occurred to non-Autistic researchers (Kaplan-Kahn & Caplan, 2023). Third, in an interpretative study working with a non-speaking young woman who did not have access to detailed expressive language to share her experience, Van Goidsenhove and De Scauwer (2020) used a variety of ways to understand the participant's world. This included: emulating the participant's interactions with objects during a planned drawing activity within a research workshop, which were "more creative" than the way those objects are conventionally handled, allowing themselves to be guided by the participant during a walk and undertaking an ongoing dialogue through filmed video clips (Van Goidsenhoven & De Schauwer, 2020).

The Development of the Autism from Menstruation to Menopause Study

We now describe the process of developing the *Autism from menstruation to menopause* study's research design and methodology *before* applying for funding. An in-depth exploration of the application for research funding is not usually taken, but the comprehensive funding, over £2.4 M, awarded to this study, allows for reflection on the earliest part of research design. First, we describe the study lead's, Aimee, background, in relation to both research focus on the maternity period and neurodivergence. This sets the stage for rounds of pre-application consultation, which are described in detail, in relation to an informal social media consultation and in-depth consultation with two Autistic community leaders (Kathryn and Willow), an Autistic midwife (Karen), an Autistic creative methods expert (Helen) and a Participatory Autism Research expert (Christina) and her AASPIRE colleague, Mirah Scharer. An overview of the process is shown in Fig. 4.1.

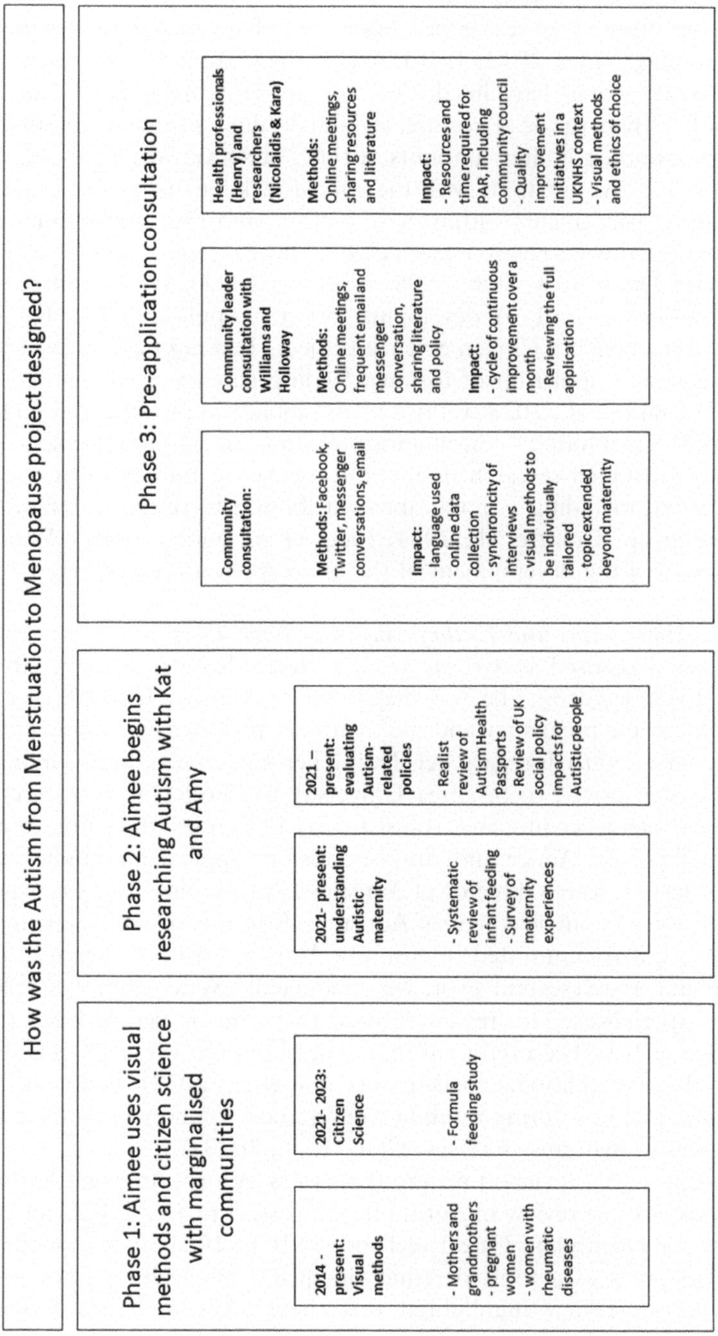

Fig. 4.1 Community involvement in the development of the autism from menstruation to menopause study

Phase 1: Aimee Uses Creative Methods and Citizen Science with Marginalised Communities
Aimee, the study's lead researcher, has focused her research on pregnancy and early parenting since 2012. In the main, these studies were not as participatory as she would have liked, due to constraints including being research staff, rather than the lead researcher, and leading studies with small grants and corresponding limited amounts of time. However, within these, she used visual and creative methods with the aim of making them more accessible to marginalised participants within these funding and time constraints (Mannay et al., 2017). Aimee's research has included those on low incomes (Grant et al., 2018) and Disabled women (Williams et al., 2022*).* We consider this early research to be "phase 1" of developing this grant application (see Fig. 4.1). In 2021, Aimee began a Citizen Science project, *Finding the Formula*, working with a general population of formula feeding parents, a stigmatised feeding method (Grant et al., 2023c). Prior to the application for this funding, Aimee undertook a community consultation, resulting in significant changes to the originally envisaged research design and increasing the depth of community participation, including a significant contribution to research design, analysis and writing up (Grant et al., 2024; Jones et al., under review). We consider this Phase 1 of the development of the grant application (see Fig. 4.1).

Phase 2: Aimee, Amy and Kathryn Begin to Work Together
Aimee was diagnosed as dyslexic as an undergraduate student in 2005, aged 23. In 2019, aged 37, she was diagnosed as Autistic. In 2020, she experienced an ectopic pregnancy and the treatment provided, including emergency surgery, appeared to disregard her sharing her Autism diagnosis (Grant, 2023). In 2021, she moved to Swansea University to develop new research in the Lactation, Infant Feeding and Translation (LIFT) research centre, directed by Amy. In this role, Aimee and Amy successfully applied for funding to work with Kathryn, research director of Autistic UK, to explore Autistic experiences of maternity. The involvement of Autistic UK in this research was invaluable, as Autistic UK is committed to ensuring Autistic voices are heard in order to shape policy and research. First, we conducted a systematic review of infant feeding experiences. This review showed that Autistic people faced the same challenges as have been reported in a general population of parents, but they also faced some additional challenges (Grant et al., 2022). Following this, we were successful in securing funding to undertake a survey of Autistic people's experiences of maternity services (Grant et al., 2023a).

In addition to a focus on maternity services, Aimee, Amy and Kathryn also undertook a realist review of Autism health passports. In the UK, the National Institute for Health and Care Excellence (NICE, 2021) recommends Autism health passports as a tool to reduce inequality of health service provision. However, our review highlighted that these tools were based on limited evidence, are under-theorised and cannot remove the barriers to equitable healthcare (Ellis et al., 2023). Aimee and Kathryn also examined official statistics, research and other reports to identify ways in which Autistic people are disadvantaged by government policy across health, education, employment, poverty and housing (Grant et al., 2023b). Throughout this time, Williams' knowledge of Autism research and policy was invaluable in Aimee's introduction to the research area, building on her existing strengths in participatory research with marginalised groups.

Phase 3(i): The Online Community Consultation

Alongside undertaking this research, Aimee planned to apply for personal fellowships to extend her fixed-term contract of employment. The fixed-term funding was intended to last for three years, although it was later reduced to 27 months. However, the full-time job it provided without the requirements of a substantial teaching load allowed sufficient time to conduct community consultation (Keating, 2021). Aimee was clear that she wanted to build on the participatory approach she had used in her UKRI-funded[1] research on infant formula feeding (Grant et al., 2023c), where a community consultation prior to submitting the funding application had resulted in a significant change in research methods. The initial concept for the fellowship application was to focus on Autistic experiences of maternity care in more detail. In terms of methods, the research would be qualitative in nature because to date there is limited evidence about Autistic people's experiences in relation to obstetric health. Furthermore, the study would be co-governed by a Community Council largely based on the AASPIRE good practice guidance for participatory Autism research (Nicolaidis et al., 2019).

Aimee knew, from the *Finding the Formula study,* that it was important to involve the community at the application stage (see also Chapter 10, 20, this volume). This involvement was not as in-depth as would have been ideal due to the relatively short time period between being given institutional approval to apply for the fellowship and the funder deadline. However, she did undertake an online consultation with 27 Autistic people who had been pregnant. This occurred through social media, including personal contacts, Facebook group conversations and strangers responding to Tweets. Different preferences were identified in different spaces. For example, within a closed Facebook group, Aimee received a **very** strong steer to use gender-inclusive ('people with wombs') and identity-first ('Autistic people') language, and this resulted in us developing the terms "Autistic People with Wombs", to represent our target population, abbreviated to APWW. Also, in conversations on Twitter, the importance of paying members of the Community Council in cash via bank transfer, rather than providing vouchers only (which is common within the UK), received strong support.

[1] UK Research and Innovation (UKRI) is a major public funder of research in the UK, across seven Research Councils which focus on aspects of arts, science and engineering.

Alongside practicalities for ethical study design, in one-to-one discussions, we identified a preference for interviews to be conducted online rather than in-person, such as at the University, the participant's home or another location the participant chose. In these discussions, Aimee was able to easily understand the discomfort arising from an unknown person coming to her home due to her positionality as an Autistic and Disabled person who has had official interviews relating to claiming benefits conducted at her home (see Grant, 2024). In these situations, it can be very hard for an Autistic person to advocate for their needs, in comparison with online encounters, when closing a window on a computer makes an interaction stop immediately. Aimee was also aware that should Autistic researchers be appointed to the researcher jobs assigned to the project, face-to-face data collection could be significantly more energy intensive, resulting in the need to reduce the number of participants who could be accommodated within the study, as the funder imposed a limit of four research assistants. Alongside this preference, the need for flexibility in relation to whether interviews should be conducted with the participant and researcher both being in the same space at the same time (synchronous) or if there should be the opportunity for participants to share information in their own time and space (asynchronous) was identified. Building on this, the need for flexibility *during* interviews was identified, to allow for a change in communication mode or synchronicity if it would be more comfortable for the participant.

When discussing the data collection/production strategies, Aimee was keen to use "creative methods" as a tool in order to allow participants to share the things most important to them and to reduce the researcher's role in shaping the data (Kara, 2020). Having described this in the pre-application consultation, responses were positive, although it was noted that sufficient time should be allocated to ensure that participants understood what was expected of them so they "did not feel stupid" for "making mistakes". The importance of creative methods meeting participants' needs has been long acknowledged (Mannay, 2016), and the same tool can be used in a wide variety of ways, changing the feel of the interview significantly. For example, the content added to timelines, designed to show changes over a particular period of time, can be entirely controlled by interviewers, or it may be done in isolation from the interview by participants who have a large degree of freedom regarding what to include (Pell et al., 2020). It was important to Aimee and those involved in the pre-application consultation that the creative methods tools could be used flexibly, potentially having unique tools for each participant. Furthermore, the way in which these tools should be used should be similarly flexible to reassure participants that they were "not doing it wrong". This included allowing participants to: do the creative methods task together with the researcher as part of the interview, or on a separate occasion; use any other way of gathering their thoughts; or not undertake a creative methods task at all, with the interviewer gently guiding the direction of the conversation using the guidance accompanying the creative methods task as a loose shared topic guide.

Furthermore, it was decided that we would not insist on creative methods materials created by participants, such as timelines or mind maps, being shared with the researcher, as their purpose was to elicit participants' narratives.

The one element that was consistently commented on across platforms was the focus of the study being limited to pregnancy only. On all three platforms, Aimee received comments of "but what about..." this other really pressing gynaecological health issue. These included specific conditions like Pre-Menstrual Dysphoric Disorder (PMDD), Polycystic Ovarian Syndrome (PCOS), endometriosis, painful periods and menopause. At this point, the suggestion to broaden the research focus away from maternity, which, Aimee knew in-depth, was very worrying, as other Autism researchers have described (Keating, 2021). For example, there would not be time to learn enough about these conditions before applying for funding, and Aimee was concerned that funders may decide that she was not a credible candidate because of a lack of publications related to wider reproductive health issues. Regardless, to be true to the ethos of being guided by the community, a complete shift in the study's boundaries occurred to cover reproductive health across the life course. To bolster her credibility, Aimee reached out to a range of individuals and organisations who had experience of undertaking research in these areas, and clearly described the expertise of these individuals on the application form.

Phase 3(ii) and (iii): Pre-application Discussion with Autistic Community Leaders (A Paid Role for "Lay" Members)

Within the UK, the research arm of the publicly funded National Health Service (NHS) states that it supports the inclusion of lay people in research related to their health, including providing funding for lay involvement in developing grant applications. Accordingly, Aimee contacted Health and Care Research Wales, Aimee's local NHS research organisation, and asked if they would fund some time for Kathryn, who Aimee had already worked with, to be involved in the development of the study. They agreed to fund her time at £25 per hour, per Involving People guidance, on Kathryn's completion of some forms. Kathryn introduced Aimee to Willow, an Autistic woman and director of many Disability organisations, including Autistic UK and Fair Treatment for the Women of Wales (FTWW). FTWW are Wales' leading organisation committed to supporting accessible and safe reproductive healthcare. In the absence of a Women's Health Plan, FTWW worked co-productively with at least 60 organisations to develop a strategy document to put to the Welsh Government. This resulted in the development of the Women's Health Plan for Wales, which has been developed co-productively with volunteers across Wales. Fortunately, Willow agreed to be involved, and Health and Care Research Wales agreed to also fund her time. Aimee had multiple conversations with Kathryn and Willow over a period of around a month, which included them making suggestions regarding research design, relevant literature and policy documents. The discussions used multiple communication platforms to ensure it met the communication preferences of all. For example, a three-way messenger conversation between Aimee, Kathryn and Willow was used to discuss small details on a frequent basis. Both reviewed the grant application in full. This would not have been possible (in an ethical way) without the funding from Health and Care Research Wales (see also Chapter 19, this volume).

Alongside this, Aimee and Karen began a conversation on social media. Karen is an Autistic Registered Midwife who was undertaking a Florence Nightingale Foundation scholarship at the time, to improve care for Autistic people during the maternity period. Due to Karen being Autistic herself, she was able to identify how inaccessible and inequitable UK maternity care can be for Autistic women and wanted to do more to rectify this imbalance. Karen shared her experience of intervention development, and barriers to implementation within a UK healthcare context and provided much needed contextual information from her role as an Autistic health professional. It was decided that Kathryn and Willow would join the study's Community Council and the study steering group, and that Karen would only join the study steering group, due to her expertise as a health professional.

Discussion

Our chapter has reinforced the importance of including Autistic people in research about Autistic health, and how that is potentially even more important when it comes to considering APWW. We outlined some existing good practice in co-produced Autism research, including referencing existing high-quality guidance in participatory Autism research (Nicolaidis et al., 2019). Following this, we described citizen science, visual and creative methods and early work on "the ethics of choice", identifying examples of good practice from Autism research. The substantive content of the chapter then focused on what later became the Autism from menstruation to menopause study, and how it began as a community consultation to ensure the involvement of Autistic people in the research design from the outset. This also fits with some of Aimee's most recent research experience using citizen science and creative methods and her focus on social justice. However, we did encounter challenges to community participation.

Autistic people were involved in the development of our grant application, primarily through two Autistic community leaders, Kathryn and Willow, and Karen, an Autistic midwife. Although we did undertake a pre-application consultation, this was somewhat limited and was mostly limited to short conversations with 27 Autistic people who had been pregnant. This interaction was limited because of the lack of funding available to pay for the time of those people involved, meaning that Aimee felt uncomfortable to ask for too much from them. The availability of funding for Kathryn and Willow's time from Health and Care Research Wales removed this barrier. However, both had to complete paperwork to register as Patient and Public Involvement representatives, and completing this paperwork could be challenging for many neurodivergent people. Regardless, we recommend that Autism researchers seek funding for pre-application consultations and support Autistic people to complete any paperwork required to access funds.

Our experience highlighted that a model of research funding that prioritises developing detailed grant applications before being awarded funding is a barrier to involving the public as partners from the outset, as has previously been reported (Keating, 2021). However, the application form completed for this study was relatively brief following the Wellcome Trust's restructuring of their fellowship schemes. We believe that reducing the length of application forms will allow researchers time to conduct greater public involvement activities. It may also be that the use of small pots of funding can be used to undertake more in-depth consultation regarding developing new research studies, although at present, there are limited funds available for this purpose in the UK, and these are not available from many major health research funders.

Within our consultation, members of the Autistic community strongly recommended some changes to our application. First, they insisted on gender-neutral language, which is appropriate in relation to Autistic people due to the higher proportion of LGBTQIA + (Hall et al., 2020). We may not have used this language in the grant application if the pre-application consultation had not been so strong on this point, for fear of a negative response from reviewers. Being able to say that the community chose this language provided a clear rebuttal in the case of a negative response. A second major change was to change the period of focus, from only obstetric health to include all gynaecological and obstetric health, "from menstruation to menopause". Aimee's research and publication history were largely concentrated in the perinatal period, and shifting the focus of the research period to a much broader area could have been considered a weakness by reviewers. Again, this was defended by making it clear that it was a community preference and ensuring that sufficient expertise was held on the study's steering committee. The peer review comments and questions at interview did not focus on either of these areas, although they were covered in the presentation required at the beginning of the interview, again providing a way to show that the decisions had been made *with* the community.

Our experiences with this project have led us to develop a checklist for involving Autistic people in pre-application consultations (Box 1). Although this may not be a perfect fit for all funding applications, we hope that this may be valuable to other research teams.

Box 1: Checklist for Involving Autistic People in Pre-application Consultations

- Identify opportunities for paying community members at the outset:

- Ensure that those who are taking part know how much funding is available and how and when this will be paid.
- Where paperwork needs to be completed to receive payment, support should be provided.
- Where payment is made through bureaucratic systems, researchers should check that people have been paid, and request updates from the funding organisation if there are delays.
- Ensure that community members' welfare benefits (where received) are not negatively impacted by being paid—provide guidance and signposting regarding how to ensure this does not happen. Likewise, do not assume that somebody cannot accept payment due to receiving welfare benefits.

• Consider the timescales available and tailor your consultation methods appropriately:
  - A social media consultation can be useful for finding answers to quick questions.
  - One-to-one discussions (using a range of communication methods) can add depth to these answers and may be particularly useful with Autistic co-researchers.
  - Group discussions require significant coordination to align availability and may result in a lot of energy being expended before considering the project design, and may not suit the communication preferences of all Autistic co-researchers.
  - Combining methods may add value but needs to be carefully planned.

• Ensure that you provide flexibility in how people can take part in the pre-application consultation:
  - Be ready to respond to interaction across multiple platforms (email, WhatsApp, social media direct messaging) whenever possible, including outside of your usual working hours, to capitalise on momentum.
  - If you are seeking a full review of your application, this may be best achieved in a collaborative meeting, for example using Zoom alongside talking through various aspects, rather than expecting written comments on the document.
  - Use straightforward language and ensure any required "jargon" or acronyms are fully explained in the first instance.

Conclusion

The "Autism from menstruation to menopause" project was shaped by a pre-application consultation, including relatively superficial conversations on social media and in-depth collaboration with three Autistic people, two community leaders and a midwife. Wherever possible, researchers should involve members of the public with lived experience in the development of their funding applications. Where researchers are not able to access funds to pay community members for this time, researchers should consider how they can involve the community in a way that is at low cost (in terms of time and effort) to the community whilst still allowing them to have a voice. Where the community suggests significant changes to a research project, research teams should consider if this is something that the team can accommodate in a high-quality way. Researchers should take advantage of funding that is available to pay those involved in pre-application consultations, from funders or health authorities. Furthermore, funders should consider making funds available for pre-application consultations to ensure that researchers are able to appropriately compensate community members for their time.

References

Akobirshoev, I., et al. (2020). In-hospital mortality among adults with autism spectrum disorder in the United States: A retrospective analysis of US hospital discharge data. *Autism, 24*(1), 177–189. https://doi.org/10.1177/1362361319855795

Asztély, K., et al. (2019). Chronic pain and health-related quality of life in women with autism And/Or ADHD: A prospective longitudinal study. *Journal of Pain Research, 12*, 2925–2932. https://doi.org/10.2147/JPR.S212422

Benson, K. J. (2023). Perplexing presentations: Compulsory neuronormativity and cognitive marginalisation in social work practice with autistic mothers of autistic children. *The British Journal of Social Work, 53*(3), 1445–1464. https://doi.org/10.1093/bjsw/bcac229

Bishop-Fitzpatrick, L., & Kind, A. J. H. (2017). A scoping review of health disparities in autism spectrum disorder. *Journal of Autism and Developmental Disorders, 47*(11), 3380–3391. https://doi.org/10.1007/S10803-017-3251-9

Brown, S. A. et al. (2016). The case for conversation: a design research framework for participatory feedback from autistic children. In *Proceedings of the 28th Australian conference on computer-human interaction—OzCHI '16. The 28th Australian Conference* (pp. 605–613). ACM Press. https://doi.org/10.1145/3010915.3010934

Brown, A. (2021). *Breastfeeding uncovered: Who really decides how we feed our babies?* Pinter & Martin.

Cage, E., Monaco, J. D., & Newell, V. (2018). Understanding, attitudes and dehumanisation towards autistic people. *Autism, 23*(6), 1373–1383. https://doi.org/10.1177/1362361318811290

Centers for Disease Control and Prevention. (2023). Autism spectrum disorder (ASD). *Data and Statistics on Autism Spectrum Disorder*. https://www.cdc.gov/ncbddd/autism/data.html (Accessed: 4 August 2023).

Collins, S., et al. (2020). Parenting science gang: Radical co-creation of research projects led by parents of young children. *Research Involvement and Engagement, 6*(1), 9. https://doi.org/10.1186/s40900-020-0181-z

Condon, L., & Ingram, J. (2011). Increasing support for breastfeeding: What can children's centres do? *Health and Social Care in the Community, 19*(6), 617–625. https://doi.org/10.1111/j.1365-2524.2011.01003.x

Corden, K., Brewer, R., & Cage, E. (2021). A systematic review of healthcare professionals' knowledge, self-efficacy and attitudes towards working with autistic people. *Review Journal of Autism and Developmental Disorders, 2021*, 1–14. https://doi.org/10.1007/S40489-021-00263-W

Crenshaw, K. (1989). Demarginalizing the intersection of race and sex: A black feminist critique of antidiscrimination doctrine feminist theory and antiracist politics. *University of Chicago Legal Forum, 1*, 139–167.

Curtis, V. (2018). *Online citizen science and the widening of academia*. Springer.

Ellis, R. et al. (2023). A realist review of health passports for Autistic adults. *PLOS ONE, 18*(9), Article e0279214. https://doi.org/10.1371/journal.pone.0279214

Emerson, L. (2021). Autistic people need a greater say in where NZ's autism research funding is spent—here's a way forward. *The Conversation*, 5 December. https://theconversation.com/autistic-people-need-a-greater-say-in-where-nzs-autism-research-funding-is-spent-heres-a-way-forward-171948

Emerson, L. M. et al. (2023). Autistic co-led community priorities for future autism research in Aotearoa New Zealand. *Autism in Adulthood* [Preprint]. https://doi.org/10.1089/aut.2022.0109

European Citizen Science Association. (2015). Ten principles of citizen science. *Ten principles of citizen science*. https://ecsa.citizen-science.net/wp-content/uploads/2020/02/ecsa_tenprinciples_of_citizen_science.pdf (Accessed: 21 February 2023).

Fletcher-Watson, S., et al. (2019). Making the future together: Shaping autism research through meaningful participation. *Autism, 23*(4), 943–953. https://doi.org/10.1177/1362361318786721

Van Goidsenhoven, L., & De Schauwer, E. (2020). Listening beyond words: Swinging together. *Scandinavian Journal of Disability Research, 22*(1), 330–339. https://doi.org/10.16993/sjdr.756

Grant, A. (2023). Miscarriage and curtains: A phenomenological autoethnography of curtains, privacy, and loss in an Early Pregnancy Unit in the UK. *Geoforum, 143*, Article 103786. https://doi.org/10.1016/j.geoforum.2023.103786

Grant, A. (2024). Scroungers, shirkers and the sick: Disability and welfare in the 21st century. In L. Gregory, & S. Iafrati. Diversity and Welfare Provision. Policy Press.

Grant, A., et al. (2019). Understanding health behaviour in pregnancy and infant feeding intentions in low-income women from the UK through qualitative visual methods and application to the COM-B (Capability, Opportunity, Motivation-Behaviour) model. *BMC Pregnancy and Childbirth, 19*(1), 1–16. https://doi.org/10.1186/s12884-018-2156-8

Grant, A., et al. (2022). Autistic women's views and experiences of infant feeding: A systematic review of qualitative evidence. *Autism, 26*(6), 1341–1352. https://doi.org/10.1177/13623613221089374

Grant, A., Griffiths, C., Williams, K., & Brown, A. (2023a). It felt like I had an old fashioned telephone ringing in my breasts: An online survey of UK Autistic birthing parents' experiences of infant feeding. *Maternal and Child Nutrition*, Article e13581. https://doi.org/10.1111/mcn.13581

Grant, A., et al. (2023b). Unmet need, epistemic injustice and early death: How social policy for Autistic adults in England and Wales fails to slay Beveridge's Five Giants. In *Social Policy Review, 35*. Policy Press, 239–257. https://bristoluniversitypressdigital.com/display/book/9781447369219/ch012.xml

Grant, A., et al. (2023c). The safety of at home powdered infant formula preparation: A community science project. *Maternal and Child Nutrition*, Article e13567. https://doi.org/10.1111/mcn.13567

Grant, A., et al. (2024). Analysing data with members of a stigmatised community: Experiences, reflections and recommendations for best practice from the finding the formula community analysis group. *International Journal of Qualitative Methods, 23*. https://doi.org/10.1177/16094069241229983

Grant, A., Mannay, D., & Marzella, R. (2018). People try and police your behavior: The impact of surveillance on mothers and grandmothers' perceptions and experiences of infant feeding. *Families, Relationships and Societies, 7*(3), 431–447. https://doi.org/10.1332/204674317X14888886530223

Gray, L. J., & Durand, H. (2023). Experiences of dysmenorrhea and its treatment among allistic and autistic menstruators: A thematic analysis. *BMC Women's Health, 23*(1), 288. https://doi.org/10.1186/s12905-023-02370-8

Hall, J. P., et al. (2020). Health disparities among sexual and gender minorities with autism spectrum disorder. *Journal of Autism and Developmental Disorders, 50*(8), 3071–3077. https://doi.org/10.1007/s10803-020-04399-2

Halladay, A. K., Bishop, S., Constantino, J. N., Daniels, A. M., Koenig, K., Palmer, K., Messinger, D., Pelphrey, K., Sanders, S. J., & Singer, A. T. (2015). Sex and gender differences in autism spectrum disorder: Summarizing evidence gaps and identifying emerging areas of priority. *Molecular Autism, 1*(6), 1–5.

Israel, B. A. et al. (2008). Critical issues in developing and following community-based participatory research principles. In *Community-based participatory research for health*. Jossey-Bass.

Jones, S. Cooper, J., Dolling, A., McNamara, T. , Dvorak, S., Sibson, V., Brown, A. Yhnell, A. Buchanan, P. Breward, S., Ellis, R. & Grant, A. (under review). The feasibility and acceptability of developing an online community of parents as community scientists to explore infant formula preparation safety in the home.

Kaplan-Kahn, E. A., & Caplan, R. (2023). Combating stigma in autism research through centering autistic voices: A co-interview guide for qualitative research. *Frontiers in Psychiatry, 14*, Article 1248247. https://doi.org/10.3389/fpsyt.2023.1248247

Kara, H. (2020). *Creative research methods: A practical guide* (2nd ed.). Policy Press.

Keating, C. T. (2021). Participatory autism research: How consultation benefits everyone. *Frontiers in Psychology, 12*. https://doi.org/10.3389/fpsyg.2021.713982 (Accessed: 6 August 2023).

Long, J., & Clarkson, A. (2017). Towards meaningful participation in research and supporting practice: Effecting change in autism services. In *Autism and intellectual disability in adults* (Vol. 2, pp. 41–45). Pavillion.

Mannay, D. et al. (2017). Negotiating closed doors and constraining deadlines: The potential of visual ethnography to effectually explore spaces of motherhood and mothering. *Journal of Contemporary Ethnography*, Online Fir. https://doi.org/10.1177/0891241617744858

Mannay, D. (2016). *Visual, narrative and creative research methods: Application, reflection and ethics*. Routledge.

Mason, D., et al. (2019). A systematic review of what barriers and facilitators prevent and enable physical healthcare services access for autistic adults. *Journal of Autism and Developmental Disorders, 49*(8), 3387–3400. https://doi.org/10.1007/S10803-019-04049-2

Miller, D., Rees, J., & Pearson, A. (2021). Masking is life: Experiences of masking in autistic and nonautistic adults. *Autism in Adulthood, 3*(4), 330–338. https://doi.org/10.1089/aut.2020.0083

Milton, D. E. M. (2012). On the ontological status of autism: The "double empathy problem." *Disability and Society, 27*(6), 883–887. https://doi.org/10.1080/09687599.2012.710008

Moyse, R. (2024). I've got some more to add to that: Using creative methods to empower autistic young people to tell their stories. In H. Kara (Ed.) T*he Bloomsbury handbook of creative research methods.* Bloomsbury.

NICE (2021) *Autism spectrum disorder in adults: Diagnosis and management. CG142.* NICE. https://www.nice.org.uk/guidance/cg142 (Accessed: 7 August 2023).

Nicolaidis, C., et al. (2011). Collaboration strategies in nontraditional community-based participatory research partnerships: Lessons from an academic-community partnership with autistic self-advocates. *Progress in Community Health Partnerships: Research, Education, and Action, 5*(2), 143–150. https://doi.org/10.1353/cpr.2011.0022

Nicolaidis, C., et al. (2013). Comparison of healthcare experiences in autistic and non-autistic adults: A cross-sectional online survey facilitated by an academic-community partnership. *Journal of General Internal Medicine, 28*(6), 761–769. https://doi.org/10.1007/s11606-012-2262-7

Nicolaidis, C., et al. (2016). The development and evaluation of an online health-care toolkit for autistic adults and their primary care providers. *Journal of General Internal Medicine, 31*(10), 1180–1189. https://doi.org/10.1007/s11606-016-3763-6

Nicolaidis, C., et al. (2019). The AASPIRE practice-based guidelines for the inclusion of autistic adults in research as co-researchers and study participants. *Autism, 23*(8), 2007–2019. https://doi.org/10.1177/1362361319830523

O'Dell, L., et al. (2016). Critical autism studies: Exploring epistemic dialogues and intersections, challenging dominant understandings of autism. *Disability and Society, 31*(2), 166–179. https://doi.org/10.1080/09687599.2016.1164026

Pavlopoulou, G. (2021). A good night's sleep: Learning about sleep from autistic adolescents' personal accounts. *Frontiers in Psychology*, *11.* https://doi.org/10.3389/fpsyg.2020.583868 (Accessed: 7 August 2023).

Pell, B. et al. (2020). Using visual timelines in telephone interviews: Reflections and lessons learned from the star family study. *International Journal of Qualitative Methods,* 19. https://doi.org/10.1177/1609406920913675

Pellicano, E., Dinsmore, A., & Charman, T. (2014). What should autism research focus upon? Community views and priorities from the United Kingdom. *Autism, 18*(7), 756–770. https://doi.org/10.1177/1362361314529627

Peuravaara, K. (2021). Emotional nuances: Critical reflections on emoji methods. *Scandinavian Journal of Disability Research, 23*(1), 305–316. https://doi.org/10.16993/sjdr.822

Phillips, R., et al. (2018). Identifying the unmet information and support needs of women with autoimmune rheumatic diseases during pregnancy planning, pregnancy and early parenting: Mixed-methods study. *BMC Rheumatology, 2*(1), 21. https://doi.org/10.1186/s41927-018-0029-4

Raymaker, D. M., et al. (2020). Having all of your internal resources exhausted beyond measure and being left with no clean-up crew: Defining autistic burnout. *Autism in Adulthood, 2*(2), 132–143. https://doi.org/10.1089/aut.2019.0079

Raymaker, D. M., et al. (2023). [I] don't wanna just be like a cog in the machine: Narratives of autism and skilled employment. *Autism, 27*(1), 65–75. https://doi.org/10.1177/13623613221080813

Samuel, P., et al. (2021). Sensory challenges experienced by autistic women during pregnancy and childbirth: A systematic review. *Archives of Gynecology and Obstetrics, 1*, 3. https://doi.org/10.1007/s00404-021-06109-4

Shepherd, J., & Shepherd, J. (2015). Interrupted interviews: Listening to young people with autism in transition to college. *Exchanges: The Interdisciplinary Research Journal, 2*(2), 249–262. https://doi.org/10.31273/eirj.v2i2.114

Smith, K. (2013). *Beyond evidence-based policy in public health.*. Palgrave Macmillan. https://doi.org/10.1057/9781137026583

Stone, B., Cameron, A., & Dowling, S. (2023). The autistic experience of homelessness: Implications from a narrative enquiry. *Autism, 27*(2), 489–499. https://doi.org/10.1177/13623613221105091

Teachman, G., & Gibson, B. E. (2018). Integrating visual methods with dialogical interviews in research with youth who use augmentative and alternative communication. *International Journal of Qualitative Methods, 17*(1), 160940691775094. https://doi.org/10.1177/1609406917750945

Wilkinson, S., & Wilkinson, C. (2018). Researching drinking "with" young people: A palette of methods. *Drugs and Alcohol Today, 18*(1), 6–16. https://doi.org/10.1108/DAT-08-2017-0036

Williams, D., et al. (2022). Identities of women who have an autoimmune rheumatic disease (ARD) during pregnancy planning, pregnancy and early parenting: A qualitative study. *PLoS ONE*. https://doi.org/10.1371/journal.pone.0263910

Wood-0, H., et al. (2021). Sex/gender differences in camouflaging in children and adolescents with autism. *Journal of Autism and Developmental Disorders, 51*(4), 1353–1364. https://doi.org/10.1007/s10803-020-04615-z

CHAPTER 5

Neurodiverse Research Collaboration Within a European Biomedical Project: Limits and Possibilities

Joke Struyf, Cecilia Ingard,
and Hanna Bertilsdotter Rosqvist

Introduction[1]

R2D2-MH is a large European multinational biomedical project on Risk and Resilience in Developmental Diversity and Mental Health. It aims to improve the mental health outcomes of young neurodivergent people across the lifespan. The project made it a priority to work with co-creation groups (CCG) and explicitly engages with the neurodivergent community, with a special focus on autism and ADHD. An adult CCG and adolescent CCG were created, made up of mainly white, middle-class people with and without intellectual disabilities from the Northern or Western parts of Europe. The aim of the groups was to (i) think about language about developmental diversity (c.f. Hens & Van Goidsenhoven, 2023) in an inclusive way, (ii) reach a

[1] The authors wish to acknowledge the important contribution of Leni Van Goidsenhoven as discussant on the chapter. In the later stage of the chapter, Kristien Hens, Thomas Bourgeron, Vincent Grimaldi de Puget, and José Miguel Carrasco have contributed to the chapter with important comments and reflections on the chapter.

J. Struyf (✉)
University of Antwerp, Antwerp, Belgium
e-mail: Joke.Struyf@uantwerpen.be

C. Ingard
University of Gävle, Gävle, Sweden

H. Bertilsdotter Rosqvist
School of Social Sciences, Södertörn University, Huddinge, Sweden

H. Bertilsdotter Rosqvist and D. Jackson-Perry (eds.), *The Palgrave Handbook of Research Methods and Ethics in Neurodiversity Studies*,
https://doi.org/10.1007/978-3-031-66127-3_5

consensus on the outcome measures that they consider acceptable and useful, and (iii) give feedback on the research progress. The authors of this chapter have different roles in the project. For a period of time, Joke facilitated the adult CCG. Cecilia is a participant in this CCG. Hanna is part of R2D2-MH's scientific advisory board.

The chapter focuses on the first of the five years of the R2D2-MH project. During this time, different project teams and CCG got to know each other and sought ways to work together. Between January and October 2023, there were eight online CCG meetings and several informal meetings between the CCG members. From June 2023, CCG members were invited to join meetings of the separate work packages (WPs) and the executive board.

We draw on a collaborative autoethnographic methodology (Chang et al., 2016). We aim to explore and reflect on whether and how this large biomedical project *is* or *could be* taking steps to centre neurodivergent participation in knowledge production. We further consider how to create a more emancipatory academic space, including and enabling different minds (Bertilsdotter Rosqvist et al., 2019). We explore the difficulties and possibilities of emancipatory co-creation from two different perspectives: *fight for co-creation* and *listen and act upon input* departing from some 'critical incidents' (Roos, 2002, see also Chapter 12, this volume) based on our experiences within the project.

Fighting for co-creation includes being the trouble shooter or a 'killjoy' (Ahmed, 2023). In an environment that 'does not sustain the shape and function of the body that enters it' (Garland-Thomson, 2014), we become misfits. By pointing out problems, we might be regarded as being the problem ourselves. While searching for co-creation opportunities we also consider how serious the people involved are about it, or whether co-creation is merely window dressing—an effort to make the project more eligible for funding. What we do is 'diversity work', which signifies both work that has the explicit aim of transforming an institution (in this case the institution of scientific research) and 'the work required, or what we do when we do not quite inhabit the norms of an institution' (Ahmed, 2014). Fighting for co-creation also means suggesting alternatives, which we will be doing in this chapter.

To listen and act upon input includes viewing trouble shooting as a valuable act of participation and acting accordingly in response to this; taking account of these suggestions to move the traditions of research into other ways of doing research. We not only consider the limits and possibilities of neurodiverse research collaboration but also pay attention to how this way of doing research can be a caring, epistemic healing experience, as well as a learning experience for all people involved.

In this chapter, we will describe some critical incidents during our experiences in R2D2-MH. These critical incidents were chosen as being particularly illustrative of situations where we were constructed as neurodivergent subjects. Each description will be followed by a 'reimagining' (c.f. Garland-Thomson, 2005). In reimagining problematic situations, we seek to explore alternative futures in which we collaborate in a more epistemically just way: 'doing neurodiverse collaborations differently'. At the end of the chapter, we have put together a check list for 'doing it differently'.

Critical Incidents: Exploring and Reimagining Co-creation

> One time, in a meeting with no CCG members present, a researcher said that they 'realised now that many of the participants are autistic'. They already knew, but only then they realised that it would have consequences, e.g. for the kind and number of questions the participants ask, the level of detail and depth they expect, and their persistent critical view (the word 'negative' was used). I thought the negative undertone of the remark was unfair, but I might interpret that unnecessarily as negative since I had encountered hesitation about co-creation at previous instances. This moment reinforced my belief that we needed to fight for co-creation because it wouldn't happen on its own. *(One of us[2])*

We have found the notion of 'fighting' useful in the context of co-creation and neurodivergent knowledge production in neurotypical-led projects. 'Fighting' stresses that the hard work is not only about trouble shooting or diversity work in the sense of letting people know about accessibility needs and stressing the importance of seeking solutions together. The assumed 'openness towards all neurotypes' is not a fixed quality of the project and needs to be negotiated for constantly. The hard work is also to engage with knowing but not knowing—as when NT researchers 'knew that, but only then they seemed to realise that it would have consequences'. To be the killjoy and not being met with 'listening and acting upon input' is hard work. It is about acknowledging the impact of inequality in power relations between neurotypes and the challenges of democratizing research processes in practice as pointed out by Pellicano and Stears (2019, p. 452).

The desire to 'democratise' any process involves recognizing that each participant deserves some equality of respect in that process and that means, in turn, that they must be granted relative equality of standing in any decision-making process. Furthermore, this requires that inequalities of power and influence, often either hidden or ignored in the research process, become the subject of considerable ethical attention (c.f. Chapter 7, this volume).

In the following section, we have chosen to focus on some illustrative critical incidents we have encountered during our first year with R2D2-MH. We

[2] Due to the sensitivity of this situation, we have chosen to refer to the one of us speaking here as 'one of us', rather as a named one of us. 'One of us' also functions as the expression of 'joint action' which feminist researchers Francis and Hey (2009, p. 231) have stressed as "core to feminist action over the years" but particularly within 'academia' where 'joint action' counter-narrates the position of academics as 'individual experts'. For further discussion and use of this term, see chapter 7, this volume.

describe and reimagine the critical incidents of *the door* (Hanna), *the consensus process* (Joke) and *the glossary* (Cecilia). All our incidents illustrate a fight for openness, where doors are to be engaged with collectively.

Doing Trouble Shooting: the Critical Incident of the Door

Hanna's first encounter with the project as part of the scientific advisory board took place during the first annual general meeting in Paris. This was also the meeting where the critical incident of the door took place. The door was physically placed in the main meeting room. The door illustrates not a singular moment happening at a meeting, but a structure underlying the whole project, exemplifying the structural in the personal. In the following, Hanna describes the critical incident of the door.

> I had deliberately chosen to sit close to the door in case I got an overload and would need to as discretely as possible get out of the space, to be able to manage my overload on my own and avoid a meltdown. The door was either closed and locked, or open. Implicitly the person sitting directly beside the door, closer to the door than I was, was assigned the role of door opener. When anyone wanted to get in after an unscheduled break, they knocked on the door and that person had to divert their attention between what was discussed and the door, get up, and open the door as friendly as ever, regardless of how many interruptions had already happened. Eventually a chair was placed in the doorway to keep the door open and enable anyone to pass freely and without disturbing. Noises from outside started to also pass freely into the meeting room. The increased sensory input of the noises put increased demand on my cognitive processing which soon evolved into an overload. I started to regularly, as in a dance with the door opener, close the door. I didn´t understand at first there was a door opener. I thought it was a matter of someone being lazy, forgetful, who didn´t think of taking away the chair in the door after a scheduled break. But after a while, I saw them in action putting the chair in the door. I asked as quietly as possible why not keep the door closed. They responded that it was better if the door was kept open if someone wanted to come in or out. I don´t know how long it took me to lose it. Fortunately, I didn´t get into a full meltdown. I just went very angry very quickly and decided my sensory needs were a legitimate reason to interrupt the whole meeting room. I am not sure how loud I cried out that the whole meeting needed to stop for a while and turn its collective attention to the door. The person who was currently presenting fell silent. I told the whole meeting room the door was a collective problem to be solved, not an individual problem of mine, the door opener and the people choosing to take unscheduled breaks. I am not sure how many really got it. A man a bit away from me suggested in a private tone that the easiest solution for the problem was for me to leave my place and take a place far away from the door. I answered him equally privately, that it was not a solution to treat this as an individual problem, and that the solution should be collective responsibility for the space. A person from the organising committee apologized for the impossibility to keep

> the door closed but unlocked, due to the restrictions of the host university. I reflected upon why it is so hard to act collectively to work with different needs in an open and pragmatic way, why it is so easy to expect some to 'move' and others to 'keep moving freely', to keep 'not doing things' until someone decides to be a killjoy? At the end of the general meeting I told the project PI I was sorry for being so 'bothersome' about the door. He replied that it was not a problem and that this was why I should be part of this, that my intervention was important.

The incident with the door made Hanna feel 'conditionally welcomed' to the general meeting in Paris. It illustrates a barrier for her to do her assigned task as a member of the scientific advisory board. It is also reminiscent of Nick Walker's question as to whether 'having a real voice within [the project] is contingent upon acting neurotypical and upon making the sort of social contributions that don't trouble the collective trance of neuronormativity' (2021, p. 76). We think about the critical incident of the door not only as a concrete sensory barrier towards participation but following Sara Ahmed, we also think about the door in a more metaphorical way. According to Ahmed, perceptions are doors which are not perceived by those who are not stopped by them. So, if you are stopped by a door, 'it can seem like you have stopped yourself' (2023, p. 165). Ahmed states that those who fit because the world has been made for them do not experience these doors, which is an energy-saving privilege. Aiming for co-creation with neurodivergent people, openness is part of the project's objectives, in the sense that the official view is that 'the door is open, anyone can enter'. Openness is assumed to be a 'quality of [the project] instead of a relation to it' (Ahmed, 2023, p. 165), ensured by the selection of neurodivergent participants and the aim of co-creation. That openness is assumed by people who do not experience the 'doors' in Ahmed's meaning.

Next, Hanna reimagines the meeting and another way of setting up the annual general meeting. What would have made her feel welcome at the meeting and enabled her to do her task as member of the scientific advisory board? In this reimagination, the door incident is not happening, as it can't happen in another structure.

> The project should be an opportunity to try out and set up **a new standard for neurodiverse research meetings**, paving the road ahead for others to follow. In this new standard, **cognitive differences are taken for granted** (where no cognitive functioning is taken more for granted than others). **Trouble shooting would be seen as natural parts of aiming at developing the meeting standard further**. Without trouble shooting we can´t see what does not work. To take cognitive differences for granted includes focusing on both the task and how to enable people with different minds to do their assigned tasks. For example, during an annual meeting of a big research project it is important to meet and discuss, share experiences of the research process so far and support each other in the different research processes. At the same time each member

of the projects needs to be steered in order to keep the project on the general track of the project aim. Focusing on the task at hand – to enable people with different minds to meet and support each other aiming at fulfilling the greater aim of the project means to think creatively about **different ways of meeting**.

The planning of the general meeting would include people with cognitive differences at the start of the planning process in order to enable trouble shooting at an early stage, not at the stage when there is not much one can do other than extinguish fires. During this planning process it is important to have an openness for different solutions to different needs, to creatively explore different ways of meeting. This would mean a point of departure without an already set meeting template to put in practice. Rather, the template is the process, where the way of meeting is to be developed as we go. This includes a process where **to learn from trying is expected**.

A way to explore different ways of meeting is to initially share a wish list-document to gather different ideas based on different needs. One can also have a brainstorming meeting in the planning group.

I can think of ways of meeting for me which would be more adapted to my processing needs. An annual general meeting can be spent over **an extended period of time, perhaps a week rather than 2-3 days**. It could consist of several shorter meetings including breaks with time allowing for processing knowledge. Meeting up could mean meeting up both **in real time and digital synchronous and asynchronous meetings**. As part of a meeting in real time one could have **walk and talks in surrounding parks** in different smaller group settings where we were going to discuss matters related to the project and our tasks in the project.

The talks and presentations at the meeting could consist of **pre-recorded talks with Q&A in real time**.

Fighting for Co-creation

When Joke started, the job with the CCG seemed to be a strictly management job. There was a protocol about communication with the participants, a provisional schedule with meeting data and an agenda with project-led themes. To keep a structure was deemed highly important. In this phase, the communication with the CCG was project-led: we provided information, asked questions upon which the CCG would comment, discuss, and give feedback.

Soon, the structure started to fall apart. The exact dates of the meetings were not decided upon, and it was not clear to Joke why she and her colleague CCG facilitators couldn't pick a date based on the availability information the participants had given prior to her joining. At a certain point, she asked to choose the dates for the next three meetings because she could not get her own agenda sorted out. The CCG members had not asked her about it, but she felt strongly that they needed this information as soon as possible.

CCG discussions took a lot more time than planned and they did not manage to finish the agenda. Without knowing about the core principles leading the project, some participants thought it not meaningful to have certain discussions. More information was needed about the project, and the role of co-creation in it, for the participants as well as for the facilitators of the CCG. They all lacked an overview. Joke felt a close connection to the participants, who were asking relevant questions, often via mail. If she couldn't provide an answer, she told them. She felt at ease being honest and at the same time curious about their questions and the project.

She started to realize that within the project, ideas about co-creation were not aligned. Co-creation involves searching, waiting, giving time to each other and the process, and it can feel a bit'messy' (see also, Chapter 8, this volume). A similar 'messiness' comes with working within a big European project involving many researchers, each with their own interests, communication style, timeline, and goals. Joke feels at ease with a certain level of messiness, but for another colleague in charge of the CCGs, it was hard. They thought the participants shouldn't question everything and that they were making it difficult. Unlike Joke, they felt uneasy with co-creation, with centring the experience and skills of the participants and trusting the process. They were focused on deliverables and a strict schedule but with the participants growing more critical and the meetings circling around the lack of information, a fixed structure was more getting in the way than helpful.

In the meantime, the participants needed to fill out a feedback form after each meeting, complete a reimbursement form and a document on their needs for the Paris meeting, etc. Joke needed to pass their information through to other people, like a hatch, which wasn't satisfying. Many participants did not reply to her emails, so she assumed that she was asking too much of them. Moreover, she could not give them the real co-creation role they had envisioned. Some chaos is to be expected in a big project, but the CCG members had fundamental questions about their own task, which remained unanswered at the time. While both Joke and the participants lacked basic information about the project, an imbalance was created in the project, asking too much while not giving them enough. In the following, Joke will describe one critical incident from her experiences of working with the CCG: the consensus process.

> Soon after me joining the project, one of the meetings considered the 'consensus process', to find out themes which the participants deem important for resilience. A colleague from another team in charge of this process was involved. She did not know the CCG participants and was initially not planning to attend the CCG meeting discussing the outcomes, but I asked her, since she could explain the process much better than I would. I did not understand her role and involvement nor what her relation would be with the CCG participants.
>
> To keep the discussion as free and open as possible, the team in charge decided to inform the participants only minimally. During the meeting I realised that was a huge mistake, which the participants confirmed in the feedback forms: they wanted to be prepared and needed more context. Because I did not really

understand the specific colleague's role in the consensus process, I had not inquired more about it.

Next, the data from the consensus discussion was put into categories, decided upon by the other team. They did not know how to interpret some of the participants' remarks but did not take initiative to contact the CCG members again to ask for clarification. They did ask me to read their list afterwards, but I felt that I was not able to interpret the participants' expressions. There was no time to discuss the categories with the participants afterwards. The other team clearly had different ideas about the participants' involvement. I wanted the CCG participants to be involved from start to end, which is how I understand co-creation, and I felt that it was becoming increasingly my responsibility to ensure that. However, at this moment, there was no time to organise their involvement. There was a sense of urgency about this consensus process but not enough transparency as to why it was so urgent and why we could not take more time to understand the context first or involve the participants more.

Although the consensus list was co-created with the participants, both the outcome of the discussion and the participants' involvement could have been much better should they have had time to prepare. Recalling Ahmed's 'doors', I believe the other team's strategy had not considered the participants' needs to get sufficient preparation material and time. The consensus meeting made me realise that I would have to take up a more active role, even towards the colleagues of the other team, to unveil the 'doors'.

These colleagues included the CCG's answers into a questionnaire to send out to all the researchers in the project. Later, CCG members 'discovered' this questionnaire at a WP meeting. They raised critical questions and when they asked to see the document, these colleagues said they needed to wait for the next version, which created frustration on the part of the participants. Next to the lack of transparency, they questioned the validity because they had not understood fully what was at stake during the discussion, and some questioned whether they were enough of a representative of their community to influence such an important questionnaire, leading to define the outcomes of the project.

The misunderstandings around the consensus process were frustrating for the CCG members but they were vocal about it. At the September meeting, another colleague from the other team took the time to explain the whole process all over again. They apologized for the way things had gone and suggested to have an extra reflection moment about the questionnaire at the Paris meeting, when there can still be changes made. The CCG accepted this suggestion.

In the following, Joke reimagines the work with the CCG as central in R2D2-MH, ways of doing it differently. In this reimagination, the current CCG work is not happening, as it can't happen in another structure.

I would **involve the participants from the beginning** in the writing of the project proposal and in all meetings. They would know the project structure and whom to address their questions and useful information to. Co-creation would be central in the project including direct communication as a norm, and the understanding that extra care is needed since we are working together as people of different neurotypes.

All project partners would discuss co-creation, both practically (about the possibilities co-creation holds and how to do it), **and on a more ethical level** (the value of co-creation in research; how to make sure the process is respectful towards everyone; issues of representation, legitimacy and power), before starting the project. It should be an informed, positive choice to use co-creation, supported by all R2D2-MH partners. **Time** should be allocated specifically towards this end. It would take a lot of time to have those profound discussions and that should happen at the beginning. There would be understanding of co-creation and trust in each other and the process. **Other researchers would be curious about co-creation**, and would spontaneously reach out to the CCG, as a reciprocal interest would mean possibilities of co-creation as a process happening between partners. All researchers would understand that some participants act like researchers due to their own need for detail and their ability to understand difficult content, and they would embrace this quality. Researchers would allow themselves to be led by CCG input, believing in the value of participants' experience and skills and the need for a different kind of research. Suggestions or questions by CCG participants would be systematically listened to by researchers, their relevance taken for granted. All researchers, not only CCG participants, would make efforts to point out and suggest solutions for situations where different needs collide.

All people should be able to get out of the co-creation process whenever they feel uncomfortable. Co-creation is a creative, messy process, especially in a big project. For some people, it is too messy. Participants can easily get out of the project, but colleagues should also be able to get out if they feel uncomfortable. It is not a failure and it does not determine the success of the project.

While there needs to be structure, it cannot be a fixed structure imposed by one party on others. Nevertheless, from the start there would be more clarity about the **organisational structure** of the project, the different studies and research questions, and the broader vision, all in layman's terms.

Allowing for dreaming and learning by doing may create the possibility to **let the CCG 'seize power' of the project**, exploring how the CCG can really make it 'their' project. We need for the CCG participants to meet, amongst themselves or with the CCG facilitators (whatever they like), and to dream. I believe they have already started to do that.

The Glossary

Cecilia is a CCG member, a researcher, and a member of the Swedish Autistic and ADHD community. She has worked with participation for persons with different disabilities. She expects to be involved in a project where the researcher regards the co-creators' experiences as necessary knowledge. The following section is her description of the first meetings of the CCG and how the group members slowly tried to figure out what the group's aim was and what the group could contribute to the project, including the critical incident of the glossary.

> The first meeting with the CCG included presenting the project and members of the CCG for them to get to know each other. However, it was also essential to understand in what context the CCG was going to work. We needed an overview of the project. This included what the project expected that the co-creators should do and how much power the researchers would be willing to give to the CCG.
>
> During the second meeting, we discussed glossaries that the projects will use. We had a meaningful discussion about the glossary. However, it was clear that even if we all came from neurodivergent communities, it differs from which country we came from and how we defined the different words in the glossary discussions. Moreover, autism and ADHD are not a homogenous group; instead, they differ regarding gender and country.
>
> The work with the glossary just came up, and I didn't understand why we had to work with those words. Before the glossary meeting, we got the list with the dictionary we should work with, a list of words that someone had identified as essential to discuss. I was surprised that work on the glossary was important before any other work in the CCG. Maybe using the right word is symbolic for some. However, discussing the different research projects is more enjoyable for me. I need the project's history, such as how they constructed the glossary. During the meeting, everyone had a lot of comments on the glossary but the meeting was too long, the participants were exhausted, and it was not a productive meeting in the last hour. Some of us experienced brain fatigue.
>
> We all needed clarification and did not understand what we participated in during the first meetings. Nevertheless, we slowly empowered ourselves and created a chat group via an app. Through that app, we discuss the project without any of the researchers and formulate our frustration about the lack of power and participation. Moreover, we discussed the glossary meeting and felt that the glossary meeting was essential, but the group deemed it of lower importance than the researchers did.

Another, similar, critical incident Cecilia recalls is the matter of the website turning into a discussion about resilient factors for mental health.

> At the third meeting we discussed what topics are crucial to research regarding resilient factors for mental health. It was a critical discussion, but I had not prepared myself for those discussions. Before the meetings, we got information that we should talk about the website, and it was that discussion I had prepared. I got a small amount of panic when I thought we should talk about another topic, especially an important one that should be a ground for some studies. We discussed many factors that affected mental health, and it was an exciting discussion. However, I was curious to know if the researcher would value our points and if they were willing to adopt our point of view in their project.
>
> The CCG members wanted to participate from the beginning in the design of the different projects. We experienced that our role now is to be an advisory

board which researchers can address whenever they need us, in which case we have no control. The researchers truly wanted to include us, but they needed to learn how, and it was a concrete task to ask us about the glossary. However, the project could improve the level of participation for the CCG members by listening to how the group members want to contribute to the project. Most of the researchers are not used to involving co-creators and did not expect that they need to give the CCG some of their power.

The CCG had an informal meeting one evening to discuss our visions without anyone working for R2D2-MH. Many of us needed to have the whole project picture, but most importantly, we demanded to be involved in the different studies from the beginning. If the researchers let us react to the design, aim, and other aspects of the studies when it is too late to affect the study, that is paternalism. More transparency is needed: during the whole project, the researchers should let the CCG know what is going on and not just call for our feedback occasionally when they need our point of view. The researchers might think that we are fragile people who should be protected from too much work, but that is also paternalistic.

A third critical incident Cecilia recalls is that of choosing members from the CCG to participate in one of the meetings with researchers.

Some people decided due to practical reasons that we must choose two of us to join the next meeting, by lottery. However, the CCG members felt we should decide how and to whom we select our representative. We must have the power to facilitate how our group should work.

Finding a process involving the co-creation group that will not be tokenism is challenging, according to Arnstein's (1969) ladder of participation, which includes eight steps that show the grade of empowerment for a group of people with less power (see also Chapters 8, 24, this volume). It is unrealistic that the CCG should have total control over the different studies. However, if R2D2-MH has high ambitions about co-creation, they must strive to reach a higher step, according to Arnstein's ladder of participation.

During a meeting with the co-creators and the leaders of the WPs, it was apparent that the WP researchers didn't expect that CCG members wanted to be involved in more than just some part of the projects. The highly qualified co-creators, some of whom are researchers in other fields, want more than just being informed. They want to contribute with knowledge during the whole process rather than only when the researchers have already decided on the design and aim for the different projects. They have knowledge of living with autism and ADHD and about the research process, which is unique.

When we told the project leaders we wanted to participate more, they invited us to many WP and executive board meetings. That was an excellent ambition, but it resulted in a lot of email invitations, which were overwhelming for some CCG members.

In the following, Cecilia will be reimagining the work with the CCG as central in R2D2-MH and ways of doing it differently.

The project should decide with the CCG **which task is the most important**, **how we should work with that task**, and how we value the glossary work next to the other jobs we identify.

The researchers could have sat down with us and planned our participation in the different WPs. Together we could figure out how we could contribute best to the project and how the project can use our knowledge best. Some of us may be more interested in glossaries, and others may prefer to participate in just one WP.

Due to nearly all CCG members belonging to the neurodiversity community, we need schedules of all meetings with aim and scope, and clear roles. I need to plan my time, otherwise I can feel stress. The CCG members should have the power and possibility to plan and structure how our group should work and which person should represent the group in different meetings. R2D2-MH must trust the CCG members.

The WP researchers must be interested in co-creation and humble towards the co-creator's knowledge. Co-creation is a complex and challenging process that takes time.

Concluding Reflections: Rethinking Co-creation

In R2D2-MH, there is enthusiasm about focusing on co-creation in the project. However, the sometimes messy practices of co-creation which can loosen and unsettle dominant and traditional ways of doing research also have been a challenge for R2D2-MH. We have used project meetings within R2D2-MH as illustrative examples of the risks of 'business as usual' (c.f. Bertilsdotter Rosqvist et al, 2020). The welcoming into a neurodiverse space (Bertilsdotter Rosqvist et al, 2019) has sometimes been marked by a 'thinness' of neurotypical-led conditionality, meaning that neurodivergent people are welcomed and invited to participate while the space remains adapted to neurotypical needs (see, e.g., Parsons et al, 2020).

Co-creation was not yet defined within the project proposal. The idea was to do this together, to craft and deepen the understanding of what this kind of research can mean in the context of R2D2-MH. However, as the project started, research labs went on to do research as usual. Members of the CCG experienced a hierarchy of knowledge, where researchers' knowledge was ranked higher than knowledge based on lived experience, illustrating encounters of epistemic injustices between groups (Fricker, 2007). In the context of R2D2-MH, epistemic injustices impact who has the priority to interpret the data, decide what is essential to research, and what the aim of the different studies is. Is the researcher willing to give some of their power to the co-creators, and if they disagree, who will then make the decisions? This became a dilemma in particular as many of the members of the co-creation group are researchers in another area. The distinction between 'researcher' and 'CCG member' was therefore not as clear-cut as commonly in research, where the researchers have research expertise and co-researchers need to learn how to do research along the way.

Co-creation and inclusive research are more than asking for feedback from the community when you present your research aims or results and doing research as usual despite changes in communication, gatherings, and presentations. Drawing on feminist theorizing, we reflect upon the possibilities of making neurodivergent participation not only an add-on to the project, but really integrate it into the research and how this can add to the debate around collaborative research. By this we want to, in line with an inclusive research discourse, 'maximise participation' (Harrington et al, 2014). At the same time, we want to stress that maximizing participation is not only about 'welcoming to participate' but a genuine interest in exploring and changing research methods in line with a more emancipatory research participation that includes hard work for all participants.

R2D2-MH is an interesting case to explore as it is the first co-creation project on such a big scale focusing on biomedical issues, which is a difference and a challenge compared to smaller projects. All our stories paint a picture of a large project with many groups who are not communicating with each other, which allows for things to be done 'business as usual'. There is a desire for this project to include genuine co-creation, but no one knows how this works in practice. Moreover, since co-creation was not defined within the project proposal, it took the first year to realise that this defining work, or 'co-creating the co-creation', would need to be done along the way.

In this final section, we collectively seek to rethink co-creation: how can it be done differently? We believe in the power of *many* good examples, of many toolkits, guides, and checklists (not *the* guide/checklist) to show how it *may* be done differently.[3] This includes the need to live with imperfections, without aiming to create crip utopias (c.f. van Goidsenhoven et al, 2022) including acknowledging that there is no 'frustration free' process. Due to its complexity and rich contradictory stances, crafting accountable, sensitive, accessible, and thus inclusive neurodiverse research spaces is not a straightforward practice with fixed methodologies. It is not about ticking boxes nor is it our aim to catalogue what is not working in neurodiverse collaborative work. We all know that solutions to a particular need do not necessarily meet the needs of others and can even create new barriers. Sometimes expected access and support do not live up to its promises. Different expectations, contradictory beliefs, confusing moments, and other elements are an inherent part of doing inclusive research. Sometimes just being there can be asking a lot and even be painful. Reconceiving collaboration and inclusive research are rather a set of large interdependencies and accountabilities. As our reflections are based on the very first year of the project, this is only the beginning and there is still time for the project to develop.

[3] For example, for other inspiring toolkits, see https://aaspire.org/

Opening doors as a process, a relationship and a task. In the language of Ahmed, doors—both in their literal/material and metaphorical sense—illustrate in our chapter a fight for co-creation which is not met up with 'listening and acting upon input'. The doors in our stories are not singular moments but illustrations of structures, experiences of the 'thinness' of being conditionally welcomed in neurotypical-led projects. Following Ahmed, we believe that real openness is more a process or a relation than a quality. It takes communication skills, patience, and time to communicate with both CCG members and researchers. We get frustrated when others do not perceive doors and we need to argue again and again for the existence of the doors, trying to make them visible, and knowing that effort will probably be unsuccessful. This 'diversity work' (Ahmed, 2014) those who do not fit have to do all the time, is exhausting. We realise that the CCG participants are doing this constantly, pointing out doors to us and others, and we acknowledge their effort by literally writing it down whenever a progress report about the co-creation is due. In the co-creation process, we wonder whether there should be someone who checks the doors, both the real and the metaphorical ones, constantly. Should this be a facilitator's task? Possibly. Regardless, this is a task to be done and the person doing it needs to be given authority and legitimacy to do it.

Co-creation as a dual not one-sided process. Next to fixed deliverables like the glossary, we wriggle in the so-called co-creation of the co-creation, to make this effort visible. However, so far, it feels as though only the participants are co-creating among themselves. There is no 'co'-creation without the involvement of researchers. Involving them takes time, because the whole research structure (both as an institution and in this specific project) is organised without co-creation in mind. In trying to transform the institution of scientific research, we encounter institutional resistance—but we try to turn the object of this resistance into a platform of institutional action (Ahmed, 2014).

Mutual efforts to create (safer) spaces for co-creation. During the process of working within the project, there has been a tension in the project between saying what the CCG needs to do—via project-led deliverables and a fixed agenda—and the often-mentioned purpose of 'co-creating the co-creation'. Co-creation cannot be just an afterthought, it needs to be integrated into the messy part of doing research, the design setting. If that is (still) impossible, the idea is to create space for the CCG to infiltrate into the different subprojects. Up until this point, this 'creating space' has been a one-sided effort from the CCG members and facilitators, fighting to get an overview, asking for co-creation opportunities, being critical about mere feedback questions, and constantly pointing out where this research is still neurotypical.

Need for control. Throughout, everyone involved wishes for more information about future plans and why certain meetings are arranged. The much-needed overview may not be possible in the first months, as the project is large and involves many research groups. However, the CCG members, as co-researchers, have a need for control; the difficulty of seeing the whole picture makes for an even more confused existence.

Respect for different knowledges. CCG members feel that their knowledge is not taken seriously and valued. The project should have involved the CCG from the beginning of the planning, even if there is yet no space or financial support for writing grant applications involving CCG. Nevertheless, we recognise that although it seems artificial to do co-creation after the design set-up, it is an important first step in discovering new ways of doing this in the future.

Requirement for efficient co-creation: time. Most fighting for co-creation involves fighting for time. Currently, no time is explicitly scheduled in the project timeline for this. Time also includes flexibility, supporting time management (sms alerts or reminder emails), keeping time during the workshops/meeting with stakeholders, time-thinking (short and efficient sessions), etc.

A Checklist: How to Do it Differently?

1. Involve co-creation from the beginning: **collectively write a project proposal** and include all the people you want to involve once the project gets funded, especially those you want to co-create with.
2. Take your time finding out what everyone involved means by co-creation. Should no one know, take initiative. **Draft a definition** and ask the others for feedback.
3. **Write down together the practical consequences** of this kind of co-creation. Schedule regular meetings to talk about co-creation: should the definition and the practice be changed?
4. Add **alternative deliverables** regarding co-creation to the official deliverables. If you don't write them down, they won't count (because they won't exist). Suggestions: 'a working definition of co-creation in this project', 'practical co-creation guidelines', 'x participants have been connected to x researchers', 'x people have worked on a working document about future collaboration', or 'CCG participants found their areas of interests within the project'. Choose realistic deadlines: this last one will probably take over a year.
5. Point out where **misunderstanding/confusion** is based on neurotypical ways of doing and thinking. You might need clarity and ask questions but keep getting different answers because of a **general vagueness. Go with the answer that you like most.** What is the worst that can happen?
6. **Read** works to remember how important and difficult it is to transform non-inclusive practices or institutions: e.g. *The Feminist Killjoy Handbook* or *On Being Included* by Sara Ahmed. They can be a comfort whenever it is too much. If you can already see a utopian horizon of research done differently, take a deep breath and **focus on the small changes**.

They need to be repeated to make a real structural change. It will be frustrating, but **don't despair**, frustration is part of it. Others will get frustrated too, and anxious about changes to their usual habits. You can **support each other** in the experience of those feelings; it might increase mutual understanding.

Acknowledgement R2D2-MH has been funded by Horizon Europe [grant agreement no. 101057385], by UK Research and Innovation (UKRI) under the UK government's Horizon Europe funding guarantee [grant no.10039383] and by the Swiss State Secretariat for Education, Research and Innovation (SERI) under contract number 22.00277.

References

Ahmed, S. (2014, June 4). *Practical phenomenology*. Feminist Killjoys. https://feministkilljoys.com/2014/06/04/practical-phenomenology/

Ahmed, S. (2023). *The feminist killjoy handbook*. Allen Lane.

Arnstein, S. R. (1969). A ladder of citizen participation. *Journal of the American Planning Association, 35*(4), 216–224. https://doi.org/10.1080/01944366908977225

Bertilsdotter Rosqvist, H., Kourti, M., Jackson-Perry, D., Brownlow, C., Fletcher, K., Bendelman, D., & O'Dell, L. (2019). Doing it differently: Emancipatory autism studies within a neurodiverse academic space. *Disability and Society, 34*(7–8), 1082–1101. https://doi.org/10.1080/09687599.2019.1603102

Bertilsdotter Rosqvist, H., Stenning, A., & Chown, N. (2020). Neurodiversity studies: proposing a new field of inquiry. In H. Bertilsdotter Rosqvist, N. Chown, & A. Stenning (Eds.), *Neurodiversity studies. A New Critical Paradigm*. Routledge.

Chang, H., Ngunjiri, F., & Hernandez, K. A. C. (2016). *Collaborative autoethnography*. Routledge.

Francis, B., & Hey, V. (2009). Talking back to power: Snowballs in hell and the imperative of insisting on structural explanations. *Gender and Education, 21*(2), 225–232. https://doi.org/10.1080/09540250802680081

Fricker, M. (2007). *Epistemic injustice: Power and the ethics of knowing*. Oxford University Press.

Garland-Thomson, R. (2005). Feminist disability studies. *Signs: Journal of women in Culture and Society*, 30(2), 1557–1587. https://doi.org/10.1086/423352

Garland-Thomson, R. (2014). The story of my work: How I became disabled, *Disability Studies Quarterly, 34*(2). https://doi.org/10.18061/dsq.v34i2.4254

Harrington, C., Foster, M., Rodger, S., & Ashburner, J. (2014). Engaging young people with autism spectrum disorder in research interviews. *British Journal of Learning Disabilities, 42*(2), 153–161. https://doi.org/10.1111/bld.12037

Hens, K., & Van Goidsenhoven, L. (2023). Developmental diversity: Putting the development back into research about developmental conditions. *Frontiers in Psychiatry, 13*, 2994. https://doi.org/10.3389/fpsyt.2022.986732

Parsons, S., Yuill, N., Good, J., & Brosnan, M. (2020). 'Whose agenda? Who knows best? Whose voice?' Co-creating a technology research roadmap with autism stakeholders. *Disability and Society, 35*(2), 201–234. https://doi.org/10.1080/09687599.2019.1624152

Pellicano, E., & Stears, M. (2019). Commentary 1: Weksler-Derri et al.'s, Ethical challenges in participatory research with autistic adults in Israel. *Journal of Empirical Research on Human Research Ethics*, *14*(5), 452–454. https://doi.org/10.1177/1556264619858524a

Roos, I. (2002). Methods of investigating critical incidents: A comparative review. *Journal of Service Research, 4*(3), 193–204. https://doi.org/10.1177/1094670502004003003

Van Goidsenhoven, L., Rutgeerts, J., & Sandahl, C. (2022). Differing bodyminds: Cripping choreography. *Choreographic Practices, 13*(2), 197–202. https://doi.org/10.1386/chor_00051_7

Walker, N. (2021). *Neuroqueer heresies. Notes on the neurodiversity paradigm, autistic empowerment, and postnormal possibilities*. Autonomous Press.

CHAPTER 6

Practicing What You Preach—A Commentary on Chapter 5

Kristien Hens

Let me start by thanking the editors of this volume for giving me the opportunity to write a comment on the thoughtful chapter by Joke Struyf, Cecilia Ingard, and Hanna Bertilsdotter Rosqvist. I also want to thank the authors of the piece for staying with the trouble of trying to make co-creation work in the context of a big EU-sponsored project, of being killjoys and benevolent gadflies. The project R2D2-MH has entered its second year as I write these words, and indeed, as the authors describe, the first year was a co-creation rollercoaster, sometimes to the point of despair for all involved. This 'all' includes me, the lead of R2D2-MH's 'Work Package 3,' the package that includes the setting up and managing of the co-creation groups. The authors rightly challenge the fact that the project proposal itself, as it was crafted in the summer of 2021, was not co-created from its inception onwards. Indeed, if there is anything we learned from this project, it is that research into neurodiversity should be co-created from the moment the proposal is written all the way to the delivery of the products or research results. In an ideal world, the topic of research on neurodiversity, or research that claims to help neurodivergent people and their families, is co-created and even instigated by the neurodivergent community itself. At the same time, the fact that this is so often not the case is not (merely) due to the unwillingness or negligence of the often

K. Hens (✉)
Department of Philosophy, University of Antwerp, Antwerp, Belgium
e-mail: Kristien.Hens@uantwerpen.be

H. Bertilsdotter Rosqvist and D. Jackson-Perry (eds.), *The Palgrave Handbook of Research Methods and Ethics in Neurodiversity Studies*, https://doi.org/10.1007/978-3-031-66127-3_6

neurotypical researchers who embark on project grant application writing. Let me shed some light on the actual inception of the project, and my role in it. Sometime in March of April 2021, I was contacted by Thomas Bourgeron, a geneticist from Institut Pasteur in Paris, to talk about a project proposal he wanted to submit to a Horizon Europe call. Such calls typically require that, even for basic science projects, there is some engagement with 'social sciences.' Typically, this means that researchers frantically look for ethicists or sociologists who are willing to write a bit of ethical reflections, or perhaps create a survey to query what 'the general population' thinks about the research that is being done or the intervention that is being created. As R2D2-MH also involves some technology, the first idea that people had was that there could be a part on the ethics of AI or on ethics of data sharing. As I am an ethicist and philosopher who has worked on autism, how to think about autism, and what the ethics of autism could be, and who had given online presentations about that in different research groups during the COVID-19 pandemic, they came to me. I was immediately charmed by the approach that Thomas and his collaborators proposed. The vast majority of research into neurodiversity over the last decades has focused on finding 'causes,' what genes 'cause' autism, for example, and what we can do about that. Neurodivergent people have rightly protested about that. Autism and ADHD are not things that you 'risk' having. They are different ways of being. At the same time, there is the acknowledgment that autism, ADHD, and being born preterm are themselves risk factors and are related to a higher risk of mental challenges. R2D2-MH wants to investigate exactly what factors (social, genetic) make it so that neurodivergent people flourish. This seemed and still seems to me a major paradigm shift in research in neurodevelopmental differences and something that neurodivergent people have been asking for a long time. At the same time, it seemed to me that flourishing is not something that can be investigated without the input of those living the experiences. This had been the very academic conclusion of my book *Towards an Ethics of Autism* (Hens, 2021), and now it was time to see whether I could practice what I preach. That, to me, sounded much more relevant than yet another project reflecting on privacy issues in research. So during the few months that we wrote this gigantic project (a crazy summer of 2021), we came up with the idea of co-creation groups. Moreover, we believed that we should especially have the input from those people who would be the future, so we included the idea of having an adolescent co-creation group. The groups would be an integral part of the project, have a say in different areas of research, and especially contribute to thinking about outcome measures and tools. It was a maverick idea: although stakeholder engagement is part and parcel of most big projects, having a group of people with lived experiences co-create fundamental biomedical research was quite new, I believe. The project was written in a short period of three months, and the deadline was September 2021. It was a lot of work, and I spent much time also rewriting parts of it that still contained contentious ideas such as 'the risk of autism'. Although this project was ultimately granted, the success rates of such projects are fairly low, so engagement in grant application writing is always a bit of a gamble. I truly believe co-creators must be involved with research from the start of grant application writing. I also acknowledge that this would mean that the way academia works right now, with high-stakes project writing, deadlines, and the risk that you gamble and end up with no money for the project, should change. I dream of a world (a university, a European Commission) where there is money set aside to pay those with lived experience to provide their time and expertise to projects. Funders could demand that co-creation be part of the project writing in order to make a project proposal eligible. It may be OK to ask professors and their staff to spend time and overtime on projects, outside of their usual duties. It is not OK to ask this from people with lived experiences outside of academia (see Chapters 4, 19, this volume). The same is true if the project is granted. We budgeted money for the co-creation groups, but we underestimated the time and effort this would take (see chapter 8, this volume). I think funders could demand that a substantial amount of grant money go to co-creation. The science and knowledge co-creation groups generate are of equal importance to the 'fundamental' medical science that is generated. It is time this importance is reflected in rewards. I also dream of centers or institutions, preferably also led by people with lived experience who are not necessarily academics and who have been part of co-creation endeavors, that can help researchers regardless of neurotype, to think about what co-creation means and how they can go about implementing it. What are fair wages for co-creators, and how much time is needed? How can the contact with the researchers be structured, and what kind of project management and communication tools are suitable for people with different abilities and/or neurotypes? How, indeed, as the authors rightly state, can we structure research environments, including conferences, so that people not only feel welcome but feel part of the research community? As the chapter explains very well, the first year of R2D2-MH co-creation was a struggle to find answers to all these questions. It was a period of trial and error, gaining trust, losing it, and regaining it again. For example, co-creators rightly demanded an umbrella vision of all that was going on in the project. They sometimes felt that things were decided beside their backs. This feeling was understandable. At the same time, as in a great many projects, also for the senior researchers, the first year is one of trying to get this umbrella vision, seeing the bigger picture and the connections between the different work packages. The messiness of big research collaborations is something that those who have experienced them before are familiar with, but it can be bewildering to those whose first time it is. We ended up inviting the co-creation group members to all meetings of all work packages, to ensure transparency. This, unfortunately, added even more to the confusion: these meetings often focused on technical aspects of genetic research or brain research, and co-creation group members felt unsure why they were even there. Starting from the second year, we have decided to implement R2D2-MH cafés, where researchers talk with each other and with co-creators about their work and about possible co-creation opportunities. This not only helps co-creators get a good picture of what is going on in the project, but it also helps people from different work packages get the umbrella vision. To me, it is a good example of how the struggles and challenges of the first year ended up leading to something that is productive, and that can inspire other projects as well. To end on a positive note, I would like to say something about the second general assembly we had. In the chapter, it is made clear that the first general assembly did not take into account the needs of every project member, and was largely 'business as usual' for researchers. However, as the authors rightly state, to have a space that is inclusive for everyone is a condition sine qua non for true co-creation and partnership to work. It goes without saying that neurodivergent-friendly conferences should be differently organized and that the usual format of days packed with PowerPoint presentations does not work. For the second general assembly, in November 2023 in Paris, the R2D2-MH project manager took many of the remarks of co-creation group members and the advisory panel into account. It was a general assembly with prominent contributions of the co-creation group members, including the adolescents. There were plenty of breaks, and people could use quiet spaces. Co-creation group members and traditional researchers got to know each other in an informal way, which changed the vibe and generated trust. Many people who were not co-creators said they really enjoyed the unconventional vibe of the meeting and the fact there was time to talk and rest. After a very difficult R2D2-MH year, also for myself (what did I embark on?), I thought the meeting was almost a healing experience. An academia that caters to everyone's needs, respecting different neurotypes, will be a better academia for everyone. I am hopeful.

Acknowledgements R2D2-MH has been funded by Horizon Europe [grant agreement no. 101057385], by UK Research and Innovation (UKRI) under the UK government's Horizon Europe funding guarantee [grant no.10039383] and by the Swiss State Secretariat for Education, Research and Innovation (SERI) under contract number 22.00277

Reference

Hens, K. (2021). *Towards an ethic of autism: A philosophical exploration*. Open Book Publishers.

CHAPTER 7

Strangers and Friends: Navigating Neurodiverse Research Collaboration

Amy Pearson, Hanna Bertilsdotter Rosqvist, Steven K. Kapp, Lill Hultman, Catherine Watson, Gemma L. Williams, Sofia Österborg Wiklund, Paul Stenner, Monique Botha, and Lindsay O'Dell

Introduction

In this chapter we will demonstrate the importance of 'being known' by others and sharing our ways of being and doing within the world when researching together in neurodiverse groups. The chapter is based on our experiences of working together in a multi-disciplinary project on the nature of neurodiversity: 'Odd ones in', also referred to in the text as the Q study (Stenner et al. in review). The Q study consisted of an exploration of neurodiversity as a concept through the lens of our lived and academic experiences, using Q methodology, or Q-Sort, a research method used to investigate the 'subjectivity' of participant's viewpoints on a specific topic. It involves generating a series of 'items' or statements about a particular phenomenon and having participants sort these statements using a ranking system (e.g., across a 5-point scale from strongly agree to strongly disagree) (Watts & Stenner, 2012). Our team is neurodiverse, including researchers of different neurotypes, and therefore, we aim to

A. Pearson (✉)
Department of Psychology, Durham University, Durham, UK
e-mail: amy.pearson@durham.ac.uk

H. Bertilsdotter Rosqvist · L. Hultman
School of Social Sciences, Södertörn University, Huddinge, Sweden

S. K. Kapp
King Henry Building, University of Portsmouth, Portsmouth, UK

H. Bertilsdotter Rosqvist and D. Jackson-Perry (eds.), *The Palgrave Handbook of Research Methods and Ethics in Neurodiversity Studies*,
https://doi.org/10.1007/978-3-031-66127-3_7

explore neurodiverse research collaboration and different forms of expertise. We are ten people with different disciplinary[1] backgrounds and different levels of previous academic experience. Some of us are from Sweden, most from the UK. Some of us have met physically, others only digitally. Some of us have worked together in several different group constellations, but we have not worked together as a collective before. We draw together both lived experience and scientific expertise (c.f. MacLennan et al, 2022).

During the project's initial phases, we reflected upon our experiences of setting up a neurodiverse research space and how we can learn from each other's ways of being and doing in order to make neurodiverse research collaboration work. This was done in the form of a collaborative autoethnographic methodology (Chang et al., 2016) as developed by Jackson-Perry and colleagues (2020). In this chapter we mingle our collaborative autoethnographic accounts referred to as "One of us". Writing with an anonymized collective voice ("One of us") brings attention to the text being written in a collective space, the collective "I" as "One of us". "One of us" also functions as the expression of "joint action" which feminist researchers Francis and Hey (2009, p. 231) have stressed as "core to feminist action over the years" but particularly within academia" where "joint action" counter-narrates the position of academics as "individual experts". The use of a collective "I" further counter-narrates the image of "the sole" individualized neurodivergent in a lonely space of one´s own or one dominated by other neurotypes (Sinclair,

L. Hultman
Department of Neurobiology, Care Sciences and Society, Karolinska Institute, Huddinge, Sweden

C. Watson
Department of Psychology and Counselling, The Open University, Milton Keynes, UK

G. L. Williams
Swansea University, Swansea, UK

S. Ö. Wiklund
Umeå Universitet, Umeå, Sweden

P. Stenner
School of Psychology and Counselling, The Open University, Milton Keynes, UK

M. Botha
Division of Psychology, University of Stirling, Stirling, UK

L. O'Dell
The Open University, Milton Keynes, UK

[1] Coming from Education, Linguistics, Neuroscience, Political science, Philosophy, Psychology, Social work, and Sociology, albeit the group is dominated by researchers from Psychology.

2010). Rather it presents a neurodiverse voice within a community of neurodiverse people; the presence of a neurodiverse togetherness. It also aims to ensure a sense of security, a wall against structural violence, acknowledging our different levels of vulnerabilities. For example, in our group, some are well-established in academia, and some are not. Some of us are on short-term contracts, some are on secure, permanent contracts. Therefore, being identified through quotes could conceivably have a negative impact on some of our careers: "one of us" is a way of 'caring for' those for whom this might be true within our team. However, the notion of "One of us" is not without tensions or challenges, as we will discuss.

We begin by critically analysing the notion of expertise and its implications for researchers (of all neurotypes) within autism research, before presenting the collaborative autoethnographic methodology. We then move to our four themes to analyse and reflect on our experiences of exploring the possibilities and challenges of 'doing research differently'.

Meanings of "Expertise" and Different Roles in Neurodiverse Research Groups

Historically, research on neurodivergence (e.g., autism research, ADHD research) has assigned neurodivergent people to the role of "participant" or "subject" (or sometimes, in participatory research for example) a "voice" to be "listened to"): the researched, and not the researcher. Where a more participatory approach was taken, neurodivergent people might have been included as "advisors" in research (e.g., as part of a reference group, who are involved in different stages of a project, commenting upon research design and analyses), rather than as leaders or equal partners (den Houting et al., 2021).

More recently, there has been a demand for neurodivergent-led or research practices based on full participation (see den Houting et al., 2022), aligning with a "nothing about us without us" (Charlton, 1998) approach. For example, some researchers have stressed the importance of autistic leadership roles in autism research (Chown et al., 2017). Others have advocated for shared leadership between autistic and non-autistic researchers in which academic researchers and community researchers collaborate "in an equitable research partnership " (Crane et al, 2019, p. 1), "as full and equal partners—during all stages (design, implementation, analysis, interpretation and dissemination) of the research" (Crane et al, 2019, p. 5). Still others have suggested partnerships where "final decision-making" rests with academic researchers (Arnold et al, 2019). There *has* been an implementation of such demands in some areas of neurodiversity research (Alcorn et al., 2021), particularly autism (e.g., Nicolaidis et al., 2019; Pellicano et al., 2022), and so we refer heavily to autism research in this chapter. However, uptake of these practices in neurodiversity research beyond autism is limited. While there is a small but growing body of participatory or neurodivergent-led research emerging with relation to ADHD (see Bertilsdotter Rosqvist et al, 2023b; Tucker & Govender, 2017) and dyslexia (Deacon et al, 2022), broader recognition of lived expertise and leadership in neurodiversity research is still uncommon.

One reason for the slow recognition and integration of the value of neurodivergent expertise is the positivist belief that researcher knowledge is unbiased and value-free (Botha, 2021) compared to "expertise through experience" or "lived experience" (Parsons et al, 2020, p. 207). There is often a devaluation of "expertise through experience" within research, and little consideration that often researchers may have both lived experience *and* be a researcher (Lonbay et al., 2021). The incorporation of neurodivergent lived experience in research not only challenges the position of the researcher as the ultimate expert, but can also challenge hegemonies of the views of neurotypical expertise (either from the standpoints of researchers, professionals or neurotypical parents/carers). Frequently neurotypicality or a neurotypical standpoint is assumed not only to be the norm for professionals or researchers (Wood et al, 2022, p. 3; see also Parsons et al, 2020), but the *ideal* position from which researchers should operate. This position has been referred to as an unreflective "neurotypical frames of analysis" (Schneid & Raz, 2020, p. 6) and the "neurotypical gaze" (McDermott, 2022); the taken-for-granted assumption that neurotypical ways of being are what all people should strive for. Historically, research into neurodiversity has taken a deficit-based approach through the neurotypical gaze, looking to either contrast the skills of neurodivergent people to those of neurotypicals, or to ameliorate supposed 'deficits' in neurodivergent people. However, researchers have stressed the importance of lived expertise in knowledge production about neurodivergence (Milton, 2014; see also Milton & Bracher, 2013), in order "for there to be a significant shift in the research agenda". Milton and Bracher (2013) advocate for "meaningful involvement of autistic people in understanding autism". This should include "the employment of appropriately trained autistic people in research teams", aiming at an autism research field which is less alienating to autistic people (Milton & Bracher, 2013,p. 66) and meets the priorities of autistic people (e.g., an improved and more accessible society, or lowered mental health difficulties, Pearson et al., 2022).

The growth in recognition of neurodivergent expertise at various levels of neurodiversity research (e.g., as principal investigators, as co-researchers) has led to the foregrounding of challenges for both neurodivergent and neurotypical researchers. For example, neurodivergent researchers have started to highlight a diverse range of barriers due to "traditional (i.e., neuroconventional) standards" ways of doing research, including neuroconventional ways of writing (c.f. Bertilsdotter Rosqvist et al., 2023a; Botha, 2021). Likewise, researchers who now view themselves as "outsider researchers" (i.e., who are not members of the group they are researching), especially from a more dominant neurotype, have had to grapple with what that means for their own research practices. For some this has meant explicitly situating oneself as neurotypical (see for example Crane et al, 2021; Anderson, 2021) and reflecting upon practical ways of meeting challenges of power imbalances. Peuravaara (2021, p. 306) has outlined how such reflections have led to questioning "the assumption of a non-autistic worldview" as their "main point of reference," and Anderson (2021) describes her own positionality as coming from "an autism informed, though fully neurotypical, lived experience" (Anderson, 2021, pp. 3–4). Beyond recognition of their outsider-researcher status, many neurotypical researchers are now concerned that their expertise and perspectives may not be respected by neurodivergent participants (Fletcher-Watson et al., 2019). Similar challenges have already been seen in research areas focused on highlighting the epistemic privileges of dominant groups from the outside (such as men in gender studies, heterosexuals in queer studies, white people in postcolonial studies or critical race studies and so on).

These examples illustrate the importance of invoking research methods that not only support inclusion (e.g., through allowing for differences in processing), but also explicitly acknowledge the different positionalities of research team members). Explicitly focusing on these challenges allows for and mitigates the 'double empathy' gap (Milton, 2012) in research. The double empathy gap refers to inter-group difficulties in empathizing or understanding another person's experience due to differing positionalities. Understanding how to balance differing sources of expertise, and recognition of varying needs[2] (see Chapters 8 and 12, this volume) in research can help us find ways to strengthen researcher relationships *and* drive innovation in how we *do* research. Not only is this more likely to build trust with researched communities (inherently addressing the issue of whether to respect and trust outsider researchers), but it may also lead to further developments of neurodivergent methodologies. By this we include the whole spectrum of methodological improvements such as tailoring of data collection tools to suit participant cognitive styles (e.g., see Nicolaidis et al., 2019) thus increasing the quality of the research itself, to development of new methods, based on neurodivergent cognitive styles.

In the following sections we outline how we explored and experienced working in a neurodiverse group through developing such a neurodivergent method: a form of collaborative autoethnography developed by neurodivergent researchers, and the key issues that emerged through our reflections on working together.

Methods: Doing Collaborative Autoethnography

When we started to write, the initial focus was to meta-reflect upon the experiences of working together in the 'Odd One's in'-project. Collaborative autoethnography has been sketched by Chang et al (2016), however here we are following the form of collaborative autoethnography which has

[2] Building upon the experiences of their collaboration in a neurodiverse group of researchers, Hultman et al. (Chapter 8, this volume) refer to the importance of navigating in relation to both "different ways of processing information and varying (different) and variable (they change from one situation to another) needs that sometimes collide with normative beliefs about the conditions under which research is conducted (i.e., "one size fits all")".

been developed for neurodivergent minds, albeit inclusive of all neurotypes by Jackson-Perry et al (2020). This involved working bottom-up, centring a detail-oriented cognitive style (Valla & Belmonte, 2013), creating a 'free text', where some group members wrote their own narratives, meta-reflecting on the project. Five group members engaged in this process, with the wider group of researchers feeding into and commenting on these reflections once the collective story had 'taken shape'. This collective storytelling process was iterative, whereby one of the group members would 'take' the text and add their thoughts before sending it on to the next group member and so forth. The individual narratives were produced in writing rounds, with narratives serving the dual purpose of writing to and being read by other authors. We also met up several times on a secure digital platform to discuss the writing process. Through the collective storytelling process, we explored our experiences of neurodiverse research collaboration from our differently situated knowledges (Harding, 1986). The iterative writing process lasted 3 months, after which we established several emerging narratives within the text and organized these into themes.

Findings

With the collaborative autoethnography we aimed at exploring the challenges and possibilities of neurodiverse research collaboration in thinking about neurodiversity. However, during early meetings, we departed from the initial focus of the project. We started to discuss the possibilities and challenges of neurodiverse research collaboration more broadly. This was primarily driven by early challenges in either advocating for, or trying to establish, a mutually accessible space for a diverse group and a desire to reflect on that together to explore possibilities for creating a more inclusive neurodiverse research space. Our reflections at this early point included the different experiences of safety between those of us who already knew each other and those of us who gradually got to know each other during the process of writing. The question of safety was also impacted by our different disciplinary backgrounds including different experiences of the "open and free" bottom-up cognitive style which was central to the writing process of this chapter, but which remains unusual in certain disciplines. This way of writing had been experimented with by the second author in different writing group constellations, including some members of the current group (Bertilsdotter Rosqvist et al., 2019, 2023a, 2023b). This was a way to both reflect upon our experiences and leave the direction of the text open so we could develop it in an inductive manner. The group brought with it challenges of planning, pacing and the multitude of voices and different needs. However, through our different perspectives, and encountering each other'ss storytelling. We were not only giving each other a space to be "listened to", but gradually started to tell our own stories differently in relation to what we read in the stories of others (c.f. Jackson-Perry et al., 2020). Therefore the method was both a way to enable storytelling in a safer space, and to enable new stories to appear; a gradual re-storying of our experiences. Here we outline our key findings, in the form of four themes.

Creating (An)Other Space: A Safer Space to Try and Research in Other Ways

Creating safer spaces for neurodivergent researchers is paramount (Bertilsdotter Rosqvist et al., 2023a, 2023b). In the context of the discipline of geography, Oswin (2020, p.14) has argued that in order to counteract marginalization, side-lining, erasure and dismissal of 'othered' people and epistemologies within geography, one needs "an alternate academy" including "a turn toward and for each 'other'". He refers to this alternative academy as "an other geography" which inhabits an "epistemological elsewhere".

The notion of an epistemological elsewhere is emergent in autism studies and has been referred to as "a quite radical form of the collaborative and co-constructed enquiry often now advocated for the field" (Murray et al, 2023, p. 221; e.g., see ex Fletcher-Watson et al, 2019; Bertilsdotter Rosqvist et al, 2019). It also represents new ways of countering research-based violence more generally. Instead of urging autistic people to distance themselves from emotions when encountering violent accounts of autism, for example, Botha (2021) stresses the "value of "leaning-in" as a radical act of dissent in the face of research-based violence".

Throughout the process of working together we had both dis/abling experiences. This highlights how neurodiverse research collaboration can feel both epistemically violent and healing (c.f. Khan & Naguib, 2019). Within (an)other space we might have to live with feeling simultaneously safe/unsafe, "leaning in" and sharing, in the hope of a gradual learning, where:

> The more experience we gain with working in more emancipatory ways, the more we can take these practices into new spaces and spread them like the seeds of a wildflower, growing new roots and germinating new ways of being and doing. (One of us).

For us three key challenges impacted the creation of (an)other space: feeling safe to 'unmask', being able to advocate for one's own needs, and having the scope to 'play with the discomfort that may arise when doing things differently'.

Feeling Safe to 'Unmask'

Fostering safety requires being able to 'drop the mask' to show up in the research space as our 'authentic selves'. Masking can be described as "the conscious or unconscious suppression or projection of aspects of self and identity, and the use of non-native cognitive or social strategies" (Pearson & Rose, 2023, p. 3).

Importantly, this means 'learning the ways of each other'. Murray et al (2023) explain how exploring neurodiverse research collaboration is reliant on recognizing both similarities and differences between and within neurotypes, as part of a "human spectrum". However, factors like masking can impact self-advocacy within collaborations, leaving people unable to show up as their authentic selves. Masking is often engrained in neurodivergent researchers who learn to mask their neurodivergence in order to manage academia, for example:

> working in ways that do not necessarily suit us but that are the expected norm (One of us).

Like Pearson and Rose (2023), we view unmasking as a dynamic and constructed process that is reliant not just upon oneself, but those we are interacting with and the safety that they do or do not provide. In order to unmask with collaborators, we have to feel that our authenticity and our needs will be respected and extend that same respect to others.

The process of 'unmasking' is a collective responsibility and is only possible under the right circumstances. It is neither a fair expectation nor the responsibility of an individual to be 'authentic' after a lifetime of ableism and trauma. Instead, we should strive to collectively identify and work through sources of discomfort, removing the demand for authenticity, but providing instead the context for it to emerge.

Being Able to Advocate for One's Own Needs

Relatedly, growing up neurodivergent can lead to a sense of hypervigilance around the needs of others and feeling responsible for another person's discomfort. This makes it difficult to state and advocate for one's needs. As One of us highlights here, being able to acknowledge the needs and motivations of others can lessen the need to internalize blame for another's discomfort that is more attributable to the situation or context.

> When I understand what it is like for somebody else and why that person acts in a certain way, thoughts about why will not distract me. I find it easier to accept a certain behaviour when I understand that a behaviour has nothing to do with me as a person, but it occurs in relation to a situation or a place. (One of us).

Access intimacy (see Chapter 13, this volume) or the sharing of our needs and ways of being formed an attempt to give each other access to our inner workings. Our colleagues might never be able to 'think' or 'feel' in the same way, but they gain valuable information about our inner experiences in the way we attempt to get to know each other or describe what is going on inside of us. One of us spoke about how they experienced this in their daily working life in a busy environment:

> When I'm trying to concentrate on writing and I'm distracted by a conversation between people nearby, I can explain to them that my head feels like it is going to explode and my irritation is rising like a volcano…but if those people are neurotypical without ever having experienced this type of feeling, I can see why they might judge this as an overreaction. (One of us).

Thus, (an)other space or an epistemological elsewhere provides a basis from which to actively consider the needs of both ourselves and those we work with, so that we can learn not just about the needs of others, but also understand what underpins those needs. Importantly, this understanding can assuage the guilt that may come from realizing that circumstances and clashing needs may prevent a collaboration from working: empathy and being authentic is not enough on its own to accomplish a smooth working relationship. Sometimes we must accept that we are differently epistemically compatible with different people, similar to all relationships.

Playing with the Discomfort

With regard to playing with the discomfort, trying new things can be challenging regardless of neurotype. It can be jarring for neurodivergent people, given that many of us may struggle with ambiguity, and are familiar with our needs being disregarded. However, it may also be difficult for those who are used to working within particular disciplinary expectations to shrug off familiar ways of doing in favour of new (unfamiliar) ones. As such (an)other space must provide safety to do the unsafe:

> This is what I would call 'low stakes' with my students- being able to try things that are a little scary, but that also do not have very much concrete 'risk' attached. I think there is something truly emancipatory about being able to be honest about what does and doesn't work for you, and know that you are safe to do so, but also to be able to formulate new spaces for working with others. I am thinking here also about the concept of neurocosmopolitanism[3], and creating safe space for us to explore ourselves and each other in ways that are accepting of both our similarities and differences. We cannot unpick a lifetime of 'mutual misunderstandings' and internalized ableism, but we can through these interactions start to provide a little bit of safety and space for exploration. (One of us).

The idea of a low stakes space for trying new things might seem incompatible with modern academic demands (e.g., a focus on 'outputs'), however, it is important not to stifle innovation by creating pressure for more immediate impact. Exploring beyond our own comfort zones can mean learning new skills as well as:

[3] Neurocosmopolitanism is defined by Walker as an acceptance and appreciation of neurodiversity, whereby different forms of communication and ways of being are viewed much in the way we may view another culture (Walker and Raymaker, 2021).

> creating space for us to explore ourselves and each other in ways that are accepting of both our similarities and differences (One of us).

We suggest that being curious in our research approach (with regard to learning about each other or learning a new skill) can facilitate finding comfort in the uncomfortable. The liminal space between known and unknown is a great place for learning to occur and can foster innovation within teams (Brandshaug & Sjølie, 2021).

Nurturing Interest-Based Research

The theory of monotropism (Murray et al, 2005) proposes that neurodivergent individuals' are more driven by "interest-based motivation and sociality" (Bertilsdotter Rosqvist, 2019) in comparison with their neurotypical counterparts. As neurodivergent methodologies are based on the workings of neurodivergent minds, a central aspect of our neurodivergent method is to nurture our interest-based motivation. During our working process we recurrently came back to our own interests as motivation in our research, and interest was a prescient feature within our storytelling. Some of us found the method (collaborative autoethnography) itself intrinsically "interesting", and others had the same feelings about the Q-Sort. For some of us Q-sort was a "totally new" method which made it into a learning process in itself, described as something "fun" that also prompted reflection about personal beliefs implicit and explicit beliefs about neurodiversity. This process of reflective and interest-based learning led to a sense of intrigue rather than panic, despite some disabling barriers during the work process, such as lack of time due to other obligations. One of us reflected on how interest often guides their approach to research, using an example from interviewing:

> I get into hyperfocus/monotropic attention tunnel, and the interview guide most times feels like something I feel I am expected to use in order to get the predefined research questions in the research project answered. (One of us).

Similarly, the experience of meeting other neurodivergent researchers was also "interesting", and for some led to a sense of acceptance not experienced before in academia:

> certainly I had felt like an odd one out all my life so it's lovely to feel genuinely like an 'odd one in' (One of us).

The collaborative process gave a sense of both work and sociality, integrating methods of working that, though only familiar to some group members (and not others) did foster "intimacy among strangers equally interested or invested in the task at hand" (One of us).

Working together through interest-based motivation creates the potential to explore flow states (McDonnell & Milton, 2014), try new things and be creative free of pressure (i.e., for the sake of it rather than output). The concept of 'serious leisure' can be helpful for understanding possibilities of exploring collaborations in this way. Serious leisure includes a

> systematic pursuit of an amateur, hobbyist or volunteer activity sufficiently substantial, interesting and fulfilling for the participant to find a [leisure] career there, acquiring and expressing a combination of its special skills, knowledge and experience (Stebbins, 2010).

One of us describes this as a process of following their "hunches", "being in deep waters, intuitively finding my way forward", which they find exciting. At the same time, they acknowledge that this process makes it hard to describe "the rules" to collaborators, outside of an invitation to share an experience of being in the deep waters (or the flow state) together. This example highlights the multidimensional nature of interest-based motivation, whereby our engagement may go beyond cognitive labelling (e.g., this task is interesting) and instead rely on embodied experience (e.g., engaging with this *feels* good). While the latter may be harder to describe to others it also provides alternative pathways for finding shared ground that transcends verbalizing our needs, thus creating alternative (and potentially more accessible) ways to build relationships with others.

Being in a Game with No Rules, Being 'Free in Chaos'

> I work with collectives of researchers rather than on my own, which I find difficult and less creative and productive. I´m really conscious of the process of deciding on a research focus and the ways in which we want to address the particular topic. This takes time and is often invisible in writing about research. For 'standard' (NT orientated) research there is, in my experience, less thought and consideration to the process of constructing research focus, contribution of each researcher to the project and methods. The research process in terms of ethical procedures are explicitly stated in my university and need to be followed. I find this really helpful in structuring the project and expectations of participants and the research team. (One of us).

Throughout our collaboration there was a tension between unplanned or flexible processes in contrast to more "explicitly stated" steps in the research process, which was compared to a "games with rules" and "games with no rules". In games with rules, the rules may be explicit, such as formalized ethics procedures, or implicit, as in "'standard' (neurotypical-led) research"-contexts, or "life in 'neurotypical society'" (One of us). Sometimes these rules may only be understood by "those 'in the know', and where there is freedom for those in the know "to play the game within the parameters of those safe boundaries" (One of us). In contrast, our own process was recurrently described in our writings as chaotic, messy and as "a game without rules".

This tension appeared to explicitly relate to previous experiences and individual needs. Some collaborators expressed that they had spent their lives feeling like they were playing a game in which others were aware of the rules and they weren't, which led to a fear of getting it 'wrong' in the research process. For some this was compared to feeling like the "messy one in a tidy society". Finding oneself in a context where it was ok to be messy, chaotic and unaware of rules led to feelings of both panic and being "free in chaos". One of us related this to Biggs' (2014) book 'Caged in Chaos' written by a young dyspraxic person about their experiences. These reflections revealed that processes which may initially look like a game with no rules, still have an underpinning ethos, here described as

> a shared understanding of equal acceptance of each of us and the validity of our contributions and a celebration of difference. (One of us).

However, creating spaces based on shared understanding alone does not automatically foster a sense of safety or accessibility, as outlined in the first theme. It takes time to establish stability and a sense of trust, learning the ways of each other and soothing tensions along the way, as One of us reflects based on their previous experience of working in 'messy' collectives:

> In the beginning, it is messy, and people are looking for someone with the rules. And after some while, "rules" appear. They have been created. We suddenly know how to do it. (One of us).

Finding our collective way through the quagmire of chaos provided the groundwork for new stories (re-storying) to begin to emerge in the text, facilitated by the interest-based research approach outlined in the previous section and the nature of the collaborative autoethnography. We began to see that our discussions were focused much more on the process of working together rather than the process of conducting a Q-Sort or our feelings about the items it included, and what grew instead was the basis for this chapter. The emergence of a more focused narrative within our discussions led to a sense of "empowerment" for those who have previously struggled with the chaos of free-writing and formed what was described as a "potentially transformative" experience, albeit one where (some degree of) comfort developed over time.

Limits of Collectivity: The Matter of 'One of Us'

Joint action—such as our use of 'One of us'—aims to level power differentials and create protective barriers against structural violence levelled at any one individual of a group. However, there is also a risk that more marginalized voices (e.g., those in more precarious (academic) positions or dissenting opinions are sidelined). As discussed throughout this chapter, a majority neurodivergent space does not equal smooth sailing and unanimity. As One of us outlines, perhaps one benefit of a space dominated by neurodivergent people is the possibility of making explicit that it is 'ok' to acknowledge a poor fit:

> In a neurodivergent space there is another way of knitting, people use different yarns. Some of them don´t work on me at all (I keep being the sole/the one). Some of them draw me into a sense of we, the togetherness, a sense of collective monotropic force (Jackson-Perry et al, 2020) which I find it hard to resist. (One of us).

Taking collective responsibility for a viewpoint, even where we may individually disagree, displays a sense of solidarity and a respect for differing perceptions of a single situation.

As such, the experience of collective knowledge production can lead us to 'unknow' what we had 'known' (see Chapter 22, this volume); we started to question our own biases, grow our knowledge and build bridges between different standpoints. This is a process which can be deeply rewarding. However, during the process of writing together and learning the ways of each other, as the initial citation implies, different tensions due to our divergence from each other became central. While the sense of almost anonymity facilitated an ability to be vulnerable with each other, sharing our inner thoughts and feelings about the working process, there was a discrepancy with how well it aligned with the different cognitive styles of the members of our group. Though this way of working may provide an emancipatory approach to unknowing (e.g., shrugging off old norms and internalized ableism), it was important to create a space where those who were uncomfortable felt able to voice their discomfort without reproach. Reflecting honestly on our feelings about the process had a therapeutic effect, but it took continuous discussion and reflection on our different "needs and functionalities" to create the space to do so.

To this end, it is essential to acknowledge the distinction between ensuring and trying to ensure that the space is welcoming and friendly and "doing it accessibly"—whereby a "welcoming space" can still be experienced as inaccessible. Even if a safer space allowed us to talk about our needs and point out problems with inaccessibility, the challenge was to go from mutual understanding ('being nice' to each other) to a change of practice; to find ways to accommodate those different needs, especially in the light of limitations of energy, time, geographic distance and cross-disciplinarity.

Among the differences which greatly impacted comfort in the writing process was whether we process in a more deductive, inductive or abductive way when we do research. In this chapter deductive processes refer to more explicit or formalized, top-down ways. Top-down (or deductive) processing (Haker et al., 2016) is commonly associated with a neurotypical cognitive style, in contrast to bottom-up (or inductive) processing, which is commonly represented as 'extreme' attention to details (Valla & Belmonte, 2013), and is often associated with an autistic cognitive style. However, in our group top-down writing was a need across neurotypes, associated with a need for clearer rules/instructions in order to feel safe enough to write, rather than the cognitive style of a particular neurotype. This may be related to previous experiences of ambiguity (e.g., being worried about 'getting it wrong' and as such wanting more concrete instruction) or could relate to specific needs for clarity that transcend a particular neurotype. However, our experiences also align with the reflections of Murray et al (2023), noting from their experiences of a neurodiverse research collaboration; "Most commonly in our enquiry, when we expected difference, we found different levels of similarity, every time we expected to find qualitatively different outcomes, we found common processes" (Murray et al, 2023, p. 228).

The "game with no rules" approach was seen to align with a more inductive or bottom-up approach. This did not feel comfortable to all group members. In contrast with bottom-up processing, some of us refer to their more abductive (mixture of bottom-up and top-down) way of processing. They associated this with their AuDHD ways of functioning; switching between more top-down to bottom-up ways of processing as a way of letting their ADHD and autism enable each other, instead of disabling;

> I wandered from the top to the bottom, and then up again, back and forth. I just had to *start* from the top. And it worked so well for me. I finally had some kind of framework to stick to when analysis becomes too scattered, fragmented and abstract. I finally could make sense of all my printed papers, colors and post-its. It was like autism and ADHD had been having a fight, and now finally could start to reconciliate, cooperate. (One of us).

Different stages (e.g., data production, analysis) of the collaborative autoethnography were more or less accessible for various members of the group based on their usual ways of thinking and working. In the following extract, two of us reflect upon the process of collectively constructing and analysing data during the process of working together in this project.

> I prefer to do 'data collection' in a way that treats the one giving the data as an equal to the one collecting the data....It was interesting having another person reflect on how they felt about my reflections and vice versa. But I found

> this very hard to convert into something clear and sensible to say in a research article. I think my neurodivergent wiring makes it hard for me to combine these things without feeling a little anxious. (One of us).
>
> As my form of neurodivergence typically lends itself to solo work, previously I've enjoyed analysing qualitative interview transcripts and/or dialogue, closely tuning in to the words on the page and noticing pattern and meaning found in and created by the words. In this current collaborative project, key words and phrases either leapt out from the page as holding significance, or I would notice a phenomenon or pattern arising and assign a new term, phrase or name to it from my previous experience or research. This collaborative project has been powerful as a process of co-creation of shared understandings through language, elaborating and expanding upon each other's definitions and linguistic representations of experience. This has led to richer data than could be achieved with only the words of one writer (One of us) and/or one interpreter of words. (One of us).

These reflections demonstrate that even within teams where we share similar research skills in social science qualitative methods, we will not always find comfort or enjoyment in the same ways of working. Being aware of such differences may provide a pathway to consider different roles and responsibilities within research teams that better align to our different ways of thinking and doing.

Concluding Reflections: Rethinking Neurodiverse Research Collaboration

In this chapter we have outlined several challenges of neurodiverse research collaboration with a particular focus on challenges we faced during the process of setting up a neurodiverse research space and learning each other's ways of being and doing. While pursuing research based on our interests was a relatively universal enabler across the group, this alone was not enough to smooth tensions in the working process. We highlighted how conflicting varying needs (see Chapters 8, 12, this volume) and different ways of thinking and doing might lead to challenges of neurodiverse research collaboration. We can be kind to each other and attempt to be inclusive, but this alone is not enough to actually create an inclusive environment. In order to address this we make several suggestions based on our experience described here:

- **It is hard for neurodivergent researchers to shrug off prior expectations that they will simply 'put up and shut up' about working practices and the importance of self and co-advocacy**

We should seek to create a compassionate and curious culture, a collaborative space that is not grounded in neuronormative expectations. This goes some way towards meeting the need for 'reasonable adjustments' in creating a more inclusive (or universal) space. It also paves the way for fostering a sense of safety among group members who may be used to suppressing their own needs in order to fit in with neuronormative expectations. We recommend that researchers be proactive in taking time at the start of a project to establish the needs of group members and look for overlapping or conflicting needs. Think about how you might deal with these, e.g., can you have online meetings where some people use spoken communication and others use written?

- **Collaborators may have conflicting needs and we cannot make assumptions about another's needs based on a shared neurotype**

As a neurodiverse group of researchers, the attempt to create (an)other space was challenging and had mixed success. Some group members discovered that the bottom-up approach did not, and was unlikely to, mesh with their cognitive styles. Likewise, we also realized that it is not always possible to remedy conflicting varying needs. Where this is the case, it is important that we are able to find solutions (even if this means deciding not to work together on a specific project) that are blame free and underpinned with compassion for each other. One suggestion is to work out where incompatibility might be most pressing, and where they might be easier to address and then map these on to who and what is needed for each stage of the project. For example, perhaps someone who struggles with large group meetings doesn't need to attend but could receive a written transcript afterwards and provide comments in a shared space. Or could the project organizer split project members into smaller working groups to address specific tasks?

- **We might need to balance stepping outside our comfort zone with creating a new comfort zone together**

Forming neurodiverse research collaboration, as reflected in this chapter, points to a very active, deliberate way of working where we must unknow (see Chapter 22, this volume) the norms of neurotypical-led research and neurotypical cognitive styles, encompassing productive, playful chaos rather than instrumental researching towards a known objective. This is a risky and potentially hazardous way of working that may include both (unintended) epistemic violence and healing. However, it offers new ways of doing neurodiversity research. We encourage researchers to be actively reflexive in their working together, encouraging them to talk openly with each other about their needs and acknowledge what is working and what isn't.

Overall, we hope that readers will recognize that neurodiverse research collaboration can be both challenging and deeply rewarding, but that to make it work we need to explicitly recognize and address both our similarities and differences.

Questions we might ask ourselves before embarking on neurodiverse research collaboration in the future:

1. How do we create a space where people feel comfortable sharing their needs from the start? If you are the project lead you might **share yours first**? Or create a shared document where people can anonymously share together.
2. What sorts of roles are there in the project? Think beyond principal and co-investigators- **where do our strengths lie**? Are some people more skilled at drafting? Do others excel at planning and organization?
3. Which parts of the project are **set in-stone** and which are **more flexible**? Can we use rules at certain points to scaffold more open and exploratory working at others?
4. How do we respond to dissent and disagreement? Are there ways that we can **encourage dissenting views** while still being **respectful and kind** to one another?

References

Alcorn, A., McGeown, S., Aitken, D., Mandy, W., Murray, F., & Fletcher-Watson, S. (2021, May). Learning about neurodiversity at school: Key concepts for communicating neurodiversity to primary school children, from the LEANS project participatory design process. In *International Society for Autism Research (INSAR) Virtual Meeting*.

Anderson, A. M. (2021). Exploring the workforce experiences of autistic librarians through accessible and participatory approaches. *Library and Information Science Research*, *43*(2), art. no. 101088.

Arnold, S., Foley, K. R., Hwang, Y. I. J., Richdale, A. L., Uljarevic, M., Lawson, L. P., Cai, R. Y., Falkmer, T., Falkmer, M., Lennox, N. G., Urbanowicz, A., & Trollor, J. (2019). Cohort profile: The Australian longitudinal study of adults with autism (ALSAA). *British Medical Journal Open*, *9*(12), e030798.

Bertilsdotter Rosqvist, H. (2019). Doing things together: Exploring meanings of different forms of sociality among autistic people in an autistic work space. *Alter; European Journal of Disability Research ;Journal Europeen de Recherche Sur le Handicap*, *13*(3), 168–178.

Bertilsdotter Rosqvist, H., Botha, M., Hens, K., O'Donoghue, S., Pearson, A., & Stenning, A. (2023a). Cutting our own keys: New possibilities of neurodivergent storying in research. *Autism, 27*(5), 1235–1244.

Bertilsdotter Rosqvist, H., Hultman, L., Wiklund, S. O., Nygren, A., Storm, P., & Sandberg, G. (2023b). ADHD in higher education and academia. *Canadian Journal of Disability Studies, 12*(3), 1–26.

Bertilsdotter Rosqvist, H., Kourti, M., Jackson-Perry, D., Brownlow, C., Fletcher, K., Bendelman, D., & O'Dell, L. (2019). Doing it differently: Emancipatory autism studies within a neurodiverse academic space. *Disability and Society, 34*(7–8), 1082–1101.

Biggs, V. (2014). *Caged in Chaos: A Dyspraxic guide to breaking free*. Updated Edition. Jessica Kingsley Publishers.

Botha, M. (2021). Academic, activist, or advocate? Angry, entangled, and emerging: A critical reflection on autism knowledge production. *Frontiers in Psychology, 12*, art. no. 727542, .

Brandshaug, S. W., & Sjølie, E. (2021). In liminality: Interdisciplinary teams learning through challenges. *Higher Education, Skills and Work-Based Learning, 11*(2), 406–419. https://doi.org/10.1108/HESWBL-10-2019-0137

Chang, H., Ngunjiri, F., & Hernandez, K. A. C. (2016). *Collaborative autoethnography*. Routledge.

Charlton, J. I. (1998). *Nothing about us without us: Disability oppression and empowerment*. Univ of California Press.

Crane, L., Adams, F., Harper, G., Welch, J., & Pellicano, E. (2019). 'Something needs to change': Mental health experiences of young autistic adults in England. *Autism, 23*(2), 477–493.

Crane, L., Hearst, C., Ashworth, M., Davies, J., & Hill, E. L. (2021). Supporting newly identified or diagnosed autistic adults: An initial evaluation of an Autistic-Led Programme. *Journal of Autism and Developmental Disorders, 51*, 892–905.

Deacon, L., Macdonald, S. J., & Donaghue, J. (2022). "What's wrong with you, are you stupid?" Listening to the biographical narratives of adults with dyslexia in an age of 'inclusive' and 'anti-discriminatory' practice. *Disability & Society, 37*(3), 406–426.

den Houting, J., Higgins, J., Isaacs, K., Mahony, J., & Pellicano, E. (2021). 'I'm not just a guinea pig': Academic and community perceptions of participatory autism research. *Autism, 25*(1), 148–163. https://doi.org/10.1177/1362361320951696

den Houting, J., Higgins, J., & Pellicano, E. (2022). From ivory tower to inclusion: Stakeholders' experiences of community engagement in Australian autism research. *Frontiers in Psychology, 13*, 876990.

Fletcher-Watson, S., Adams, J., Brook, K., Charman, T., Crane, L., Cusack, J., Leekam, S., Milton, D., Parr, J. R., & Pellicano, E. (2019). Making the future together: Shaping autism research through meaningful participation. *Autism, 23*(4), 943–953.

Francis, B., & Hey, V. (2009). Talking back to power: Snowballs in hell and the imperative of insisting on structural explanations. *Gender and Education, 21*(2), 225–232.

Haker, H., Schneebeli, M., & Stephan, K. E. (2016). Can Bayesian theories of autism spectrum disorder help improve clinical practice? *Frontiers in Psychiatry, 7*, 185543.

Harding, S. G. (1986). *The science question in feminism*. Cornell Univ.

Jackson-Perry, D., Hanna Bertilsdotter Rosqvist, H., Kourti, M. & Annable, J. L. (2020). Sensory strangers: travels in normate sensory worlds. In *Neurodiversity Studies. A new critical paradigm*, H. Bertilsdotter Rosqvist, N. Chown, & A. Stenning (Eds.). Routledge.

Khan, F. R., & Naguib, R. (2019). Epistemic healing: A critical ethical response to epistemic violence in business ethics. *Journal of Business Ethics, 156*, 89–104.

Lonbay, S., Pearson, A., Hamilton, E., Higgins, P., Foulkes, E., & Glascott, M. (2021). Trauma informed participatory research: Reflections on co-producing a research proposal. *Gateways: International Journal of Community Research and Engagement, 14*(1), 1–8.

MacLennan, K., O'Brien, S., & Tavassoli, T. (2022). In Our own words: The complex sensory experiences of autistic adults. *Journal of Autism and Developmental Disorders, 52*(7), 3061–3075.

McDermott, C. (2022). Theorising the neurotypical gaze: Autistic love and relationships in The Bridge (Bron/Broen 2011–2018). *Medical Humanities, 48*, 51–62.

McDonnell, A., & Milton, D. (2014). Going with the flow: Reconsidering 'repetitive behaviour' through the concept of "flow states." Published online 2014. Accessed June 29, 2022. https://kar.kent.ac.uk/62647/

Milton, D. E. (2012). On the ontological status of autism: The 'double empathy problem.' *Disability and Society, 27*(6), 883–887.

Milton, D. E. M. (2014). Autistic expertise: A critical reflection on the production of knowledge in autism studies. *Autism, 18*(7), 794–802.

Milton, D. E. M., & Bracher, M. (2013). Autistics speak but are they heard? *A Journal of the BSA MedSoc Group, 7*(2013), 61–69.

Murray, D., Lesser, M., & Lawson, W. (2005). Attention, monotropism and the diagnostic criteria for autism. *Autism, 9*(2), 139–156.

Murray, D., Milton, D., Green, J., & Bervoets, J. (2023). The human spectrum: A phenomenological enquiry within neurodiversity. *Psychopathology, 56*(3), 220–230.

Nicolaidis, C., Raymaker, D., Kapp, S. K., Baggs, A., Ashkenazy, E., McDonald, K., Weiner, M., Maslak, J., Hunter, M., & Joyce, A. (2019). The AASPIRE practice-based guidelines for the inclusion of autistic adults in research as co-researchers and study participants. *Autism, 23*(8), 2007–2019.

Oswin, N. (2020). An other geography. *Dialogues in Human Geography, 10*(1), 9–18.

Parsons, S., Yuill, N., Good, J., & Brosnan, M. (2020). 'Whose agenda? Who knows best? Whose voice?' Co-creating a technology research roadmap with autism stakeholders. *Disability and Society, 35*(2), 201–234.

Pearson, A., & Rose, K. (2023). *Autistic masking: Understanding identity management and the role of stigma*. Pavilion Publishing and Media Ltd.

Pearson, A., Surtees, A., Crompton, C. J., Goodall, C., Pillai, D., Sedgewick, F., & Au-Yeung, S. K. (2022). Addressing community priorities in autism research. *Frontiers in Psychology, 13*, 1040446.

Pellicano, E., Lawson, W., Hall, G., Mahony, J., Lilley, R., Heyworth, M., Clapham, H., & Yudell, M. (2022). "I knew she'd get it, and get me": Participants' perspectives of a participatory autism research project. *Autism in Adulthood, 4*(2), 120–129.

Peuravaara, K. (2021). Emotional nuances: Critical reflections on emoji methods. *Scandinavian Journal of Disability Research, 23*(1), 305–316.

Schneid, I., & Raz, A. E. (2020). The mask of autism: Social camouflaging and impression management as coping/normalization from the perspectives of autistic adults. *Social Science and Medicine, 248*, art. no. 112826.

Sinclair, J. (2010). Being autistic together. *Disabil Stud Q, 30*(1). Accessed June 29, 2022.

Stevenson, R. A., Sun, S. Z., Hazlett, N., Cant, J. S., Barense, M. D., & Ferber, S. (2018). Seeing the forest and the trees: Default local processing in individuals with high autistic traits does not come at the expense of global attention. *Journal of Autism and Developmental Disorders, 48*(4), 1382–1396.

Tucker, L. A., & Govender, K. (2017). Ethical considerations for research involving boys diagnosed with attention-deficit/hyperactivity disorder. *Early Child Development and Care, 187*(7), 1147–1156.

Valla, J. M., & Belmonte, M. K. (2013). Detail-oriented cognitive style and social communicative deficits, within and beyond the autism spectrum: Independent traits that grow into developmental interdependence. *Developmental Review, 33*(4), 371–398.

Walker, N., & Raymaker, D. R. (2021). Toward a Neuroqueer future: An interview with Nick Walker. *Autism in Adulthood, 3*(1), 5–10.

Watts, S., & Stenner, P. (2012). *Doing Q methodological research: Theory, method and interpretation*. Sage.

Wood, R., Crane, L., Happé, F., & Moyse, R. (2022). Learning from autistic teachers: Lessons about change in an era of COVID-19. *Educational Review*, 1–23.

CHAPTER 8

Co-creation in a Neuromixed Space

Lill Hultman, Mikael Åkerlund, Fredrik Sandman, and Jeanette Nelson

Introduction

Involving people with lived experience of acquired brain injury (ABI) in the development of community services may contribute to services that better meet their needs (Mockford et al., 2012; Bombard et al., 2018). This chapter draws on experiences of co-creation in a neuromixed space, consisting of people differently situated on the neurological spectrum. Our experiences come from working together in the research project "Together—for participation and belonging in society for people with acquired brain injury". The aim of the research project was to increase knowledge about experiences

L. Hultman (✉)
School of Social Sciences, Södertörn University, Huddinge, Sweden
e-mail: lill.hultman@ki.se

Department of Neurobiology, Care Sciences and Society, Karolinska Institute, Huddinge, Sweden

M. Åkerlund
Vällingby, Sweden

F. Sandman
Bandhagen, Sweden

J. Nelson
Bromma, Sweden

H. Bertilsdotter Rosqvist and D. Jackson-Perry (eds.), *The Palgrave Handbook of Research Methods and Ethics in Neurodiversity Studies*,
https://doi.org/10.1007/978-3-031-66127-3_8

of being involved and feeling a sense of belonging in society among adult people with ABI in Sweden. The larger research group consists of seven academic researchers from different disciplines and three co-researchers with lived experience of ABI. The core research group, referred to as the co-research group, consists of five permanent members: two academic researchers and three community researchers with own experience of ABI. The community researchers live with the consequences of ABI, which include brain fatigue and aphasia. Being a neuromixed group entailed having different ways of processing information and *varying* (different) and *variable* (they change from one situation to another) *needs* that sometimes collide with normative beliefs about the conditions under which research is conducted (i.e., "one size fits all"). The research project was initiated in January 2020 and was fully financed for three years with an additional fourth year (2024) without further financing. The project entailed different study designs, such as inductive content analysis, collective autoethnography and photovoice methodology. Initially, the community researchers were involved in the project plan that was approved for funding. Later, when the funding application was approved, the community researchers were employed as research assistants. With limited time and variable interest in the different studies, the involvement of the community researchers gradually shifted from full participation at the beginning, to more limited participation both concerning the overall planning of the project and with the different sub-studies.

The chapter is written by one of the academic researchers (an ADHD:er) and the community researchers. All are part of the co-research group. We will discuss our co-creation experiences and reflect upon the reciprocal learning processes over time. We consider unevenly distributed hermeneutical resources as well as experiences of cognitive difference and varying needs. Different parts of the research processes are used to describe how we worked. All parts underline the benefits of integrating experience-based knowledge and the challenges of doing research in a neuromixed space. We focus particularly on situations in the data collection and the analytical processes where the community researchers' own experience of ABI brought forward alternative forms of data collection and analyses. Finally, we will discuss the challenges of building a community of practice based on a joint enterprise with shared knowledge, artifacts, and trust. Unlike most qualitative research, where participants' identities are anonymized, auto ethnographers can use their own names in a published research paper (Lapadat, 2017). In this chapter, drawing upon our own experiences, the community researchers wished to be acknowledged by their real names when quoted in the text, as in our previous published paper (Hultman et al., 2022). In our previous paper we focused on experiences of collaboration between academic and community researchers and analyzed challenges in relation to participation and epistemic injustice. In this chapter we explore challenges and opportunities of collaboration in relation to unevenly distributed hermeneutical resources, and experiences of cognitive difference and varying needs. The first part of the chapter introduces the work process in a more general sense. The second part, which is illustrative of the first part, provides a detailed description of how we collaborated with data collection and analyses in a Facebook community for stroke survivors.

The Development of the Project, "Together for Participation in Society"

Designing this project was a collaborative effort between academic researchers and people living with the consequences of an acquired brain injury (ABI): the co-research group. We named the project "Together for Participation in Society" to underline the partnership between academic researchers and community researchers. In our collaboration, full participation has entailed being involved in contributing input to the overall purpose of the project as well as participating in all parts of a specific sub-study: formulating the purpose, discussing the study design, participating in data collection and analysis, and disseminating results. Limited participation entailed only taking part in specific parts of a sub-study, i.e., data collection or validating research results.

Our co-research process has been influenced by Participatory Action Research (PAR). The goal of PAR is to 'bring together action and reflection, theory and practice, in participation with those providing the service to pursue practical solutions to those of pressing concern to people, and more generally the flourishing of the individual persons and their communities' (Reason & Bradbury, 2008). Community-Based Participatory Research (CBPR) involves a partnership between academic researchers and community stakeholders (Springer & Skolarus, 1970/2019). When we discussed our work in terms of collaboration, we did it with reference to Greenhalgh et al's. (2016) description of co-creation as "the collaborative generation of knowledge by academics working alongside stakeholders from other sectors" (p. 393).

The Gradual Formation of "the Co-research Group"

Our collaboration started by gradually forming what later became known as "the co-research group". In January, Jeanette Nelson, and Mikael Åkerlund were employed as community researchers, and in February Lill Hultman, alongside the principal investigator,[1] Malin Tistad, became the other academic researcher. The last person to join the co-research group was Fredrik Sandman who also was a community researcher. By March 2020 the recruitment of the members of the co-research group was completed. It was decided that stability and continuity were more important than adding more community researchers or possibly replacing the existing community researchers and

[1] In this context, the principal investigator acted as the formal research group leader.

having some as more permanent members and others as participants in a reference group. Altogether, the co-research group consists of five members with different neurological functioning: neurotypical (NT), innate neurodivergent (ND) (ADHD), acquired ND (ABI).

When we embarked on this journey none of the researchers had any previous experience of this kind of collaboration. The community researchers' previous experience had been as research participants in other studies. Two of them had also been involved in a regional supportive network that consisted of patient organizations, rehabilitation services, and care providers. Two of the community researchers became involved when the draft for the research project was formulated, and when the project received financial support, they became part of the "co-research group". Two of the community researchers already knew each other since they had attended an education specifically designed for people with aphasia, at a "Folk high school".[2]

Working together we considered ourselves as co-researchers and colleagues. Initially, a lot of effort was put into simultaneously getting to know each other, discovering what our joint enterprise would be like, developing working methods, and gradually creating a shared repertoire. This included a shared understanding of research vocabulary and producing artifacts such as manuscripts and presentations at conferences (Hultman et al, 2022). Working regularly together eventually transformed us into a community of practice (CoP), defined by Lave and Wenger (21, p. 98) as "…an emergent activity system within which participants develop and share understandings concerning what they are doing and what that means".

The Integration Between Experience-Based and Scientific knowledge

An important point of departure for the overall research project was to combine experience-based knowledge with scientific knowledge and thereby, integrate an insider perspective in all parts of the research process: research design, data collection, analysis, and dissemination. However, since we all had different amounts of time allocated to the project, we had to decide in which studies the community researchers could have full participation, and in which they should not participate in or have more limited participation. During a study's start-up phase, we made decisions about degrees of involvement, from 'full participation' to participating in certain phases of the study. We decided to aim for full participation in three studies: an autoethnography with a focus on our research process; a study focused on interactions in a specific Facebook community for people with stroke, and a photovoice study for people who had aphasia because of a stroke. At the time of writing this chapter we have finalized the data collection in the Facebook study and work on writing reports

[2] "Folk High School" is a Swedish form of adult education. One of the central features that distinguishes "The Folk High School" from other adult education is conversation and joint learning in group. It is common to work both practically and theoretically in projects.

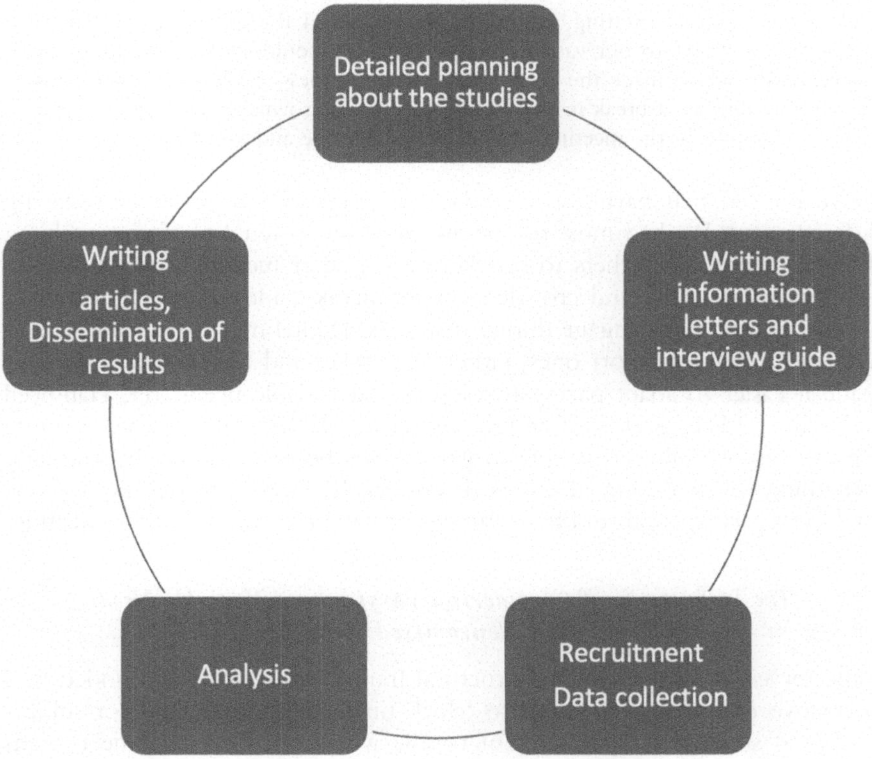

Fig. 8.1 What full participation in one study entailed

from the data. Figure 8.1 illustrates what we understood as 'full participation' in a sub-study.

Time and Energy Management

Time and energy management is something that all of us struggled with in different ways and to a different extent. For Fredrik, the most difficult part was to work in between meetings. Mikael found it more challenging when the structure changed, i.e., when we swopped meeting days, or the frequency of meetings was altered since data collection or analysis demanded a higher intensity. In the data collection, however, we also had to consider the research participants' needs as well. Sometimes this resulted in overriding the community researchers' preferred time limits (in opposite to what is discussed in terms of prioritizing participants' comfort over the researcher´s in ch. 12, 19). In a neuromixed space, one needs to learn to live and work with broken temporalities (Bertilsdotter Rosqvist et al., 2023a). As Lill wrote:

> At times it was frustrating, to know that we lacked the full flexibility that is needed to attend to our variable needs, either we could shorten meetings for everybody, which made the community researchers less involved in the process or we could take a break in the middle of a meeting which mostly was helpful if the duration of the meeting was no more than one and a half hour.

We noticed that having a fixed day for meetings was better than having different days. We had most of our meetings on a digital platform, enabling the community researchers to rest before and after meetings. Being able to avoid traffic or noisy and crowded environments made it possible for all to have more energy to engage during meetings. Digital meetings enabled us to have more frequent short ones, rather than fewer and longer meetings. This made it easier to adapt pacing to varying and variable needs, (cf, Hallowell and Ratey, 2021; Bertilsdotter Rosqvist et al., 2023c). Being able to mute the microphone allowed people to get a quick boost of energy by grabbing something to eat during meetings. It also made it easier to join even if one could not participate from the beginning or stay until the end of the meeting.

The Importance of Considering Varying and Variable Needs in a Neuromixed Space

Time serves as an overarching structural framework which is grounded in a normative temporal orientation to 'clock time'. Without regard for subjectivity and situated temporalities of people who both have and meet needs (Lanoix, 2013), normative time management may exacerbate inequities in the distribution of temporal and other associated resources (Katzman et al., 2020). According to Kafer (2013) 'crip time' accounts for natural variations in the temporal and other related resources that people need and can access to accomplish everyday activities. Kafer describes crip time as:

> 'a reorientation to time' that: requires reimagining our notions of what can and should happen in time or recognizing how expectations of 'how long things take' are based on very particular minds and bodies. We can then understand the flexibility of crip time as being not only an accommodation to those who need 'more' time but also, and perhaps especially, a challenge to normative and normalizing expectations of pace and scheduling. Rather than bend disabled bodies and minds to meet the clock, crip time bends the clock to meet disabled bodies and minds. (Kafer, 2013, p.27).

Crip time is a relational phenomenon that arises as non-normative bodies and minds interact with social worlds that anticipate a particular version of 'normalcy' (Kafer, 2013). This became salient in our neuromixed research space, or as formulated by Samuels (2017, p. 2),

> *Crip time is time travel.* Disability and illness have the power to extract us from linear progressive time with its normative life stages and cast us into a wormhole

of backward and forward acceleration, jerky stops and starts, tedious intervals and abrupt endings

In our co-research group, we all have varying needs that are affected by how we experience and manage time in relation to variable amounts of energy. Our needs are both varying and variable. With variable needs and functionalities, we had to create methods for collaboration that balanced being efficient and being fair. Nevertheless, being task- and goal-oriented sometimes made us transgress time and energy limits. Without having experience of brain fatigue, it can be "easy to forget", that an additional 15 minutes does not have the same consequences for everybody. Participating in a meeting, analyzing data, or discussing methods for data collection does not require the same amount of energy for everybody. The "easy to forget" can be seen as micro expressions of disabling practices. This is related to the notion of everyday ableism that can include assuming people have to have a visible disability to be disabled. It could also imply making casual remarks that trivialize disability, i.e." we all have a whiff of ADHD". Central in energy management strategies is finding a rhythm that makes it possible to manage tasks without being exhausted afterward (Bertilsdotter Rosqvist et al., 2023b). Common among autistic people is sensory overload which implies hypo-or hypersensitivity toward auditory tactile stimuli in the environment (Siegelman, 2005). Having brain fatigue or ADHD can also result in an overload if one does not listen to the body's signals and allow oneself to take a break or regulate pacing. Hence, it requires the ability to recognize warning signs (e.g., a heightened sensitivity to noise, being less patient, experiencing an itching body, etc.), in time to prevent oneself from becoming exhausted (Bertilsdotter Rosqvist et al., 2023b). A central aspect of the balancing act of maintaining a functional energy level is other people's abilities to validate, recognize, and be sensitive to the different energy levels of others (Bertilsdotter Rosqvist et al., 2023b). When others do not acknowledge varying and variable needs and different patterns of interaction, it can be difficult to attend to varying and variable needs. Maintaining a functional energy level can be particularly challenging if one's functionality has had a drastic change. For example, having an acquired neurodivergent functionality such as living with the consequences of ABI or being an ADHDer whose energy levels may change over time. Sometimes the drain of energy does not become salient until after meetings, when "the cost of" participating takes its toll.

I get tired no matter what I do- but it's fun. I'm exhausted afterward... For me... other people must understand what it is like to live with brain fatigue. It's tiresome to commute and it's important to understand how it feels. Suddenly, all energy runs out; to think, talk, or move. I often rest but simple things become difficult. I can be finished in the evening, and it can continue the next day. It has consequences for what you can do the next day, e.g., social activities must be canceled. Usually, I rest after meetings. Thursday: meeting and cooking. You must plan your time; you must take it easy on Wednesday and Friday. It's good to have a fixed meeting time. You need to prioritize your time and not do too many activities that take a lot of energy. "Ordinary people" sleep 8 hours, have 8 hours of leisure time, and work for 8 hours, but I need 14 hours of rest. Thus the number of daily active hours varies (Fredrik).

I have flexibility- I can plan my time. I go to bed when I'm tired. I wake up when I wake up. I'm not governed by time. An airy schedule is a prerequisite, like being able to rest on Fridays (Mikael).

To pay attention to variable needs was something we 'learned along the way'. In our group, variable needs had to do with researchers' different pacing and energy levels. They are affected by micro-processes (such as group interaction, the task at hand) and structural barriers (such as delayed special transportation and inaccessible meeting rooms). Variable needs can also be affected by what had happened earlier that day, or the day before, which implies that energy levels and the ability to stay focused shift from day to day.

Communicative Skills and Challenges

Within the co-research group, we have different abilities to express our opinions. Due to the aphasia, two of the community researchers are not able to talk as fast as the rest of the group members. This makes it important to allocate time for everyone to express their opinions. For some of us, communicating can be much more demanding than for others,

I often choose to be quiet if someone else talks- I do not have the same opportunity to do something fast. I wait for my turn, but it depends on whom you talk with... (Fredrik).

In opposition to this, for others it was challenging to not talk too much or too fast when one gets excited about a topic.

I was very fast- I want it to go fast- I used to be 'Turbo Micke'. In the past, I managed to be fast- being fast is highly valued in society. It feels like a hidden requirement, a stressor. It is better to take it easy... You did not get any diagnosis when you were young. I've had to think about not talking too much, to remember to hold back. (Mikael).

Both for ADHDers and people with ABI, being able to communicate at the right pace is considered valuable, a social skill that one is expected to know. For stroke survivors who have aphasia, it can be frustrating that it takes time to make oneself understood, and even at times not being able to make oneself fully understood.

This type of experience-based knowledge was valuable concerning the photovoice study, where the participants shared similar experiences as researchers who had their own experience of aphasia. Specially in neuromixed encounters, there is a need for new translation methods, as Bertilsdotter Rosqvist, Hjorth and Nygren (2023, p. 409), notes:

> When writing is traveling from one researcher/storyteller to another, knowledge production takes place as a constantly changing exchange. At the same time, there are moments of response all the way during the process, cultivating intersubjective reflections

This becomes particularly important when people depending on their functionality utilize and understand language in different ways, which is expressed by one of the community researchers:

> We "talk/write" together. You come up with things you hadn't thought of... group conversations are valuable. You cannot have a conversation with a book. [It is important] to break things down- to do limited tasks, so you don't get overwhelmed. There are many ways to reach the same goal. It can be challenging if you are a larger or more spread-out group. (Mikael).

When we had digital meetings in the larger research group it was more difficult for the community researchers to become involved in the spontaneous discussions that occurred between academic researchers. Although we had the same amount of time as in our co-researcher meetings (one and a half hours) we often spent less time on contextualizing the topics that were being discussed. Shifting focus was both perceived as exciting and challenging which Buckle et al (2021) refer to as *autistic inertia*, which implies resistance toward a change of state, i.e., difficulties in starting, stopping, and changing direction.

Challenges of—and Solutions to—Varying and Variable Needs

Attending to variable needs can be a challenge when people in the neuromixed space have varying needs. For example, sometimes what is beneficial for one person could be experienced as a barrier for another person.

> It's nice to be together, you get to know each other better by seeing each other physically. You see body language and by seeing other people's reactions you become more present. You are more sensitive to noticing each other's needs and it is easier to take breaks or walk and move around (Jeanette).

To ensure that everyone could be involved during meetings we decided that one of the academic researchers should be responsible for screen-sharing and making notes on PowerPoint during the meetings. This served both as memory support and transformed meetings into a more inclusive practice, where everyone could alter or add meaning to what was written down during the meetings. The collective responsibility for content and form of notetaking made us aware that we on a micro level had to adapt the research process to our variable needs. Adapting the pace allowed everybody to become engaged in all parts research process, such as formulating a text, providing input to a study design, adding a question to an interview guide, having suggestions regarding data collection sites. The need to adapt the pace made visible the clash between normative ideas about how long different parts of the research process should take, and under what conditions research should be conducted. Among such normative ideas are that formally funded project should not be carried out during your leisure time, rather it should be done within your employment within a certain time frame. In line with the idea that research can be conducted according to "one size fits all", adapting the pace may be considered as "slowing down" rather than viewing it as a tool intended to democratize the research process (Pellicano & Stears, 2019). In our research process adapting the pace was a necessary measure to ensure that community researchers would be enabled to fully participate in the research process.

Working closely together made us aware of many challenges and opportunities we could not have foreseen and as a response to those challenges we had to come up with solutions. "Talk/writing" (making notes while talking, which allowed everyone to give their input and alter the meaning of the text, see also Chapter 12, this volume) during meetings made it possible for us to discover a common language. The community researchers' questions enabled us to identify language barriers and to a certain degree simplify language by exchanging academic language for words that lay people use. Participating in collaborative "talk/writing" processes helped us to clarify our thoughts and it enabled us to shorten sentences.

In the start-up phase of the research project, there was a dual process of making sense of what we were doing and being comfortable in each other's company. A lot of our meetings revolved around getting to know each other and being safe with expressing different opinions and perspectives. Initially, we felt a bit lost since there was this simultaneous task of getting to know each other and finding appropriate methods for co-creation. After a while our common quest for meaningful collaboration developed into mutual learning processes which Jeanette explains in the following manner:

> There was a lot to do- it created focus. You get excited. It creates commitment and things happen! It got better, we had a better structure. One could be prepared before the meeting, knowing what to talk about and being able to read beforehand if you wanted to. There was time to think before [the meeting]. Sometimes it is important to be able to think ahead and sometimes thoughts arise in the moment (Jeanette).

Becoming a tight-knit group made us interested in reflecting on our research process. Lill suggested that we transform our experiences into two autoethnographic studies of our research process (Hultman et al., 2022). The first paper scrutinized challenges related to the legitimacy of different types of knowledge (academic and experience-based knowledge) and degrees of participation in the overall research process. This chapter focuses on opportunities and challenges with co-creation in a neuromixed research group where there exist both varying and variable needs. Being able to add parts, like this new sub-study to the overall research project, is an example of our organic research process which allowed for a flexible approach that enabled creativity, inclusion, and mutual engagement in the research process.

An Illustration: Data Collection in a Facebook Community

Early on, the community researchers had identified Facebook as an important environment to explore experiences of participation and sense of belonging. The community researchers' own experience of living with acquired brain injury (ABI) enabled them to identify and suggest relevant Facebook communities that we could approach for data collection. To cover different aspects of living with ABI, we approached two different Facebook communities, one with a particular emphasis on living with brain fatigue and another group whose members had suffered from a stroke or were significant others to people with stroke. In this study, we wanted to explore what topics members of the groups wrote about, and how they wrote about those topics.

Initially, the Facebook group's administrators were approached by the academic researchers, to ask for permission to explore which type of topics were discussed within the group. The group administrators for both Facebook communities allowed us access to the Facebook community. To gain trust among the Facebook community's members we decided that it was important that both an academic and a community researcher presented the study to the members of the selected Facebook group. Having their own experience of stroke gave the community researchers the legitimacy to approach the members of the group since they had their own experience of ABI and were familiar with many of the topics being discussed in the group. In that sense, the community researchers' involvement in the co-research group gave all of us the legitimacy to conduct research within the Facebook community (c.f. Chapter 10, this volume). Before providing information about the study, we discussed what we should include in the information about the study, when we should provide information, and where we should provide the information. The community researchers' previous experience of being research participants as well as having their own experience of ABI was valuable when we discussed which information to provide about conditions for participation in a research study and information that could be difficult to understand.

We provided information about the study in three different ways (see Fig. 8.2).

Before the digital information meetings, we prepared a PowerPoint presentation, where we shared information about the research project and the specific study design of the Facebook study. For those meetings, we paired up one academic researcher and one community researcher and decided which parts of

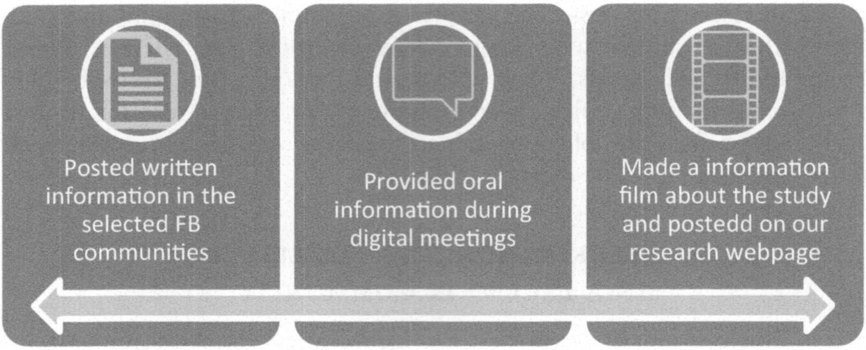

Fig. 8.2 Dissemination of information about the Facebook study

the presentation an academic researcher should do and which parts a community researcher should be responsible for. In one of the Facebook groups, the group administrator was involved and contributed to the discussion.

Holding information meetings was a learning process. One of the challenges was to find "the best time" for having digital information meetings since many of the members in the studied Facebook communities, like some of the community researchers, suffered from brain fatigue which made it difficult to know whether one could participate depending on variable needs, i.e., how you are feeling that day. Finally, we ended up having the information meeting on several occasions, during different times of the day, and both during weekdays and weekends. This turned out to be the best solution since it allowed both community researchers and Facebook members to attend to their variable needs. The members who attended those meetings often wanted to share their experiences of living with stroke, which enabled us to recruit informants for interviews. Even though we had put much effort into reaching all members we were not sure if we had managed to make information accessible to everyone interested in how the study was conducted, which type of data was collected, etc. We therefore decided to record the information meetings and make them available on the project homepage[3] so that people could watch them at a time that felt most convenient to them, and contact us if they had any questions.

During data collection, we worked in different constellations depending on study design. In the Facebook study, we utilized the same data for different types of analysis. For data collection we developed a protocol to extract information about posts and comments in Facebook groups. We followed and noted all the posts and comments for six months. Extracting data proved to be quite difficult since we had permission to describe the content in the blog posts without copying the blog post itself. Writing without interpreting

[3] Karolinska Institutet. (4 april 2024). *Ihop—för delaktighet och tillhörighet i samhället vid förvärvad hjärnskada*. https://ki.se/nvs/ihop-for-delaktighet-och-tillhorighet-i-samhallet-vid-forvarvad-hjarnskada.

the text proved to be quite difficult. A text written in a phenomenological spirit is characterized by the fact that it invites the reader to wonder, to put previous knowledge aside and experience the text with an open mind (van Manen, 2002). For us this meant to take notes, without adding or removing anything from the original blog posts. Therefore, we did data extraction together, looking at the same blog posts and holding discussions about how to describe the content concerning the level of detail. For three consecutive months, each blog post was noted in a separate protocol. All protocols were compiled monthly and provided with a unique number and a code, i.e., (0101 = protocol 1, January, stroke anniversary) and a list that specified the number of blog posts.

When we initiated the Facebook study, we did not realize how much time and energy it would take from us which eventually affected our view on how to think about full participation in our last study.

> The hardest part was with the Facebook study- it was a lot. It was very intense when there was too much time spent and work to do in the evening. Commuting, and making changes made it messy. I felt that one must be able to cancel when it is too much. (Mikael).
>
> Working with the Facebook study was challenging. From our previous experiences of doing analyses of our research process, we wanted to continue to work together with data collection and analysis, especially since this was grounded in the community researchers' interest-based motivation. (Lill).

Working intensively with the Facebook study had put other processes on hold, and afterward we realized that during the most intensive phases of data collection and data analysis we had not managed to attend to our varying and variable needs.

Modified Inductive Content Analysis

Embedded in our research process was a commitment to meaningfully integrate the processes of participatory data analysis into the research process. In the literature on PAR research, some scholars argue that the processes of data analysis are often not fully integrated in PAR projects, leaving many 'analytical' tasks left up to the academic partners (Fricker & Nixon, 2015; Jackson, 2008).

To work with the data generated from posts in the Facebook community we decided to do a modified inductive content analysis. Working with the data analysis online required that everybody was comfortable with coding and categorizing separately. Since the community researchers lacked previous experience in data collection and data analysis, we decided to meet up in a conference room and do it together. When we sorted the data into codes and themes we worked in pairs– one academic researcher and one community researcher. We coded part of the data together and made a matrix with

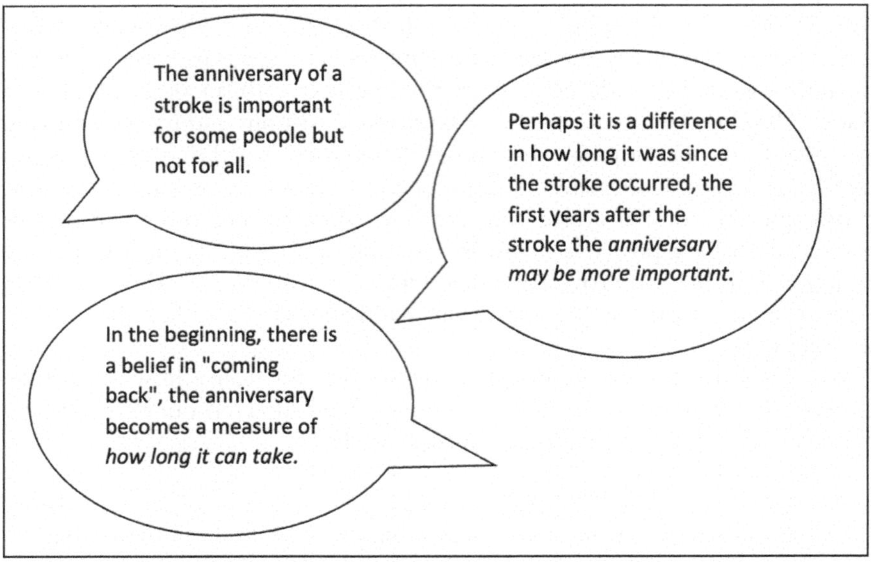

Fig. 8.3 Examples of reflections on Facebook posts about "Anniversary of stroke"

sub-themes and themes. We started with the theme which was identified as significant and important by all the researchers: the personal stroke narrative. In a later stage, we worked on the three identified themes utilizing "reflections":

(I) The personal stroke story.
(II) The way back- Rediscovering oneself.
(III) Everyday life with a stroke.

We discussed emerging themes based on reflections from fellow community researchers. Reflections are based on associations, own experiences, and thoughts. The purpose of the reflections was to nuance meanings, provide more perspectives, and broaden and deepen the understanding (Fig. 8.3).

The Ethical Dilemma of Full Participation

Our collective writing processes revealed that the community researchers were less interested in writing scientific papers than texts that were going to be published in patient organizations' magazines. One of the main reasons for the community researchers' reluctance to co-write scientific papers was the unevenly distributed hermeneutical resources. Since English is not our first language there was therefore a dual language barrier for certain members of the research group (English and academic language).

Our first scientific text written in Swedish was about starting up the research project, which initially made us believe that it would be easier for everyone to be involved in all parts of the study. But even though we tried to simplify language, we still had to write a text that would fit into an academic journal. To solve this dilemma, we did two separate but intertwined analyses, which enabled the community researchers to contribute their experience-based knowledge of living with an acquired brain injury which was filtered through theoretical concepts such as Arnstein's (1969) ladder of participation (see also Chapters 5, 24, this volume) and Fricker's concept of epistemic injustice (Hultman et al., 2022).

In hindsight, we should have put more effort into discussing our varying and variable needs more frequently and integrated those in working methods. In their recommendations for participatory research, MacLeod and colleagues (2014) underline the importance of identifying and adapting the level and type of participation that are appropriate and possible given the expectations and limits of a specific research project. Nevertheless, the amount of time spent together enabled us to learn and adapt along the way, and we have managed to develop into a community of practice, with a joint interest in doing participatory research with people with own experience of ABI. After working together for several years, it reinforced our mutual engagement and joint interest.

Conclusions: Wrapping It Up

With less than a year left in the "Together for participation project", we end our chapter with some lessons learned from our co-creation process. The lessons are summarized under the following subheadings: challenges and mitigating challenges.

Challenges

- All parts of the research process take more time than you think.
- In a neuromixed group we can keep our focus for different lengths of time, our ability to focus is both varied and variable.
- It can be difficult to change meeting times.
- It was difficult to do data collection in the Facebook group.
- It was difficult to find a balance between what we want to achieve and what we can do both as a group and as individuals.

Mitigating Challenges

- Discuss different group members' varying and variable needs before starting up the research process since it affects planning and time management.

- Be honest with each other. Everyone can do an equal amount. Sometimes someone can do more.
- Be patient with each other and see challenges as a learning process.
- Communicate with each other, i.e., dare to ask questions.
- It is good to have fixed meeting times—the same day and same time. It makes it easier to remember and to plan one's time.
- Try things out and be willing to change the structure if it does not work.
- Have digital meetings—it saves energy and can prevent a sensory overload.
- Think about the balance between physical meetings and online meetings. Online meetings can save time and energy consumption, but physical meetings can make it easier to get to know each other.

References

Arnstein, S. R. (1969). A ladder of citizen participation. *Journal of the American Planning Associa-Tion, 35*(4), 216–224.

Bertilsdotter Rosqvist, H., Hjort, E., & Nygren, A. (2023a). Meeting up in broken word/times: communication, temporality and pace in neuromixed writing. *Medical Humanities, 49*, 407–415. https://doi.org/10.1136/medhum-2022-012384.

Bertilsdotter Rosqvist, H., Hultman, L., Hallqvist, J. (2023b). Managing vocational work performance: Support and self-management amongst young autistic adults in the context of vocational support interventions in Sweden, *The British Journal of Social Work, 53*, 258–275.

Bertilsdotter, Rosqvist H., Hultman, L., Österborg Wiklund, S., Nygren, A., Storm, P., Sandberg, G. (2023c). Intensity and variable attention: Counter narrating ADHD, from ADHD deficits to ADHD difference. *The British Journal of Social Work, 53*(8).

Bombard, Y., Baker, G. R., Orlando, E., Fancott, C., Bhatia, P., Casalino, S., Onate, K., Denis, J.-L., & Pomey, M.-P. (2018). Engaging patients to improve quality of care: A systematic review. *Implementation Science, 13*(1), 98. https://doi.org/10.1186/s13012-018-0784-z

Buckle, K. L., Leadbitter, K., Poliakoff, E., & Gowen, E. (2021). "No way out except from external intervention": First-hand accounts of autistic inertia. *Frontiers in Psychology, 12*, 631596.

Flicker, S., & Nixon, S. (2015). The DEPICT model for participatory qualitative health promotion research analysis piloted in Canada, Zambia and South Africa. *Health Promotion International, 30*, 616–624.

Greenhalgh, T., Jackson, C., Shaw, S., & Janamian, T. (2016). Achieving research impact through co-creation in community-based health services: Literature review and case study. *The Milbank Quarterly, 94*(2), 392–429.

Hallowell, E. M., & Ratey, J. J. (2021). *ADHD 2.0: New science and essential strategies for thriving with distraction-from childhood through adulthood*. Ballantine Books.

Hultman, L., Sandman, F., Nelson, J., Åkerlund, M., von Koch, L., & Tistad, M. (2022). Som erfarenhetsforskare, då är man med och bestämmer i forskningsprojektet: En autoetnografisk studie om en gemensam forskningsprocess. *Socialvetenskaplig Tidskrift, 29*(3–4), 305–324.

Katzman, E. R., Kinsella, E.A., & Polzer; J. (2020). 'Everything is down to the minute': Clock time, crip time and the relational work of self-managing attendant services, *Disability and Society*, *35*(4), 517–541. https://doi.org/10.1080/09687599.2019.1649126

Lanoix, M. (2013). Caring for money: Communicative and strategic action in ancillary care. *International Journal of Feminist Approaches to Bioethics*, *6*(2), 94–117. https://doi.org/10.3138/ijfab.6.2.94

Lapadat, J. C. (2017). Ethics in autoethnography and collaborative autoethnography, *Qualitative Inquiry*, *23*(8), 589–603. https://doi.org/10.1177/1077800417704462

MacLeod, A. G., Lewis, A., & Robertson, C. (2014). CHARLIE: PLEASE RESPOND! Using a participatory methodology with individuals on the autism spectrum. *International Journal of Research and Method in Education, 37*(4), 407–420.

Mockford, C., Staniszewska, S., Griffiths, F., & Herron-Marx, S. (2012). The impact of patient and public involvement on UK NHS health care: A systematic review. *International Journal for Quality in Health Care, 24*(1), 28–38.

Pellicano, E., & Stears, M. (2019). Commentary I: Weksler-Derri et al.'s "Ethical challenges in participatory research with autistic adults in Israel", *Journal of Empirical Research on Human Research Ethics*, *14*(5) 452–454.

Reason P., & Bradbury, H., (Eds.). (2008). *The Sage handbook of action research: Participative inquiry and practice*. Sage Publications.

Samuels, E. (2017). Six Ways Of Looking At Crip Time, *Disability Studies Quarterly, 37*(3). https://doi.org/10.18061/dsq.v37i3.5824

Springer, M. V., & Skolarus, L. E. (1970/2019). Community-based participatory research partnering with communities, *Stroke*, *50*(3), 48–50.

van Manen, M. (2002). *Writing in the dark*. Althouse Press.

CHAPTER 9

Neuromixed Participatory Action Research: A Dialogical Method

Marjorie Désormeaux-Moreau and Isabelle Courcy

Introduction[1]

A growing number of openly Autistic[2] researchers are conducting studies that focus on autistic people's and communities' perspective and life conditions, helping to redefine the boundaries of dominant biomedical research

[1] Acknowledgement: We extend our gratitude to all partners, citizens, and researchers involved in the research collective that inspired this study and the conceptualization of Autisme Soutien, a social support service "by and for" Autistic individuals in Canada. We are particularly thankful to Catherine Aubry, Lucila Guerrero, Mathieu Giroux and Richard Marcotte for their valuable involvement as well as for their role as Autistic citizen partners in the research project that led to the launch of Autisme Soutien's online social support service. We wish to emphasize the quality and richness of their involvement. We would also like to extend our gratitude to the book's editors, Hanna Bertilsdotter Rosqvist and David Jackson-Perry. Their feedback throughout the writing of this chapter allowed us to refine and clarify our ideas, thereby significantly enhancing its quality.

[2] In this chapter, we alternate between the terms "autistic" and "Autistic" when referring to individuals. This practice acknowledges the diversity within autistic communities: i) some autistic individuals perceive their autism (lowercase 'a') as a state of being or a condition, while ii) others identify as Autistic (uppercase "A"), viewing it as an integral part of their identity and culture. Although we recognize that this alternation might initially seem confusing, we believe it is important to reflect these identity positionings, which are central to the support service discussed in this article.

H. Bertilsdotter Rosqvist and D. Jackson-Perry (eds.), *The Palgrave Handbook of Research Methods and Ethics in Neurodiversity Studies*,
https://doi.org/10.1007/978-3-031-66127-3_9

programs (Chown et al., 2017; Nuwer, 2020). Their social and political positioning, as reflected in their work, represents a break with the dominant institutions whose practices confine them to the identities of "sick", "disordered", or "psychiatric patients" as well as with neuronormative practices and culture (Dwyer et al., 2021) shaped by neuronormalized people. In parallel, academic and scientific research is increasingly conducted by teams rather than by individual researchers (National Research Council, 2015). In this context, what are the possible collaborations for Autistic researchers (and neurominorized researchers more broadly) given that they remain underrepresented in academic circles? More specifically, what about collaborations and working relationships between neurominorized and neuronormalized[3] researchers? Is there such a thing as equitable collaboration in a neuromixed research team where some members belong to neuronormalized and others to neurominoritized groups? How do these collaborations influence the research process and the academic knowledge that is produced? Specifically, how does the interplay between neuronormalized and neurominoritized researchers in a neuromixed research team shape both the process and the outcomes?

To reflect on these questions, we—the authors of this chapter—build on the co-analysis of our experience, as co-researchers engaging in a study concerning the development of an online social support "by and for" Autistic adults. In this chapter, we begin by providing context regarding both the online social support that was launched and its co-development, embedded in a research process. We then discuss the dialogical method we implemented in a qualitative and interpretative approach to reflect on and share our interpretation and understanding of several issues that emerged during the course of the project and more broadly over our collaboration, including in the writing of this chapter. In doing so, we address three key lessons that emerged regarding

M. Désormeaux-Moreau (✉)
Université de Sherbrooke, Sherbrooke, Québec, Canada
e-mail: Marjorie.Desormeaux-Moreau@USherbrooke.ca

I. Courcy
Université de Montréal, Montréal, Québec, Canada
e-mail: isabelle.courcy@umontreal.ca

[3] In this chapter, the terms "neuronormalized" and "neurominorized" refer to neurocognitive profiles respectively described as "neurotypical" and "neurodivergent" (sometimes also referred to as "neuroatypical"). As explained by Catala (2023), these terms highlight the power relations on which the production and differentiation of neurocognitive profiles categories are based, as well as the dimensions of privilege and marginalization they bring into play, which all refers to neuronormativity.

the participatory approach that was used within our neuromixed[4] research collective: (1) valuing the diversity and plurality of knowledge; (2) reflecting on our respective standpoints and developing a joint political engagement; and (3) rebalancing power relations in the collective. We will conclude that strategies such as those we have implemented can foster the emergence of research knowledge while promoting the agency of Autistic people.

A Citizen Initiative: Launching Autisme Soutien

In December 2018, Autisme-Écoute (which can be translated as Autism Listening and which later became Autisme Soutien) was launched in Québec, a francophone province within a mostly anglophone country (Canada). It was spearheaded by Catherine Aubry, an Autistic woman and mother of two autistic sons to address a lack of adequate social support resources for autistic adults (Cordeau, 2019; Désormeaux-Moreau & Aubry, 2020). The initiative derived from Aubry's own experience, as she felt she would have needed social support following her sons and her own "autism diagnostic" assessments (as referred to in a medical model), and yet this was difficult if not impossible to find (Désormeaux-Moreau & Aubry, 2020), especially as an Autistic adult. In Québec, very few support services are available to autistic adults and most of them still target so-called deficits and undesirable behaviors. This reflects both the prevailing pathologizing understanding of autism (Chapman, 2019; Evans, 2013) and the powerful long-standing influence of the medical and clinical paradigms on support services as well as on community-based services intended for Autistic people (Long, 2023).

Autisme-Écoute came about as a citizen's initiative in Québec (Canada), at a time when counter-narratives to medicalized, pathological, and deficit understandings of autism were only beginning to spread throughout francophone autistic communities. Autisme-Écoute offered free, confidential, and geographically accessible social support to both Autistic people and parents of Autistic kids (whether Autistic themselves or not) via an online platform. Recognizing that social exclusion and disenfranchisement have the most negative impact on their mental health (Milton & Sims, 2016), it provided synchronous (via chat) and asynchronous (via e-mail) social support which was initially intended to be provided by Autistic adults but ended up being offered by autistics and allies,[5] due to recruitment and financial constraints.

[4] The term neuromixed reflects the fact that our team was mixed in terms of neurocognitive profiles and identities. However, the team was also mixed in terms of various other social attributes (e.g., gender, profession, language, cultural background, age).

[5] To be an ally means to be united with and in support of another person's cause (Duopress Labs, 2022). Being an ally more specifically means to offer one's "support to a particular group of people in society that they themselves do not belong to, in order to help ensure their basic rights and ability to be happy and successful in society" (Oxford Advanced Learner's Dictionary, nd). While some allistic people are allies to autistic individuals, this is not always the case. Indeed, since neuronormalized privilege is most

In fact, this was a citizen-led initiative, carried by too few people, making the workload too heavy to sustain over time. Furthermore, it was an independent initiative, unaffiliated with any structures or organizations, without funding. The lack of resources (human and financial) adversely affected the initiative's credibility and recruitment process, leading to a lack of training and support for volunteers. Finally, it should be noted that involvement in such an initiative requires one to be in a stable emotional and psychological state. Consequently, it was not possible to recruit and involve all autistic individuals who expressed interest in the initiative. Despite being a groundbreaking initiative, it was discontinued six months after its implementation. However, a great deal of research has shown that social peer support, which is neither clinical nor institutional, is a valuable aid and protective factor for mental health (Cyr et al., 2016), while contributing to empowerment and a sense of community belonging (Crompton et al., 2022; Milton & Sims, 2016). Therefore, the first author of this chapter (MDM)—who is openly Autistic and used to volunteer in Autisme-Écoute, suggested that a research process could help relaunch the initiative. This led to a reevaluation of the service's conceptualization and a name change, from Autisme-Écoute to Autisme Soutien,[6] marking the rebirth of the initiative and the transformations brought about by the research process.

A Participatory Action Research: Co-Developing Autisme Soutien and Its Online Social Support Service

The participative action research (Greenwood et al., 1993; Reason, 1994) that guided the Autism Soutien project was grounded in an emancipatory approach (Oliver, 2002) that aimed at shifting power and control (Oliver, 1997; Walmsley, 2001) from (mostly allistic[7]) autism researchers to Autistics. Based on the premise that both the needs that gave rise to the study (i.e., the lack of social support that would be aligned with autistic adults' needs) and its potential solutions would be drawn from and located in Autistic people's day-to-day concerns and experiences, this research can also be seen as a situated practice research, as described by Long (2023).

Such paradigmatic and epistemological anchoring is not trivial, given that most of what is thought to be known about autistic experiences and needs is derived from knowledge of autism which inspires most services designed for (not with or by) Autistic people. Hence, core to our epistemological stance is our recognition of the historical privileging of certain types of knowledge within the design of therapies and interventions aimed at autistic people, and the consequent silencing or invisibilization of autistic knowledge. Therefore,

often imperceptible, unconscious, and taken for granted, allistic people, as a group, are at the root of the oppressive and stigmatizing practices experienced by Autistic people.

[6] Autisme Soutien could be translated as "Autism Support".

[7] In this chapter, we refrain from capitalizing the word "allistic" when referring to people, as it does not denote a collective culture or shared historical narrative.

to challenge and to break the cycle of such epistemic injustice (Fricker, 2007), while keeping with the spirit that had driven the founder of Autisme-Écoute, it was decided from the outset, even before the formation of a research team and the design of the study, that the target service would be led by Autistics.

Since the goal was to develop a formal social support service that had hitherto been non-existent in Quebec, the research accompanying the Autism Soutien project relied on a developmental research design (Contandriopoulos et al., 1990; Loiselle & Harvey, 2007; Van der Maren, 1995), with a pragmatic focus. As this design focuses on problem-solving through the search for functional solutions (Van der Maren, 1995) it was deemed ideally suited to this project, which aimed to use scientific knowledge and research data to develop a social support service by and for Autistic people. The participatory action research had three objectives:

1. To assess the perceived social support needs of autistic adults from Quebec, and their willingness to use an online social support service by and for Autistic adults;
2. To document the process by which the online social support service by and for autistic adults would be conceptualized and implemented;
3. To develop and implement a prototype of the social support service.

Led by a researcher who identifies as Autistic, this research project was based on the interweaving of experiential, professional, and scientific knowledge—and it is therefore important to unpack these different types of knowledge and how they intermingle, sometimes *within* a body, sometimes *between* bodies. The project was propelled by a research collective that initially comprised five Autistic citizens and two allistic researchers,[8] in addition to the lead Autistic researcher. It's worth noting that three of the Autistic members of the initial collective had an experience of the first initiative (two of them as volunteers). The other members had been invited to join the collective because of their interests, commitment, and work in autism research or within autistic communities. Autistic members all occupied positions within autistic communities in Québec, by either regularly taking public positions, having a media presence, occupying various forums or participating in different community initiatives, such as consultation tables, or exercising roles as consultants or experts. This research project ran from summer 2020—starting with the formulation and the design of the research project—to fall 2022, when Autisme Soutien was formed as a registered nonprofit organization. To foster the participation of members who are geographically distant from each other, all activities conducted by the research collective were carried out virtually, through e-mail, chat sessions, or virtual meetings, using Microsoft Teams. This eventually facilitated the continuation of the project as planned, even after the emergence of

[8] We acknowledge that our collective includes people who are not Autistic but are otherwise still minorized based on other social identities.

the COVID-19 pandemic and the enforcement of health measures designed to restrict in-person gatherings.

In the following sections, instead of presenting the study's results, we move to our reflective process of both carrying out that research and writing the current chapter. What follows is a dual-layered description: one layer detailing our use of the dialogical method during our research, and the other layer reflecting on our collective thought process during the writing of this chapter. This approach allowed us to engage in meta-thinking and collective reflection on our collaborative process.

Reflecting on the Research Process Using of a Dialogical Method

Dialogism can be defined as "the capacity of the human mind to conceive, create and communicate about social realities in terms of alter, i.e., in relation to other individuals, groups, communities and cultures" (Marková & Orfali, 2005, p. 27, our translation),[9] regardless of the forms or mode of communication (e.g., verbal, written, conceptual, gestural, technology-mediated, etc.). The dialogical principle lies at the heart of our reflections on the research process that we collectively experienced. This principle is based on the idea that in any communication, social practice or discourse, we are from the outset in relation with others. The concept of dialogism is generally attributed to the Russian literary theorist Mikhail Bakhtin (Todorov & Bahtin, 1981). Bakhtinian "dialogism" refers to how an individual elaborates, formulates, and expresses their personal position vis-à-vis themselves and others (Peytard, 1995). In expressing their worldview, the individual calls upon the points-of-view and conceptions of the world they perceive in others, contrasting them, and confronting them with their own personal position. Therefore, the relationship with others is thus omnipresent in our personal conception of the world (social, physical, sensory, cultural).

This epistemological stance becomes political when the dialogue involves protagonists with respectively normalized (e.g., allistic) and minoritized (e.g., Autistic) statuses and perspectives. Autistic individuals have historically been, and remain, restricted in certain institutionalized spaces, in their capacity to elaborate, formulate, and express their personal position vis-à-vis themselves and others, which can be understood as a form of epistemic injustice (Fricker, 2007). Epistemic injustice refers to "a specific type of injustice that affects an individual in their capacity as a knower or epistemic agent, i.e., as a person who produces, uses, or transmits knowledge—including knowledge regarding their personal or social experiences" (Catala et al., 2021, p. 9017, see also Chapter 24, this volume). Such injustice often arises when Autistic people

[9] The French original citation: « la capacité de l'esprit humain de concevoir, créer et communiquer au sujet des réalités sociales en termes d'alter, c'est-à-dire par rapport à d'autres individus, groupes, communautés et cultures» (Markovà & Orfali, 2005, p. 27).

and autistic knowledge are confronted by dominantly situated knowers (e.g., allistic researchers, allistic clinicians, allistic parents of autistic children) who refuse to make the effort to understand realities that can only be known from a position other than their own. Remedying such injustice guides our joint and respective research activities. Starting from the dialogical premise that we form our ideas in contact with (or in anticipation of) those of others, it becomes possible to enrich perspectives on both sides by capitalizing on this productive dialogism of shared meaning.

In the following section, we address three specific issues that occurred in our collaborative process. We also discuss how we established, applied, and negotiated the way we work together. To do so, we will illustrate the dialogical method in four steps, as we have experienced them:

(1) Cross-fertilizing knowledge and establishing equitable collaborative relationships between Autistic and allistics;
(2) Valuing the diversity and plurality of knowledge;
(3) Reflecting on our respective standpoint and developing a joint political engagement;
(4) Rebalancing power relations in the collective.

Although these steps are presented sequentially for editorial purposes, they occurred iteratively in practice (see Fig. 9.1).

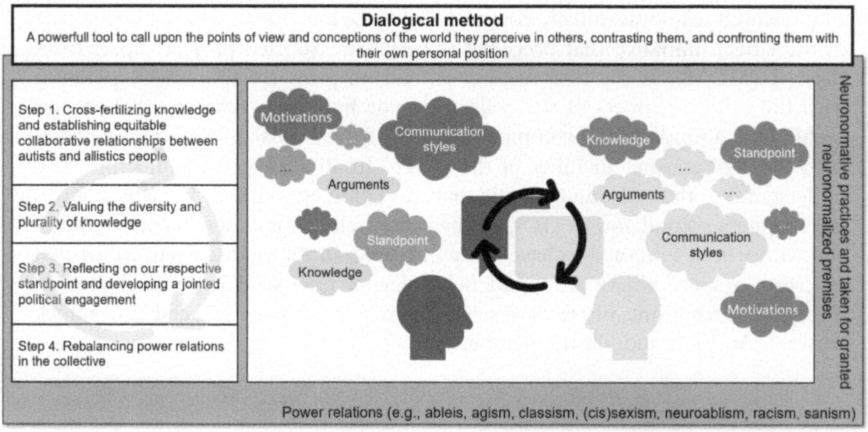

Fig. 9.1 Schematization of the iterative four steps dialogical method as we (the authors) experienced it

Step 1—Cross-Fertilizing Knowledge and Establishing Equitable Collaborative Relationships

A growing number of authors have discussed and explored how and why autistic ways of being, acting, and communicating tend to be misinterpreted by allistic individuals, leading to unfavorable perceptions of Autistic people and communities (see Milton, 2012 and Mitchell et al., 2021 for a critical review). This presents a major challenge for neuromixed collaboration among teams consisting in autistics and allistic people—a challenge that is essential to overcome in order to avoid maintaining or perpetuating epistemic injustices. The key questions we grapple with are: How can we facilitate interactions and mutual understanding? How can we identify misunderstandings? How can we ensure that the research process leads to the emancipation of autistic people?

Through debriefing meetings, we[10] sought to delve deeper into these issues, taking into account our localized experiences and our different communication styles. The conversations that transpired during these meetings ended up enriching and validating our perspectives and interpretations of the interactions within our research collective. They also enabled us to engage in a metacognitive[11] process that allowed us to iteratively reflect and analyze our experience, both together and within the research collective, thereby contributing to the cross-fertilizing of our knowledge as well as to the establishment of equitable collaborative relationships (see Box 9.1).

Box 9.1—The Starting Point of Our Dialogical Journey

The research team was initially composed of three individuals who had been part of the citizen initiative that preceded the Autisme Soutien project. Apart from these, I (MDM), as the lead researcher, had no prior collaborative experience with the other members of the collective. The first few weeks of collaboration highlighted a fundamental incompatibility of my positions and working methods with those of another member of the group. In an attempt to gain support in understanding the working dynamic that was being established, I reached out to the chapter's second author (IC), with whom I felt it was easier to communicate and with whom I seemed to have more in common. As we discussed, we wished to create a space where we could delve deeper into such issues and prevent divergent perceptions, interests or needs from becoming an intractable deadlock between Autistics and allistic members.

[10] In this chapter, "we" refers to the authors.

[11] Metacognition refers to the ability to reflect on one's own cognitive processes. It is at work in all learning, which is seen as an active process rather than a simple retention of information. Metacognition enables us to identify our mistakes as well as our successes, to understand why we achieved a given outcome (whether positive or negative), and to adjust if necessary. The development of metacognitive skills thus contributes to improving the quality of learning.

To achieve this, we sought to create a space for dialogue in the form of regular virtual meetings, scheduled every three weeks throughout the duration of the project. Additional meetings were arranged as needed and spontaneous discussions also took place in response to specific circumstances. Deliberation and validation practices have been at the heart of our implementation of the dialogical principle. For instance, we engaged in reflection and discussion until we reached consensus and collaboratively constructed a stance on emerging issues that arose during the research process or other events related to the project. Deeply anchored in knowledge cross-fertilizing, most of our meetings were devoted to validating our interpretations of various situations experienced within the research collective. In doing so, we sought to gain a shared understanding of the collaborative research process at stake as well as of any problems and their consequences in terms of epistemic (in)justice which constitutes a step towards establishing equitable collaborative relationships.

Step 2—Valuing the Diversity and Plurality of Knowledge

Feminist standpoint theory and other work related to epistemological points-of-view have highlighted the situated character of all knowledge as being inscribed in a historical time, localized and rooted in a particular world-view and material conditions (Flores Espínola, 2012; Harding, 2004). This suggests that all individuals' knowledge is necessarily unique and colored not only by formal knowledge they might have acquired, but also by their whole background and experiences. Therefore, it is crucial for research teams to respect and value the diversity and plurality of knowledge whether, experiential, professional, or scientific knowledge[12]. It must also be acknowledged that these forms of knowledge are not interchangeable and none is superior to another. While recognizing that all forms of knowledge are important, we must avoid the relativism that would imply all knowledge is equal in all circumstances. It is important to acknowledge that some forms of knowledge will provide more relevant insights in certain contexts. Simply put, it is crucial to recognize that each member brings a unique global knowledge that is the result of their accumulated experience, but also to accept that they might bring different types of knowledge (see Box 9.2). While working in neuromixed groups, it is consequently crucial to consider the arguments made, rather than the people making them or the type of knowledge on which they are based (see Chapters 8, 10, this volume).

[12] While in this chapter our primary focus is on experiential, professional, and scientific knowledge, we also acknowledge the existence and importance of other forms of knowledge, such as pragmatic and traditional knowledge.

Box 9.2—Knowledge in our Neuromixed and Interdisciplinary Collective
Our research collective initially comprised five Autistic citizens, an Autistic researcher and two allistic researchers. Various types of knowledge intersected within the collective, but also within individual members. The knowledge held by the various members was as follows:

- All Autistic citizen partners had experiential autistic knowledge. In some, this knowledge was interwoven with professional knowledge (i.e., as trained and practicing peers in a mental illness context) and scientific knowledge (i.e., acquired through experiences as a research collaborator). Being Autistic, the lead researcher and first author of this chapter (MDM) had experiential autistic knowledge, interwoven with their professional knowledge (i.e., as an occupational therapist) and their scientific (methodological) knowledge.
- The other researchers had professional (i.e., as health professionals) and scientific autism knowledge rooted in health or social sciences. Such knowledge was, in one, interwoven with the social experience of being the parent of an autistic child.

Acevedo argues that Autistic researchers have an invaluable role to play in enabling autistic people to reclaim and actualize their own stories and thus de-center the epistemic authority that allistic professionals have historically held over autistic experiences (Acevedo, in Dwyer et al., 2021). As knowledge producers, Autistic people and Autistic professionals with critical and experiential knowledge offer valuable insight regarding subjects and realities that concern and affect them. By way of example, StEvens' (2022) research demonstrates that the experiential and professional knowledge of autistic teachers is essential for thinking about and implementing inclusive teaching methods. Yet, autistic researchers often face a lack of credibility as producers and transmitters of knowledge. Their positioning places them under the suspicion of violating a supposed epistemic neutrality, dear to the defenders of a positivist vision of science (Botha, 2021). Despite benefits associated with dual positioning (particularly in terms of decreasing the risk of bias (Raymaker, 2017), Autistic researchers face a dual risk of rejection, both by the autistic community and by the scientific community (see Box 9.3).

Box 9.3—To be Autistic While Being a Researcher or to be an Autistic Researcher
At first, I (MDM) did not feel legitimized to lead this study, because my research program was not related to autism. Following discussions related to what it means to be an "expert", I eventually came to terms with the idea of the interweaving of my autistic experiential, my professional and my scientific knowledge representing an added value, both for the team and the project.

While I had hitherto considered that being Autistic was something that belonged to my personal life and had to remain distinct from my professional life, I engaged in a reflexive process that led me to recognize that the two were, in fact, indissociable. I therefore came to acknowledge that the knowledge acquired through my lived experience, as well as the knowledge thoroughly acquired over time to learn to better understand and know myself, intersected with my scientific knowledge and expertise. Ultimately, this project led me to completely reorient my research program.

Still, questioning from outside of our collective about my positionality as an Autistic researcher has proven to be an issue. Perceived as a political standpoint, such involvement implied a disclosure among scientific peers, whereupon some collaborators and other players questioned the legitimacy of my involvement in such a project, because of my belonging to the autistic community. My involvement in this project also led to an unplanned unveiling, as a researcher, within the autistic community. This led to a noticeable change in relationships, particularly in relation to my positioning in the academic field and my understanding of the higher education environment, as well as my teaching experience. For instance, I began receiving requests to provide my perspective on some individual's post-secondary education choice; to offer my advice and support concerning academic challenges, particularly those related to interactions with teachers/ professors; as well as to assist in clarifying ambiguous instructions in academic assignments in classes (other than those that I taught).

Step 3—Reflecting on our Respective Standpoint and Developing a Joint Political Engagement

Reflecting on our respective positions as allistic and Autistic researchers appeared to us to be essential, as our respective standpoints (see Box 9.4) influenced both what, in our eyes, might be problematic and what should be the aim of our research. Reflecting on our standpoints also led us to question the social and political implications of our involvement in the project.

Box 9.4—Our Understanding of Our Political Engagement

Over time, we came to acknowledge that we are both researchers that have found understanding and allyship through and with each other in recent years. We are both white and French-speaking Quebecers/ Canadian who have undertaken higher education and research, and we have the privilege of holding academic positions.

We also recognize our specific experiences and motivation. For instance, the first author (MDM) identifies as a non-binary Autistic. They are also aphant and prosopagnosic[13], and consider themselves as having a lived experience of disability. Their interest and motivation to research the experiences of Autistic people is intertwined with their own experiential knowledge as well as with their activism. They consider that critical disability, critical autism studies and neurodivergent perspectives are vital to the understanding of human occupations, interactions, and experiences. They have experienced othering, tokenism, exclusion, and denial of their perspectives, voices, identities, and minds in their personal, academic, and professional lives.

The second author (IC) is a sociologist and an allistic person. She has a keen interest in the notion of everyday life and the social experience of Autistic people. Engaged in projects aiming for greater social justice, she has been working for several years with groups and communities on issues that particularly impact women and people concerned with gender diversity/plurality (e.g., access to care and social support, street harassment, violence and stigmatization, mental load, cognitive and emotional aspects of care work).

In research collectives such as ours, the cross-fertilization of disciplinary knowledge is also noteworthy, as "[t]he move to broaden autism research from neuroscience, psychology, and clinical studies to include social sciences and humanities [opens] up a space for critical interdisciplinary debate" (Long, 2023, p. 27). This is of great importance in the critical development of new autistic and autism knowledge. What is more, and as pointed out by Schneid and Raz (2020), emancipatory participatory research offers the possibility of deconstructing neuronormalized premises that are taken for granted in academic research, by questioning and enriching the formulation of concepts that may be central to a given discipline, in sociology, for example.

It is likely that the understanding of autism and what it means to be Autistic will vary among neuromixed research teams where some members belong to neuronormalized and others to neurominorized groups. Indeed, Autistic people's view of themselves—a view that often deviates from a pathological model (Bagatell, 2007; Kapp, 2020)—is frequently in tension with that of researchers, particularly those coming from clinical disciplines. Based on stereotyped social representations of what autism "is" (or rather, "autism spectrum disorder" as it is still considered within many professional and scientific circles), there has historically been an important artificial divide between, on the one hand, caregivers, and service providers (allistics or presumed to be), and on the other, those that need help and service recipients (Autistic people). This dichotomy reinforces neuronormative practices or those based on "pity-charity" logic. Consequently, support initiatives developed by Autistic people

[13] The term aphant is used to describe people who do not experience voluntary imagery, while the term prosopagnosic—sometimes said face-blindness, describes people who do not distinguish the characteristic features of a face or associate them with a known person and therefore, who do not use faces to recognise people.

for Autistics people risk being confronted with judgments, questioning—insidious or explicit—and even pressure to surrender to allistic and medical authority. Despite these tensions being abusive, offensive, and disabling, they can still have a catalytic effect on several levels. When dialogue is possible, it can enable groups and individuals to discuss the detrimental effect of the commodification of medical authority in community services and social support. Doing so can challenge, question, or even trouble the traditional (and assumed to be accurate) idea that support must be provided by allistics to autistics people (see Box 9.5) (Bertilsdotter Rosqvist et al., 2023).

Box 9.5—Two Incidents That Proved to Be Catalytic

From the beginning, we faced a clash between the emancipatory social framework and the biomedical one held by some members. The capacity of autistic people to efficiently organize themselves to provide a social support service was discussed, as well as the widespread view that biomedical professionals' implication was essential to safely provide such a social support. In the end, that clash proved to be the catalyst that consolidated the relevance of a "by and for" autistics perspectives as well as a participatory and citizen approach (c.f citizen science as discussed in Chapters 4, 14, this volume). Similarly, concerns later arose regarding the organizational structure of the service (e.g., under which organization? Based on which coordination model?) causing another clash. In this regard, researchers within our collective argued that the service provision needed to be overseen by physicians and that it should be established through an autism organization run by allistics. Again, these concerns — rooted in both medical authority and neuronormativity[14] bias, ended up reinforcing the relevance of our "by and for" autistic perspective.

[14] Such biases are rooted in neuronormativity, which refers to "the societal assumption and expectation that all individuals' neurocognitive functioning aligns with a perceived 'normal' standard" (Catala, 2023).

Such questioning of autistic autonomy and emancipation proved to be emotionally taxing for most of the autistic collective members. Still, they ended up being catalytic and reinforced both autistic leadership of both the project and the initiative and the service's community-based approach. The dialogical method we developed and put into practice came in the wake of these incidents. Indeed, it was important to us to be able to overcome the dualism of our respective positioning (Autistic/allistic) and set the scene for authentic and equitable working methods.

To mitigate frictions, tensions, and clashes between how to conceive the nature of perceived support and how to offer support, we argue that it is crucial to develop a joint political engagement that does not erase stakeholders' respective standpoints (Step 4 delves more into this). This is all the more important as not everyone has the same opportunities or advantages (privileges), given the social relations that run through society and by which we are situated. Despite this, all stakeholders are endowed with agency and reflexive skills, enabling them to envision and attempt to transform the social relations in which their relationships are embedded. Given this point, it is unrealistic and impossible to totally avoid all tensions. When they occur, they should serve as a reminder that, in addition to the power relations induced by being neuromixed, other power relations (e.g., classism, agism (cis)sexism, racism, ableism) intrude into research collectives and can manifest themselves in interactions.

Step 4—Rebalancing Power Relations in the Collective

Contemporary societies produce inequalities that discriminate against neurominorized people—including Autistics—to the advantage of neuronormalized people. It would be naive to believe that any interpersonal relations could be magically exempt from neuroableism. Setting up research teams, collectives, and groups where Autistic members form the majority can contribute to a power shift (see Box 9.6).

Box 9.6—There Is Power in Numbers

The formation of a research collective in which there would be more Autistic than allistic members was a non-negotiable condition that I (MDM) laid down at the very beginning of the project's formation. I believed that such a group composition would ensure and defend, if needed, the decisions and self-determination of the Autistic members. For similar reasons, our research collective decided, as a group, that decisions would be taken by consensus or, if not achievable, by vote (i.e., majority plus one). The fact that Autistics members were highly represented ensured that no decision could ever be imposed by allistic members.

However, as we aimed to adopt an over-representation of autistic experiential knowledge in our neuromixed research teams, it was important to ensure that Autistic members' participation was not inequitable or economically detrimental. We recognize that Autistic citizen research partners are often not compensated for their work while university researchers are paid for their research activities: it is therefore important to change this. We believe that budgetary rules must be collectively established to provide adequate and sufficient compensation to non-academic members who give their valuable time and expertise to the project.

Researchers who are part of a neuronormalized group benefit from dominant (and most common) communication both in everyday interactions and collaborative research practices. For example, face-to-face work, which is often valued in neuronormative research settings, can prove incapacitating and hinder some autistic people's participation. The structure of meetings is another example: while meetings with no structure other than an agenda are frequent among research teams, they can be unsatisfactory for some, hence the importance of formalizing research group or teams functioning. Similarly, communication modes should take into account each member's optimal participation and avoid perpetuating neuronormative collaborative practices. Neuronormalized researchers are invited to reflect on their "taken for granted" and "common sense" practices by questioning their usual working methods and recognizing that these may limit other members' participation (see Box 9.7). To this extent, establishing working and collaborative methods that suit everyone is essential to formalize new and inclusive practices.

Box 9.7—Allistic Practice of Self-Disempowerment

In neuromixed research teams, allistic researchers are invited to do research "differently" and under conditions that are not solely decided by them. It is in terms of rebalancing powers that I (IC) have reexamined my role as an allistic researcher in the project. It felt crucial to me to be conscious of my own privilege and to adjust my conduct and interactions accordingly, so that dominant norms were not mirrored in our group. To do so and act with critical awareness, I attempted to reduce my power as an allistic and my ties of complicity and connivance with the neuronormalized majority by practicing self-disempowerment.

This idea of self-disempowerment might seem pejorative. However, a meaningful and equitable redistribution of power requires relinquishing any surplus privilege. **In fact, a privilege that is not redistributed is kept.**

Through my involvement in the collective and following various incidents, I tried to act in solidarity with the denunciations of the appropriation of Autistic people's voice or their tokenism in events or research programs entirely designed and organized without autistic people. While doing so, I was careful not to take on a "saviour" or "protector" role that would have undermined the rebalancing of power, or even reversed it in my favor by possibly making me gain symbolic capital. Allistic self-disempowerment is also about understanding that certain contexts or situations are particularly sensitive for some members. For example, I either withdrew or was not invited to certain discussions held between Autistic members. In fact, allistic self-disempowerment practice focuses on being supportive and being present at the right time.

Within our collective—and in terms of responsibilities and roles, a self-disempowerment practice implied leaving, as much as possible, the leadership of the project in the hands of autistic people. This meant making a significant contribution while not seeking individual recognition. In other words, the practice of allistic self-disempowerment means not losing sight of the collective work, which might pose an additional challenge for both Autistic and allistic researchers at an early stage in their careers, as the academic world often values individual achievement and success.

The ethical importance of not speaking for others is a relatively well-known principle and most researchers are familiar with it. Enabling people to act on their own terms, unless explicitly requested to do so, is another equally important principle, but perhaps less widely conveyed. From our dialogical perspective, it is also important that the implemented self-disempowerment practices are discussed and agreed upon between autistic and allistic collaborators to reach a common understanding.

Conclusion

Building on our shared experience, this chapter aimed at expanding on issues related to participatory action research. We specifically aimed to explain and illustrate how reflecting on our respective positioning allowed us to establish an open dialogue that enabled us to validate and to counter-validate subjective interpretations of certain situations as well as to unravel anticipated imbroglios. Rooted in a reflective stance, this dialogical method allowed for cross-checking information, making inferences, and deepening reflections, while contributing to strengthening our relationships.

Our experience has highlighted the importance of creating time and space for a broader shared reflection—an invisible yet essential labor of establishing relationships and making sense of the process as well as of what might have happened between proceeding meetings. Such time and space have been crucial in the process of exploring and confronting our respective understandings, be it in relation to types of knowledge, or to the idea of expertise and to research conducts—thereby highlighting the richness of the diversity and plurality that characterizes our respective knowledge. In doing so, we reflected on our respective standpoints and developed a joint political engagement. This allowed us to get to know each other better, build trust, and identify shared values as well as a common understanding of what constitutes knowledge and how we produced it. This has also proven to be a lever for strengthening our collaboration and carrying out more projects together. Through our dialogue, we have learned, unlearned, and grown. More importantly, we have reoriented ourselves as researchers—as well as people with personal lives—in a direction that leads us to do research differently, in a more sensitive and culturally informed way, and in a way that promotes Autistic people's agency both in the process of knowledge production and in their applications.

Finally, the writing of this chapter benefited our dialogical journey, as we revisited our thoughts during the co-writing process and deepened our understanding of some aspects of the project. We are looking forward to the continuation of this dialogical process over the coming years, with the advancement of autism support-related knowledge, critical autism studies, and, more widely, neurodivergent perspectives. By sharing our journey, we hope to inspire other teams to lead projects in a neuromixed context to produce knowledge and promote the epistemic agency of neurominorized people.

References

Bagatell, N. (2007). Orchestrating voices: Autism, identity and the power of discourse. *Disability and Society, 22*(4), 413–426. https://doi.org/10.1080/09687590701337967

Bertilsdotter Rosqvist, H., Milton, D., & O'Dell, L. (2023). Support on whose term? Competing meanings of support aimed at autistic people. In D. Milton & S. Ryan (Eds.), *The Routledge international handbook of critical autism studies* (pp. 270–276). Routledge.

Botha, M. (2021). Academic, activist, or advocate? Angry, entangled, and emerging: A critical reflection on autism knowledge production. *Frontiers in Psychology, 12*, 1–12. https://doi.org/10.3389/fpsyg.2021.727542

Catala, A. (2023, April 11). Understanding Neurodiversity, Unlearning Neuronormativity. *Blog of the APA*. https://blog.apaonline.org/2023/04/11/understanding-neurodiversity-unlearning-neuronormativity/

Catala, A., Faucher, L., & Poirier, P. (2021). Autism, epistemic injustice, and epistemic disablement: A relational account of epistemic agency. *Synthese, 199*(3–4), 9013–9039. https://doi.org/10.1007/s11229-021-03192-7

Chapman, R. (2019). Autism as a form of life: Wittgenstein and the psychological coherence of autism. *Metaphilosophy, 50*(4), 421–440. https://doi.org/10.1111/meta.12366

Chown, N., Robinson, J., Beardon, L., Downing, J., Hughes, L., Leatherland, J., Fox, K., Hickman, L., & MacGregor, D. (2017). Improving research about us, with us: A draft framework for inclusive autism research. *Disability & Society, 32*(5), 720–734. https://doi.org/10.1080/09687599.2017.1320273

Contandriopoulos, A. P., Champagne, F., Potvin, L., Denis, J. L., & Boyle, P. (1990). *Savoir préparer une recherche. La définir, la structurer, la financer*. Presses de l'Université de Montréal.

Cordeau, M. -J. (2019, April). Présentation d'Autisme Écoute par Marie Josée Cordeau. Retrieved August 6, 2023, from https://mjcordeau.com/blogue.cfm?bId=2021

Crompton, C. J., Hallett, S., McAuliffe, C., Stanfield, A. C., & Fletcher-Watson, S. (2022). A group off fellow travelers who understand": Interviews with autistic people about post-diagnostic peer support in adulthood. *Frontiers in Psychology, 13.* https://doi.org/10.3389/fpsyg.2022.831628

Cyr, C., Mckee, H., O'Hagan, M., & Priest, R. (2016, July). Le soutien par les pairs: Une nécessité. Comité du projet de soutien par les pairs en matière de santé mentale de la Commission de la santé mentale du Canada. Retrieved January, 23, 2023, from https://www.mentalhealthcommission.ca/wp-content/uploads/drupal/2016-10/Making_the_Case_for_Peer_Support_2016_Fr.pdf.pdf

Désormeaux-Moreau, M., & Aubry, C. (2020, November). Croisement des savoirs et partenariat citoyen: L'exemple de la mise en place d'une plateforme de soutien en ligne conçue par et pour les personnes autistes. Midi-Conférence de l'Institut universitaire de première ligne en santé et services sociaux. Retrieved August 6, 2023, from https://www.youtube.com/watch?v=v3MXKNTDXPI&t=180s&ab_channel=Institutuniversitairedepremi%C3%A8religneensant%C3%A9etservicessociaux-CIUSSSdel%27Estrie-CHUS

Dwyer, P., Acevedo, S. M., Brown, H. M., Grapel, J., Jones, S. C., Nachman, B. R., Raymaker, D. M., & Williams, Z. J. (2021). An expert roundtable discussion on experiences of autistic autism researchers. *Autism in Adulthood: Challenges and Management, 3*(3), 209–220. https://doi.org/10.1089/aut.2021.29019.rtb

Evans, B. (2013). How autism became autism: The radical transformation of a central concept of child development in Britain. *History of the Human Sciences, 26*(3), 3–31.

Flores Espínola, A. (2012). Subjectivité et connaissance : réflexions sur les épistémologies du 'point de vue.'*Cahiers Du Genre, 53*(2), 99–120. https://doi.org/10.3917/cdge.053.0099

Fricker, M. (2007). *Epistemic injustice: Power and the ethics of knowing.* Oxford University Press.

Greenwood, D. J., Whyte, W. F., & Harkavy, I. (1993). Participatory action research as a process and as a goal. *Human Relations, 46*(2), 175–192. https://doi.org/10.1177/001872679304600203

Harding, S. G. (2004). *The feminist standpoint theory reader: Intellectual and political controversies.* Routledge.

Kapp, S. K. (2020). *Autistic Community and the Neurodiversity Movement: Stories from the Frontline.* https://doi.org/10.1007/978-981-13-8437-0

Duopress Lab. (2022). Ally. In *The young activist's dictionary of social justice* (p. 3).

Loiselle, J., & Harvey, S. (2007). La recherche développement en éducation: Fondements, apports et limites. *Recherches Qualitatives, 27*(1), 40–59. https://doi.org/10.7202/1085356ar

Long, J. (2023). Critical approaches to autism support practice. Engaging situated reflection and research. In D. Milton & S. Ryan (Eds.), *The Routledge international handbook of critical autism studies* (pp. 270–276). Routledge.

Van der Maren, J. -M. (1995). *Méthodes de recherche pour l'éducation.* Les Presses de l'Université de Montréal.

Marková, I., & Orfali, B. (2005). Le dialogisme en psychologie sociale. *Hermès, La Revue, 41*, 25–31. https://doi.org/10.4267/2042/8948

Milton, D. E. (2012). On the ontological status of autism: The 'double empathy problem.' *Disability and Society, 27*(6), 883–887. https://doi.org/10.1080/09687599.2012.710008

Milton, D., & Sims, T. (2016). How is a sense of well-being and belonging constructed in the accounts of autistic adults? *Disability and Society, 31*(4), 520–534. https://doi.org/10.1080/09687599.2016.1186529

Mitchell, P., Sheppard, E., & Cassidy, S. (2021). Autism and the double empathy problem: Implications for development and mental health. *British Journal of Developmental Psychology, 39*(1), 1–18. https://doi.org/10.1111/bjdp.12350

National Research Council. (2015). *Enhancing the effectiveness of team science.* National Academies Press. Retrieved August 6, 2023, from https://nap.nationalacademies.org/catalog/19007/enhancing-the-effectiveness-of-team-science

Nuwer, R. (2020, June). Meet the autistic scientists redefining autism research. *Spectrum.* Retrieved August 6, 2023, from https://www.spectrumnews.org/features/deep-dive/meet-the-autistic-scientists-redefining-autism-research/

Oliver, M. (2002, December). Emancipatory research: A vehicle for social transformation or policy development. In *Inaugural NDA Disability Research Conference, Dublin, 3 December. Conference proceedings: Using emancipatory methodologies in disability research* (pp. 15–23). Retrieved August 6, 2023, from: https://disability-studies.leeds.ac.uk/wp-content/uploads/sites/40/library/Oliver-Mikes-paper.pdf

Oliver, M. (1997). Emancipatory research: Realistic goal or impossible dream. In C. Barnes & M. Mercer (Eds.), *Doing disability research* (pp. 15–31). The Disability Press.

Oxford Learners' Dictionary. (n.d.). Ally. In *Oxford Learners' Dictionary.* https://www.oxfordlearnersdictionaries.com/definition/english/ally_1

Peytard, J. (1995). *Mikhaïl Bakhtine. Dialogisme et analyse du discours.* Bertrand-Lacoste.

Raymaker, D. M. (2017). Reflections of a community-based participatory researcher from the intersection of disability advocacy, engineering, and the academy. *Action Research, 15*(3), 258–275.

Reason, P. (1994). Three approaches to participative inquiry. In N. Denzin & Y. Lincoln (Eds.), *Handbook of qualitative research.* Sage Publications.

Schneid, I., & Raz, A. E. (2020). The mask of autism: Social camouflaging and impression management as coping/normalization from the perspectives of autistic adults. *Social Science and Medicine (1982), 248*, 112826. https://doi.org/10.1016/j.socscimed.2020.112826

StEvens, C. (2022). The lived experience of autistic teachers: A review of the literature. *International Journal of Inclusive Education.* https://doi.org/10.1080/13603116.2022.2041738

Todorov, T., & Bahtin, M. M. (1981). *Mikhaïl Bakhtine, le principe dialogique.* Éditions du Seuil.

Walmsley, J. (2001). Normalisation, emancipatory research and inclusive research in learning disability. *Disability and Society, 16*(2), 187–205. https://doi.org/10.1080/09687590120035807

CHAPTER 10

Innovative and Neuro-affirming Autistic Approaches to Autism Research

Sandra Thom-Jones and *Jennifer Lowe*

Introduction

The notions of an autistic approach, autistic lens, autistic vantage point, autistic point-of-view and adopting an autistic gaze in autism research are slowly and steadily gaining recognition across the disciplines of psychology (Botha, 2021), sociology (Milton, 2014), anthropology (Schneid & Raz, 2020), education (StEvens, 2022) and media studies (McDermott, 2022). However, these terms have different meanings and interpretations across disciplines and between individuals.

Although prior attempts at participatory approaches in autism research may have served as a façade for the tokenistic inclusion of autistic people by

S. Thom-Jones (✉)
University of Wollongong, Wollongong, Australia
e-mail: autisticprofessor@gmail.com

Melbourne, Australia

J. Lowe
La Trobe University, Melbourne, Australia

H. Bertilsdotter Rosqvist and D. Jackson-Perry (eds.), *The Palgrave Handbook of Research Methods and Ethics in Neurodiversity Studies*,
https://doi.org/10.1007/978-3-031-66127-3_10

non-autistic autism research teams, a truly *autistic-inclusive approach*[1] highlights foundational changes that can be made towards inclusion.

As autism researchers move towards partnering with autistic people as collaborators, rather than the objects of research, autistic autism researchers are increasingly identifying current research paradigms, processes and practices as barriers to fulfilling autistic needs. Despite widespread recognition of the 'Double Empathy Problem' existing as a mismatch between autistic and non-autistic communication styles (Bertilsdotter Rosqvist et al., 2023; Milton et al., 2022), these discrepancies have not been adequately considered by autism researchers. The development, conduct, interpretation and outcomes of autism research are all impacted by the complicated combination of autistic and non-autistic actors on either side of the research exchange.

This chapter highlights methods developed and adopted by autistic academics as a shift towards foundational, meaningful and sustainable change to participatory practices in the field of autism research. The authors have collated examples from eight autism studies in which adopting an autistic approach at each stage of the research process proved valuable. Figure 10.1 provides a roadmap for the reader to follow as we describe aspects of the various case studies. For us, as for all researchers, autistic research methods organically grow and develop over time, building upon lessons learned and improving with each autism study. For postgraduate students undertaking autism research in fulfilment of a masters or doctorate, the small qualitative studies may appear more relevant than the large-scale online surveys due to the financial and physical resources required but there are valuable learnings from each study that can be applied across topics and methodologies.

These eight case studies conducted across seven years focus on innovative, creative and adaptive 'work around' ideas born largely out of necessity in the face of common challenges.

In sharing these examples, the authors recognise that the field of autism research is in a constant state of flux and these autistic methods may well be superseded in months or years ahead. Indeed, the older of us looks back with dismay on aspects of the research she undertook in the decade or two prior to these case studies. There is a desperate need for autism researchers to strike a balance between maintaining the status quo as a means of adhering to academic rigour and developing autistic-inclusive methods in the pursuit of meaning, relevance, participation, inclusion, applicability and accessibility.

[1] In this chapter, we use the term 'autistic-inclusive approach' to refer to a way of undertaking research that is founded in understanding and prioritising the needs, perspectives and understandings of autistic people. This includes involving autistic people in all stages of the research process. It also includes acknowledging that many of the methods currently followed, and tools currently used, in research were designed from a neurotypical perspective; and adjusting these to suit the autistic community context to ensure that research is respectful and valuable to our community.

Case Study 1. Autism Inclusion Monitor (2017)

Case Study 2. Australian Autism Alliance (2020)

Case Study 3. Autism Inclusion Monitor (2021)

Case Study 4. Autism Programs in Schools (2022)

Autism Community Collective

- Autism Organisation Led
- Large Scale Studies
- Collective Unit of Analysis
- Mixed Method Surveys
- External Funding

Case Study 5. Autistic University Students (2021)

Case Study 6. Autistics in Academia (2022)

Case Study 7. Autistic Experiences of Eating (2022)

Case Study 8. Autistic Experiences of Motherhood (2023)

Everyday Autistic Experiences

- Autistic Researcher Led
- Small Scale Studies
- Individual Unit of Analysis
- Qualitative Interviews
- Passion Projects

Fig. 10.1 Case studies—roadmap for reader

Autism Community Collective: Macro Studies

The autism studies we have undertaken with a focus on collective community experiences and priorities have largely been commissioned by autism-specific organisations, motivated by the prospect of securing funding and grant opportunities through state and federal government initiatives designed to improve the quality of life of autistic people. At each stage of the research process, autistic-inclusive approaches to inquiry were carefully considered and curated based on the context of the study undertaken.

Case Study 1: Autism Inclusion Monitor (2017)

In 2017, 'Amaze', based in Melbourne, Australia, commissioned a study to understand both non-autistic Australian's attitudes towards autism and autistic people's lived experiences. This was designed to be a multi-wave longitudinal study. In this first wave, the majority of survey respondents were non-autistic Australians. Key concerns within the autism community, defined as autistic people and parents/carers of autistic people, included widespread stigma and discrimination. See the following publications related to this study (Jones et al., 2021, 2022, 2023).

Case Study 2: Australian Autism Alliance (2020)

In 2020, the Australian Autism Alliance commissioned a national online survey of autistic adults, and the family members of autistic children and adults, to inform their submission to the Federal State Senate Select Committee on Autism in support of a National Autism Strategy. Priority areas were identified as physical and mental health, education, employment, justice, housing, the National Disability Insurance Scheme (NDIS), social participation and advocacy.

Case Study 3: Autism Inclusion Monitor (2021)

In this second wave, Amaze commissioned a second national study which sought not only to replicate the first, albeit with unavoidable changes due to the COVID-19 pandemic, but also to explore new topics of importance to the autistic and autism communities. In 2021, our learnings from 2017 were applied, resulting in a dramatic increase in autistic participation.

Case Study 4: Autism Programmes in Schools (2022)

Victoria's Department of Education and Training commissioned an independent evaluation of Autism programmes run by the 'I CAN Network' in primary schools, high schools and an extra-curricular option via zoom.

An Autistic-Led Research Team

These four successful national case studies clearly share the common priority of ensuring the project is driven by an autistic-led research team. Representing 12 autism organisations, the Australian Autism Alliance specifically sought Sandra's expertise as a senior autistic academic researcher with a professional and personal passion for autism advocacy to inform policies. In case studies of similar size, driven by the need for autism organisations to secure funding and state-based government grants to improve the lives of autistic people, Amaze and the autistic-led 'I CAN Network' also sought specific autistic expertise. Although an autistic-led research team is recommended, and autistic researchers should be active and equal participants in all autism research, we are not suggesting that non-autistic perspectives should be excluded. We often have non-autistic members in our research team who provide a different perspective and bring diverse skills (see Chapter 9, this volume).

Community Informed Research Objectives

With our understanding that research topics deemed worthy of investigation should and can only be identified by members of the autistic and autism communities, the formation of an expert autism panel and autism inclusion advisory group were integral to the success of our studies. The expert autism panel initially formed in 2017 also provided consistent and longitudinal community input during the following Autism Inclusion Monitor study in 2021. Led by Sandra, and convened by autism advocate Shadia Hancock, members of the expert autism panel collectively represented a wide range of the intersectional characteristics present within the Australian autistic and autism communities. Similarly, members of the Autism Inclusion Advisory Group, formed to advise on student programmes, were instrumental in ensuring the independent evaluation of autistic-led programmes facilitated by 'I CAN Network' mentors in primary and secondary schools secured funding for meaningful change to occur.

Engaging in meaningful, sincere and ongoing collaboration with autistic people as equal partners in research is a necessary approach for autistic and non-autistic autism researchers. For academic researchers who already have autistic colleagues, existing synergy and mutual interests may provide a natural starting point to establish an ongoing collaboration across multiple disciplines. For postgraduate student researchers undertaking an autism study for the first time, or a student project with scant financial support and limited resources, adopting a bottom-up approach to identifying research priorities that align with community interests remains entirely possible.

Those new to undertaking autism research, and in particular non-autistic autism researchers, may find it more difficult to navigate the saturated and complicated landscape of autism organisations, advocates and individuals, all of whom purport to represent the shared and collective best interests of the autistic community. For postgraduate and early career autism researchers, peer-reviewed published papers about community priorities are one of the recommended starting points to understanding the potential benefit of a new project (Benevides et al., 2020; Frazier et al., 2018; Roche et al., 2021). Autism advocacy groups and their position statements may also provide guidance and direction in planning.

In initiating new collaborations, it is imperative for autism researchers to approach an autistic academic or autism advocacy group with authenticity and transparency in relation to your intentions. Autistic academics are all too familiar with the proposition of an informal zoom meeting to discuss autistic perspectives about a study that is already underway, being later positioned as a participatory approach or autistic community consultation. Non-autistic autism researchers must recognise the shared responsibility to ensure autistic insights are integrated in a way which positions research findings as accessible, feasible and available to the autistic community.

Investigators leading large-scale autism studies are encouraged to involve autistic academics in their research design, processes and practices from the inception of the study through to the dissemination of findings. Failure to do this often results in autistic academics engaging in unpaid labour under the guise of ethical or moral obligation to ensure acceptable practices.

Autistic academics are in a unique position possessing discipline-specific expertise and the lived experience of an autistic person across various domains. An autistic autism researcher's professional and personal perspective provided in the course of an autism study should be recognised as holding a dualistic value worthy of recognition in financial terms, rather than the opportunity to share the success through study citations. For non-autistic autism researchers, a list of autistic academics can be found on Sandra's site, autisticprofessor.com. For autistic autism student researchers, the Autistic Researchers Researching Autism (ARRA) Facebook community facilitates collaboration and mentorship.

Accessible Autistic Research Design

Studies designed to understand the collective experiences of large numbers of autistic people, their families, friends and wider communities are generally undertaken using online surveys, which can be easily and rapidly distributed among existing autistic networks. Both quantitative data and qualitative data can be elicited through the inclusion of established and adapted scales as well as strategically worded open-ended questions pertaining to known autistic research priorities.

In autism research, the importance of taking the time to plan and develop appropriate, relevant and respectful instruments to collect and measure data cannot be overstated. Whilst questionnaire development in our survey studies was led by the aforementioned expert autism panel and autism inclusion advisory group, the ongoing processes involved a reiterative and reflexive approach to refine the specific language and terminology used, clarify the intended meaning, and anticipate potential errors in data analysis due to misinterpretation. Particularly where autism research informs policy, additional attention must be taken to adopt respectful and neuro-affirming language.

One of the major barriers for autism researchers who wish to adopt an autistic approach to the design of online survey instruments is the underlying assumption in academia that any alterations made to pre-existing and established scales of measurement may limit the validity, reliability and therefore value of any significant research findings that follow. If autistic approaches to research cannot incorporate adjustments to the language used and intended meaning of existing instruments based on well-established evidence, the prospect that future autism research findings will directly benefit autistic people appears bleak at best.

As it stands, autism research continues to study important aspects of autistic people's lives using instruments developed by and for non-autistic researchers and survey respondents.

It is no longer acceptable for autism researchers to make minor modifications to the façade of pre-existing scales without addressing the foundational intention of the instrument in the context of the study. Continuing to measure the wrong constructs, albeit validly and reliably, does not serve the future of autism research, nor the potential to improve autistic lives.

Ethical and Meaningful Measures

There are also ethical considerations in determining meaningful measurement, particularly in studies involving autistic children with co-occurring intellectual and/or language impairments. In conducting an independent evaluation of autistic-led programmes in primary and high schools across Melbourne on behalf of the Victorian Department of Education and Training, concerns raised by the ethics committee prior to the project's approval were clearly counterproductive. Permission seeking procedures carried out by a predominantly non-autistic ethics committee for autistic autism researchers to engage with autistic children represents a paradoxical irony.

Ethics committees often incorrectly assume that autistic individuals, regardless of their age and capacity, are 'vulnerable' and 'mentally ill' based on the algorithm behind the pre-populated online ethics forms based on misinformation about autistic participants as a homogenous group (for detailed discussion, see Chapter 18, this volume, see also discussion on 'situational vulnerability' in Chapter 19, this volume). If autistic autism researchers evaluating school programmes designed for autistic children, some of whom have co-occurring intellectual and/or language impairment, are not granted ethics approval due to the project being perceived as 'too risky', when will autistic perspectives be considered as valuable in the evaluation of existing supports specifically designed for them?

Autistic Methods of Measurement

Designing an online questionnaire to meaningfully, validly and reliably capture the views of autistic students who attended autistic-led programmes at school to inform funding decisions initially seemed an almost insurmountable task. Armed with an array of pre-existing scales as potential suitors for inclusion in the online survey, a consultation with the Autism Inclusion Advisory Group quickly established that some of the standardised words, phrases and concepts were difficult to interpret as intended, even for autistic adults. There was a clear consensus as to which scale items and instructions needed to be amended in order to be perceived as originally intended. In some cases, only relatively minor changes to the wording of questions and/or responses were required to provide a significant increase in clarity, with an example provided below.

Adjusting Instruments for Autistic Participants

One of the key evaluation criteria related to assessing the impact of autistic-led programmes in schools relied on a measurable improvement in autistic students' perception of themselves, comparing scores before and after their

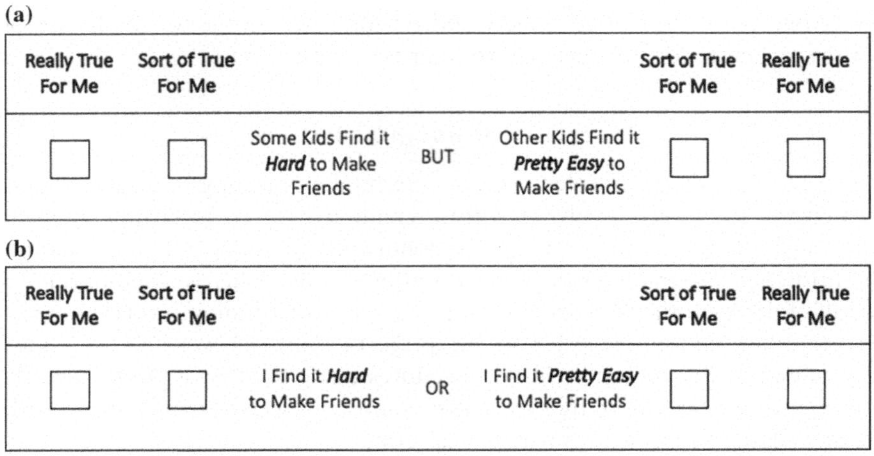

Fig. 10.2 **a** Original wording. **b** Adjusted wording

engagement in the programme. Harter (2012) 'Self-Perception Profile for Children' instrument was presented to the Autism Inclusion Advisory Group for consideration. As shown in Fig. 10.2(a), participants are asked to choose between two descriptive statements, indicating which resonates most with their own experiences. Multiple members of the advisory group reported difficulty interpreting and responding to these statements based on the impossibility of knowing 'some kids' or 'other kids' views. When these descriptive statements were adjusted to 'I' statements shown in Fig. 10.2(b), advisory group members perceived the question as intended and responded accordingly. Pilot testing these statements with another sample of young autistic people resulted in the same outcome. A non-autistic autism researcher and registered psychologist was consulted to ensure the validity of the measures was not compromised using the pilot data collected.

Participant Recruitment

Although the national survey commissioned by Amaze in 2017 engaged with an established and reputable research agency, the response rate for autistic participants was low relative to non-autistic participants. Changing tack to ameliorate this, autistic organisations and autistic researchers shared an online link to the national survey via their personal and professional social media networks. Autistic participant numbers increased from 57 in 2017 to 671 in 2021, providing detailed insights into the experiences of autistic Australians across interactive community contexts. As a broad generalisation, autistic-led autism studies often yield greater responses from the autistic and autism communities.

The potential significance of an autistic participant's contribution as a means of improving autistic quality of life has also been a clear source of motivation.

The Australian Autism Alliance and its member organisations promoted the opportunity for autistic participants to share their experiences to shape the first National Autism Strategy. Despite the limited timeframe which autistic participants had to respond as a result of the senate submission deadline, the initial target sample of 100 + was far surpassed with a total of 3,884 online surveys completed by self-reporting autistic people, and parents and carers.

The resulting dataset constitutes the largest and most comprehensive collation of autistic perspectives, not only within Australia but on an international scale.

Where large-scale collective autism studies are advertised via social media networks to recruit autistic participants, autistic academics have been likened to gatekeepers of their personal and professional networks. There is often a sense of safety, trust and approval among potential study participants when a recognised autistic academic confirms that an autism project is 'safe' to engage in and worthwhile due to anticipated community benefits.

As a student researcher or novice autism researcher, utilising social media platforms in an attempt to recruit autistic participants must be approached with authenticity, clarity and an accurate representation of yourself as a researcher and what your study aims to achieve.

Autistic Input on Analysis and Interpretation

Whilst following up with individual participants for further information or clarification about their responses is not an option in large-scale collective studies, researchers should instead engage with autistic community representatives and conclude the reiterative process that began with autistic perspectives. In our large-scale survey studies, the same expert autism panel that collaborated on the development of the survey instrument was involved in an iterative process of reviewing and providing feedback on the draft analysis and interpretation. This included aspects such as potential (mis)interpretations of findings, further analyses to undertake, and appropriate ways of presenting results. Importantly, these individuals were paid for their time and expertise.

Deliverables and Industry Outputs

The set of four case studies presented here as examples of large-scale collective autistic community projects conducted by autistic autism researchers and commissioned by autism organisations is merely a snapshot of what is possible if autism researchers commit to autistic-inclusive approaches to autism research and the development of methods and tools that are appropriate for autistic study participants. Autism researchers are also responsible for proactively identifying individuals and groups whose ethos towards autism and the autistic community does—or does not—promote a message of safety, acceptance and inclusion.

As a result of the Autism Inclusion Monitor research conducted with Amaze in 2017 and 2021, 'Autism: Change Your Reactions' was the first public education campaign to be aired via mass media platforms in Victoria. Key longitudinal research findings comparing 2017 and 2021 results post-COVID-19 were presented at the first 'Autism@VicParliament' event in 2022. In addition to academic journal articles, we have produced lay summaries, newsletter articles and other dissemination tools. Whilst not all autism research outputs are directly available to the public, their impact can be visible. For example, our independent evaluation of I CAN Networks' autistic-led programmes in schools contributed to the Victorian Department of Education's decision to commit $5.7million to expanding their programmes beyond 2022.

Everyday Autistic Experiences: Micro Studies

In contrast to Case Studies 1–4 which sought to gather brief data from large cohorts, Case Studies 5–8 focus on examining in depth the lived experiences of autistic individuals in an attempt to understand how their inner worlds intersect with activities of daily living and everyday experiences. In-depth qualitative autism studies are less likely to attract external funding opportunities and, as such, are often pursued by autistic academics as passion projects or with students whose research interests align with their own.

Case Study 5: Autistic University Students (2021)

Conducting multiple studies involving autistic university students over many years has revealed an inherent and pervasive systemic source of misinformation about what it means to be an autistic person studying at a tertiary education institution in Australia. Despite the standardised requirements to be accepted into an Australian university, predominantly non-autistic ethics committee members consistently raise concerns surrounding autistic students' impairments and vulnerabilities, suggesting that this cohort may not be able to provide informed consent and/or that 'simplified' documents should be made available. This commonly encountered and deeply offensive assumption about our competence requires systemic change. Refer to O'Connor et al. (2022) for an example from this body of work.

Case Study 6: Autistics in Academia (2022)

Whilst anecdotal evidence suggests there is an overlap between autistic strengths and the skills necessary to succeed as an academic, including an attention to detail and specific areas of interest, empirical studies about autistic academics are scant. This qualitative study was conducted over a 12-month period with 37 autistic academics from around the world, to understand barriers and facilitators to a successful career as an autistic academic. For key publications from this body of work refer to Jones (2023a, 2023b).

Case Study 7: Autistic Experiences of Eating (2022)

Qualitative autism studies detailing the seemingly mundane everyday lived experiences of an autistic woman's life—such as food and eating—offer a fortuitous opportunity for autistic academics to explore unfunded passion projects that similarly resonate with postgraduate student studies. Adopting an exploratory qualitative approach, this multi-faceted study was undertaken by an honours student and a Master's student from the discipline of psychology to normalise autistic women's experiences of eating as distinct from many pathologised eating disorder diagnoses. The initial findings are available in Stephenson et al. (2024).

Case Study 8: Autistic Experiences of Motherhood (2023)

As the majority of autism studies involve mothers of autistic children as proxy reporters who provide responses to researchers on their child's behalf, autistic mothers' lived experiences of motherhood from their own personal perspectives have largely been overlooked by researchers. This study seeks to understand the experiences of autistic women as mothers, adopting a staged developmental approach based on life stages. Our interest in this area began with a research project exploring autistic mothers' reflections on Reddit (Jones et al., 2024). Autistic women are provided a rare opportunity to share their lived experiences of motherhood throughout pregnancy, and their children's childhood, adolescence and adulthood.

Autistic-Led Research Teams with Lived Experience

Although autistic autism researchers have always experienced their everyday lives through an autistic lens they may or may not have been aware of, it is important for autistic academics to be cognisant of their knowledge limitations outside of their specific discipline of expertise.

Despite both authors' extensive lifelong subjective experiences with food and eating, first as undiagnosed autistic girls and then as late-diagnosed autistic women, as researchers they recognised a need to collaborate with colleagues from the discipline of psychology, who were similarly interested in exploring this range of phenomena. As a result of carefully curating an autistic-led interdisciplinary research team, autistic women were able to share, some for the first time, their subjective experiences of food and eating.

The research findings suggest that autistic girls and women would benefit from a normalised, rather than pathologised understanding of their specific food and eating preferences outside the medicalised parameters of diagnosable eating disorders. Lived experiences of autistic women's relationships with food and eating were established as normalised and in many cases, adaptive behaviours, challenging the long-held assumptions underpinning a disproportionate number of autistic girls and women diagnosed with an eating disorder.

More recently there have been an increasing number of autistic-led research teams setting out to collaboratively explore and develop new concepts in the field of autism research that existing concepts in non-autistic literature cannot adequately capture. Specific examples of constructs under investigation by these strategically assembled autistic-led research teams are non-autistic people's autism acceptance (Kim, 2020) and autistic burnout (Arnold et al., 2023; Mantzalas et al., 2022; Raymaker et al., 2020).

Ethics Committee Assumptions

Having addressed ethical considerations relating to large-scale collective autism studies, smaller qualitative studies involving one-on-one interviews may require more detailed planning to ensure autistic participants feel safe and comfortable throughout the semi-structured social interaction, whether the interview takes place online or face-to-face.

In Australia, National Health and Medical Research Council (NHMRC) guidelines govern all research efforts undertaken within universities, overseen by an institutional review board, more commonly referred to as an Ethics Committee. In the context of NHMRCs statement, Chapter 4.5 refers to 'People with a Cognitive Impairment, an Intellectual Disability, or a Mental Illness', and Sect. 4.5.3 reiterates persons are 'entitled to participate in research, and to do so for altruistic reasons'. Although all reasonable measures should be taken to protect autism study participants, acknowledging their vulnerabilities should not prevent their participation in research that may potentially benefit them and their communities (see Chapter 8, Munday, this volume for comparison with the UK ethics processes).

Having conducted several studies involving autistic adults as participants, we find ethics committee members often make incorrect assumptions about autism and autistic participants. One of the questions non-autistic ethics committee members consistently recommended for revision asks, 'Does your research include anyone who is intellectually, mentally or physically impaired'? In addition, autism researchers are often asked to reconsider whether autism studies include 'People who may be vulnerable or unable to give fully informed consent'.

Adopting a proactive approach to pre-emptively address the concerns consistently raised by non-autistic ethics committee members, a proforma statement outlining our position on the ethical conduct of autism research guided NHMRC guidelines has proven to be useful. The proforma statement developed specifically for autism research ethics applications is shown below in Fig. 10.3. The statement was attached to the ethics application submitted

Statement Regarding Research Involving Autistic People

The following statement was prepared by Professor Sandra Jones, PVC (Research Impact); who is an autistic person and an advocate for the autistic community.

As a researcher, I recognise the importance of ensuring the ethical conduct of research, and the need to comply with the NHMRC National Statement. As an autistic person, I recognise the autonomy and perspectives of autistic people.

Fig. 10.3 Ethics Application Statement from Autistic Experiences of Motherhood

for the autistic motherhood study, which was approved without additional revisions from ethics.

B.2.20: Anyone who is intellectually, mentally or physically impaired

As autistic people, we reject the common perception of autism as an 'impairment'. We see autism as a neurotype not an illness, a difference not a disability. For this reason, our response to this question will in most cases be 'no'. For example, in this instance research with autistic parents. We recognise that many autistic people have co-morbidities that include intellectual or physical impairments. However, this is also the case for non-autistic people. Participants in this study are (a) parents; (b) active on social media; and (c) self-selecting to respond to an online invitation, visit the study platform and complete the consent form. Consistent with any other population, and respecting their dignity and autonomy as autistic people, capacity is assumed—given that they are over the age of 18. We would also like the committee to note that the chief investigator and co-investigator are autistic researchers and autism advocates; and the peer reviewer is an autistic autism researcher, advocate and member of Australia's largest autism service provider.

A.4.1: Vulnerable Groups

There is a diversity of views on whether autistic adults constitute a 'vulnerable group'. We acknowledge that, given the high occurrence of anxiety as a co-morbidity of autism, participants in our studies may be considered 'more-than-usually' vulnerable. For this reason, all of the research we undertake is reviewed by and/or developed in collaboration with autistic people. For example, all surveys and interview guides are reviewed by a group of autistic people (to ensure that the questions are appropriate and respectful, and that the potential for distress is minimised. The two interviewers are an experienced autistic autism researcher (Lowe) and an experienced non-autistic researcher who has completed numerous qualitative studies with autistic participants. The aim of this study is to hear—and share—the voices of a population that is currently under-represented in the literature and under-served by health and social service agencies. This study seeks to support and empower vulnerable people, it is being conducted 'by, with and for' this group not 'on' this group.

Autistic Approaches to Research Design

One of the most crucial, yet undocumented, phases of an autistic-inclusive approach to research design involves gaining an understanding of the research team dynamics based on the personal and professional relationships between autistic and non-autistic academics and finding common ground with colleagues and collaborators whose expertise resides in a discipline outside your own (see Chapter 5, this volume). Developing an open and honest working relationship among all members of a research team is a critical factor to success, and even more so for autistic academics who must actively observe their peers to formulate a basic understanding of others communication preferences and styles of working, which may not always be accurate, requiring an additional but invisible process of ongoing and conscientious awareness.

For autistic autism researchers who are taking a lead role in guiding the direction of a study derived from their 'wish list' of passion projects, providing others with an authentic, personal and transparent vision for the research is worthwhile. Ensuring all members of the research team are privy to the 'why' which drives your pursuit of the passion project is an additional consideration in the early stages of the study but offers worthwhile return on investment in the analysis and interpretation of the findings. Making the decision to share personal details with colleagues about your vested interests in a specific study should be carefully considered on an individual basis, as this approach may not be appropriate and indeed be harmful in some contexts.

Adopting an autistic-inclusive approach—considering the cohort of research participants, desired outcomes and available resources—each stage of the research process can be customised to optimise participant experience. Following the collective fatigue experienced as a result of the COVID-19 pandemic, it was essential to remove the rigid restraints typically associated with research participation. To minimise potential barriers to participation, studies were designed to maximise flexibility.

Example (a): Autistics in Academia

Key considerations in the design of this study included the breadth, depth and nature of topics to be covered; the range of communication preferences among participants; and the risk of low response rates due to competing demands for participants time and energy.

The study was designed to be conducted over a 12-month period, with participants invited to reflect on a new topic each month and submit their written or audio-recorded responses via email. Participants were provided with topic specific questions and thought prompts but were free to write on any aspect of the topic that was of interest or relevance to them.

Participants were given a month to submit responses via email, and further flexibility was available to those who planned to submit a previous topic reflection at an agreed upon date. The submission of a reflection each month remained voluntary and allowed participants to prioritise the topics most important to them. Despite the duration of the study and the impact of COVID-19, 72% completed all 12 monthly reflections and 90% at least half.

Example (b): Autistic Experiences of Motherhood

This exploratory study sought to explore the experiences of autistic mothers of both autistic and non-autistic children. Understanding the importance of flexibility to autistic mothers, the potential participants were instructed to choose the most convenient day, date and time for their interview to take place, effectively offering an 'on call' autistic autism researcher to conduct interviews via phone, zoom or email, allowing participants to dictate the use of audio and/or video and/or chat settings available via zoom teleconferencing software.

Participant Recruitment

For postgraduate student or novice autism researchers, it is important to understand what potentially participating in a qualitative autism study may mean to an autistic participant as the recruitment process commences. Unlike completing an anonymous online survey from the comfort of home, engaging in qualitative research methods such as an interview may be a particularly daunting prospect for autistic people, due to a range of factors, some of which may not be apparent or even perceived as a credible consideration to non-autistic people. Pragmatic deterrents to participating in an interview may include the additional time and resources required, the process of negotiating and committing to a specific date and time based on another person's availability, access to the required technology and appropriate levels of digital literacy to utilise them as required by the researcher.

As autistic autism researchers who have conducted and participated in autism research interviews, a personal stock take of sensory, emotional, cognitive and physical resources available must be considered prior to making a commitment. It can be confronting to be faced with the reality of what it means to be an autistic person in circumstances such as these. Regardless of planning, preparation and practice, the ever-present possibility of an unanticipated emotional upheaval resulting in a state of dysregulation that continues to reverberate for hours, days or even weeks after, depending on the nature of the study.

Setting the potential for emotional dysregulation aside, it is well established that autistic people perceive and process cognitive and sensory information in different ways. This may include taking longer to process audio or visual information, asking the researcher to repeat an interview question due to an unreliable working memory and a diminished ability to retain information or maintain cognitive thought processes required to make a worthy contribution.

When faced with a new or anxiety provoking social exchange such as an online interview, the intensity of involuntarily physiological responses may require the use of medication prior to the interview, allowing the participant to maintain access to cognition and word retrieval.

Autism Research Advertising Etiquette

As autistic autism researchers, we are familiar with the autistic community's need for safe online social media spaces that are subject to membership criteria and group moderation.

Social media platforms such as Facebook and LinkedIn facilitate closed groups which are often highly valued by autistic individuals for the sense of belonging and connection they provide. Sustained online social interactions among autistic people with shared interests support the development of autistic community belonging and larger social networks.

Closed autistic community groups on social media platforms should not be joined by non-autistic researchers for the purposes of promoting their study. In the first instance, group moderators should be approached to inform them of the study and seek approval to advertise the opportunity to participate in autism research to the members of their group. When inviting autistic people to participate in autism research, it is important to provide a clear, detailed and logical rationale for why their contribution to the production of autism knowledge is valuable and how it will be used to improve the lives of autistic communities. It is also important to explicitly state whether the research is autistic-led and the extent to which autistic researchers are involved in the design and conduct of the research.

Interviewing Autistically

Whilst neither author has conducted nor participated in an interview as a non-autistic person, there are noteworthy differences when applying an autistic-inclusive approach to interview research. Sharing these insights may assist non-autistic academics and student researchers to be increasingly receptive and responsive to the needs of autistic interview participants.

As an autistic interview participant, it is comforting to know whether the researcher conducting the interview is autistic so that expectations can be adjusted accordingly. Having supervised non-autistic postgraduate students conducting qualitative studies involving interviews with autistic participants, anxiety is often experienced due to the complexity of navigating the nuance of 'politically correct' autistic terminology. Fear of offending an autistic interview participant is a common experience for all researchers; however, autistic academics are generally held to higher expectations and standards.

As autistic academics, it is important to foster a positive culture of patience, tolerance, respect and understanding to ensure the longevity and value of autism research.

Autistic Input on Analysis and Interpretation

There are benefits and risks associated with the use of Interviewee Transcript Reviews (ITR) as a tool to improve data quality and reassure participants who have engaged in an interview pertaining to a study involving highly personal or sensitive subjects (Rowlands, 2021). Where potentially identifying details could compromise interview participants' safety or well-being, sharing the interview transcript and providing them with an opportunity to make changes as they deem necessary serves to alleviate feelings of uncertainty and improve confidence. (see Chapter 13, this volume).

Adopting an autistic-inclusive approach to research design and methods, there are clear benefits to providing autistic participants with a transcript of their interview. Whilst some may benefit from additional time to process questions and formulate answers, others may appreciate the opportunity to reflect on their responses retrospectively. Considering differences in the ways autistic people perceive, process and respond to written and verbal information, providing an additional opportunity for the expression of thoughts and experiences promotes inclusion.

Student researchers and non-autistic academics may wish to factor in the use of Interviewee Transcript Reviews during the initial research design, offering autistic participants the opportunity to confirm or clarify non-autistic researchers' understanding of their intended meaning. This is pertinent when non-autistic researchers interpret responses provided by autistic participants.

Our autistic interview participants were provided with their interview transcript, offering an opportunity to correct errors, include additional comments, edit or delete any information they wish to retract. Many commented that they were delighted to receive a copy of their responses for their own reference and appreciated an opportunity to review the transcript in their own time.

Autistic Approaches to Dissemination and Integration

As autistic academics, the process of dissemination and integration of research findings into practice can be experienced as a conflict between professional and personal perspectives. For employed academic research professionals, publications in highly regarded peer-review journals are valued by the institutions paying our salaries, and career progression is the reward for following this well-worn path. However, access to journal articles and databases has traditionally been reserved for corporations and institutions who can afford to pay the hefty price tag to gain access; and thus, out of reach for those who stand to gain from the research. Whilst publishers have more recently taken steps towards addressing this imbalance of power and privilege, open access publication remains beyond the financial means of most researchers. The introduction of lay summaries and open access options are to be commended, offering autistic people who are not themselves researchers an opportunity to benefit from findings.

Where autism studies are funded by external third-party organisations or government departments, considerations surrounding how the research findings will actually reach autistic people often sits outside the scope of the project brief and thus outside of the budget. This often leaves autistic autism researchers with rich, insightful datasets that hold endless potential for improving the lives of autistic people, without a tertiary education institution, private organisation or government department willing to support such an undertaking. Autistic academics then have to choose between 'letting it go' or a 'labour of love' involving the creation of multi-media community-based resources such as YouTube videos, media articles, books, industry events, conference presentations, public speaking, social media content, factsheets and infographics. (c.f. Chapter 13, this volume).

In an ideal world, funded autism studies would incorporate the dissemination and integration of research findings in formats accessible to community members through online platforms. There should be at least as much time and energy dedicated to sharing the knowledge gained with the community of autistic people who contributed to the research, and stand to benefit from it, as there is dedicated to sharing it with other academics.

Autistic Comradery in Research

Despite the lack of financial support from governments and corporations to ensure autistic people are directly benefitting from the research they are generating through participation, there is a strong spirit of comradery between autistic researchers and autistic participants. To those who espouse the sentiment that autistic people do not want to, or cannot or will not participate in research, as autistic autism researchers we have evidence to the contrary. When autism research is designed by, with and for the benefit of all autistic individuals, the autistic and autism communities participate enthusiastically and benefit from the process.

The following statements were made by autistic participants who took part in our research, in unprompted feedback emails to the senior researcher.

> "I'm so glad that you are conducting this study and are openly autistic, it's very encouraging and such valuable research, I'm so happy to be a part of it."
>
> "Thank you for doing all you can to help my community. I cannot tell you how much it means to me."
>
> "Thanks for including me in this study – I've really enjoyed it, and I hope you get some really useful data out of it."
>
> "It's such a great project - the method in particular. I have a soft spot for good qualitative methods and am really enjoying participating in this. I've been part of a few research studies, and this is by far the best!"

Concluding reflections: An Autistic Way Forward

Every step of the research process offers an opportunity for all autism researchers to think beyond the perceived limitations of current practices and directly contribute to improving the lives of autistic individuals, their families and the wider autism community. There are simple steps that can be taken to incorporate autistic-inclusive research designs and methods in autism research, even where little time and few resources are available.

As autistic autism researchers who also live their everyday lives as autistic adults, autism research is encountered in a myriad of ways, some positive, some negative, and some outright confusing. But then there are those occasions that inspire innovation and hope that autism research itself can move towards adopting an autistic-inclusive approach.

These eight case studies represent a culmination of years spent re-imagining, developing, and trialling innovative ideas intended to improve autistic people's experiences of participating in autism studies and pave the way for autism researchers to adopt an autistic-inclusive approach to their work.

This approach is grounded in ensuring that autistic people are meaningfully involved in all stages of the research process, from inception to dissemination. It requires careful consideration of the extent to which existing research paradigms, processes and tools are appropriate for exploring issues of importance to autistic people, and for collecting meaningful data from autistic participants.

Table 10.1 provides an overview of some of the autistic-inclusive approaches to research design and methods that have been outlined in this chapter. We are certain that readers will build on this list in their own research.

Table 10.1 An autistic approach to the research process

Research team	*Autistic-led research team* *Interdisciplinary collaboration as required*
Research Objectives	Autistic community consultation, adopt a bottom-up approach Expert panel or advisory group—factor this process into the timeline
Research design	Customised for the autistic community Consider specific aspects unique to autistic participant cohort
Ethics application	Actively address common concerns Challenge incorrect assumptions Autistic input on the development of proforma statements
Online surveys	Adapt measures, scales and instruments for an autistic audience Autistic and academic community feedback Analyse pilot data to ensure validity and credibility maintained

(continued)

Table 10.1 (continued)

Research team	*Autistic-led research team* *Interdisciplinary collaboration as required*
Conducting interviews	Provide appropriate supports and or resources if required Interview experience led by the autistic participant
Participant recruitment	Minimise barriers, maximise flexibility Accept autistic participants' personal preferences to engage Communication preferences may vary, all data is valuable Adhere to autistic community advertising etiquette Contact moderators prior to accessing autistic groups online
Data quality and integrity	Autistic input on data analysis and interpretation Provide autistic participants with opportunity to review their responses Liaise with participants and/or autistic colleagues unfamiliar with the study
Dissemination and integration	Identify autistic people who benefit most
Deliverables and industry output	Advocate for funded design and distribution to autistic people and groups

References

Arnold, S. R. C., Higgins, J. M., Weise, J., Desai, A., Pellicano, E., & Trollor, J. N. (2023). Towards the measurement of autistic burnout. *Autism: The International Journal of Research and Practice, 27*(7), 1933–1948. https://doi.org/10.1177/13623613221147401

Benevides, T. W., Shore, S. M., Palmer, K., Duncan, P., Plank, A., Andresen, M.-L., Caplan, R., Cook, B., Gassner, D., Hector, B. L., Morgan, L., Nebeker, L., Purkis, Y., Rankowski, B., Wittig, K., & Coughlin, S. S. (2020). Listening to the autistic voice: Mental health priorities to guide research and practice in autism from a stakeholder-driven project. *Autism : The International Journal of Research and Practice, 24*(4), 822–833. https://doi.org/10.1177/1362361320908410

Bertilsdotter Rosqvist, H., Hultman, L., & Hallqvist, J. (2023). Knowing and accepting oneself: Exploring possibilities of self-awareness among working autistic young adults. *Autism: The International Journal of Research and Practice, 27*(5), 1417–1425. https://doi.org/10.1177/13623613221137428

Botha, M. (2021). Academic, Activist, or advocate? Angry, entangled, and emerging: A critical reflection on autism knowledge production. *Frontiers in Psychology, 12*, 727542–727542. https://doi.org/10.3389/fpsyg.2021.727542

Frazier, T. W., Dawson, G., Murray, D., Shih, A., Sachs, J. S., & Geiger, A. (2018). Brief report: A survey of autism research priorities across a diverse community of stakeholders. *Journal of Autism and Developmental Disorders, 48*(11), 3965–3971. https://doi.org/10.1007/s10803-018-3642-6

Harter, S. (2012). *Self-perception profile for children: Manual and questionnaires (Grades 3–8).* University of Denver. Retrieved 7 August from https://www.apa.org/obesity-guideline/self-preception.pdf

Jones, S.C., Lowe, J., Edwards, N., & Maloney, J. (2023). Autism: Change your reactions. In G. Hastings & C. Domegan (Eds.), *Social Marketing: Principles and practice for delivering global change*. (4th ed., pp. 321–329). Taylor & Francis. https://doi.org/10.4324/9781003200086-28

Jones, S. C. (2023a). Advice for autistic people considering a career in academia. *Autism: The International Journal of Research and Practice, 27*(7), 2187–2192. https://doi.org/10.1177/13623613231161882

Jones, S. C. (2023b). Autistics working in academia: What are the barriers and facilitators? *Autism: The International Journal of Research and Practice, 27*(3), 822–831. https://doi.org/10.1177/13623613221118158

Jones, S. C., Akram, M., Gordon, C. S., Murphy, N., & Sharkie, F. (2021). Autism in Australia: Community knowledge and autistic people's experiences. *Journal of Autism and Developmental Disorders, 51*(10), 3677–3689. https://doi.org/10.1007/s10803-020-04819-3

Jones, S. C., Gordon, C. S., Akram, M., Murphy, N., & Sharkie, F. (2022). Inclusion, exclusion and isolation of autistic people: Community attitudes and autistic people's experiences. *Journal of Autism and Developmental Disorders, 52*(3), 1131–1142. https://doi.org/10.1007/s10803-021-04998-7

Kim, S. Y. (2020). The development and pilot-testing of the autism attitude acceptance scale: An instrument measuring autism acceptance. *Autism in Adulthood, 2*(3), 204–215. https://doi.org/10.1089/aut.2019.0066

Mantzalas, J., Richdale, A. L., Adikari, A., Lowe, J., & Dissanayake, C. (2022). What is autistic burnout? A thematic analysis of posts on two online platforms. *Autism in Adulthood, 4*(1), 52–65. https://doi.org/10.1089/aut.2021.0021

McDermott, C. (2022). Theorising the neurotypical gaze: Autistic love and relationships in The Bridge (Bron/Broen 2011–2018). *Medical Humanities, 48*(1), 51–62. https://doi.org/10.1136/medhum-2020-011906

Milton, D. E. M. (2014). Autistic expertise: A critical reflection on the production of knowledge in autism studies. *Autism: The International Journal of Research and Practice, 18*(7), 794–802. https://doi.org/10.1177/1362361314525281

Milton, D., Gurbuz, E., & López, B. (2022). The 'double empathy problem': Ten years on. *Autism: The International Journal of Research and Practice, 26*(8), 1901–1903. https://doi.org/10.1177/13623613221129123

O'Connor, M., Jones, S., Gordon, C., & Joosten, A. (2022). Exploring environmental barriers and facilitators to inclusion on a university campus for autistic students. *Autism in Adulthood*. https://doi.org/10.1089/aut.2022.0053

Raymaker, D. M., Teo, A. R., Steckler, N. A., Lentz, B., Scharer, M., Delos Santos, A., Kapp, S. K., Hunter, M., Joyce, A., & Nicolaidis, C. (2020). Having all of your internal resources exhausted beyond measure and being left with no clean-up crew: Defining autistic burnout. *Autism in Adulthood, 2*(2), 132–143. https://doi.org/10.1089/aut.2019.0079

Roche, L., Adams, D., & Clark, M. (2021). Research priorities of the autism community: A systematic review of key stakeholder perspectives. *Autism: The International Journal of Research and Practice, 25*(2), 336–348. https://doi.org/10.1177/1362361320967790

Rowlands, J. (2021). Interviewee transcript review as a tool to improve data quality and participant confidence in sensitive research. *International Journal of Qualitative Methods, 20*, 160940692110661. https://doi.org/10.1177/16094069211066170

Schneid, I., & Raz, A. E. (2020). The mask of autism: Social camouflaging and impression management as coping/normalization from the perspectives of autistic adults. *Social Science and Medicine, 1982*(248), 112826–112829. https://doi.org/10.1016/j.socscimed.2020.112826

Stephenson, H., Thom-Jones, S., Lowe, J., & Milne, L. (2024) "I Know It's Very Spectrum-y": Autistic women reflect on sensory aspects of food and eating. *Autism in Adulthood*; https://doi.org/10.1089/aut.2023.011

StEvens, C. (2022). The lived experience of autistic teachers: a review of the literature. *International Journal of Inclusive Education*, 1–15. https://doi.org/10.1080/13603116.2022.2041738

Thom-Jones, S., Melgaard, I., & Gordon, C. (2024). Autistic women's experience of parenting: A Qualitative analysis of Reddit posts. *Journal of Autism and Developmental Disorders*. https://doi.org/10.1007/s10803-024-06312-7

CHAPTER 11

Fostering Ethical and Caring Fieldwork Practices in ADHD Research

Shruti Taneja-Johansson

Introduction

Even when the voices of people with ADHD are included in research on education, they tend to be framed within a certain discourse—medical and psychological—with a limited focus on experiences of the diagnosis, medication, and intervention (Eccleston et al., 2019; Ringer, 2020). The restrictive questions posed, and the research design selected exclude a space for certain experiences to be heard (Pellicano & den Houting, 2022). Epistemic injustice (Fricker, 2007) occurs when the capacity of people with ADHD as 'knowers' is negated, and when they are not considered bearers of valid and credible knowledge. However, there is a growing recognition of the value of participatory and inclusive approaches as well as the importance of the 'ADHD voice' in research (see Chapters 3, 12, 23, this volume). Still, the question here is: What does it mean to truly listen as a neurotypical researcher? What does ethical and meaningful inclusion of neurodivergent participants mean? In this chapter, I reflect on some of these questions within the context of fieldwork. In the process, I touch upon the ethical and practical possibilities of doing ADHD research fieldwork in ways that counter the epistemic injustice (Fricker, 2007) that currently holds sway in the existing research.

S. Taneja-Johansson (✉)
Department of Education and Special Education, University of Gothenburg, Gothenburg, Sweden
e-mail: shruti.taneja.johansson@ped.gu.se

H. Bertilsdotter Rosqvist and D. Jackson-Perry (eds.), *The Palgrave Handbook of Research Methods and Ethics in Neurodiversity Studies*,
https://doi.org/10.1007/978-3-031-66127-3_11

This text is a personal reflection on the decisions I took during my fieldwork and on the challenges, opportunities, and dilemmas I faced. It is also a realization of my own growing confidence as a researcher to differentiate between considering the quality of fieldwork to be limited to procedural ethics, methodological stringency and knowledge focus and perceiving quality as an ethical and caring research practice, central to which is *listening*. I am not a neurodivergent researcher, and in this chapter, I attempt to initiate a self-reflexive discussion on researching a neurotype other than my own. To do this, I draw on my own research journey as experienced during the extensive fieldwork I conducted with young adults with ADHD as part of a broader study looking at the educational pathways of persons with functional disabilities. In recent years, some neurotypical researchers have briefly reflected on their neurotypical positionality in relation to their interactions with autistic participants in their research (Crane et al., 2021; Peuravaara, 2021). Others have chosen to disclose their non-autistic status in their published article to acknowledge both possible power imbalances and their own neurotypical lived experience in the data analysis (Andersson, 2021; Martin, 2021).

Overview of the Study

This chapter draws on fieldwork I conducted during 2017 and 2020 as part of a much larger project[1] financed by the Swedish Research Council. Prior to data collection, ethical approval was obtained for the larger study from the Regional Ethical Review Authority in Linköping.[2] An ethnographic case study design (Merriam, 2016) was used to follow young adults with ADHD, looking at their retroactively constructed pathways to transitions from schooling through adult education, higher education, employment, and/or other domains of adulthood. Retrospective, current, and prospective qualitative data were collected over an average period of two years. This included the participants' narratives as gathered through numerous formal and informal conversations, text messages, photographs, certificates, and existing documentation from schools, institutional settings, social services, and health services. The young people were also shadowed (Czarniawska, 2007) in their everyday lives, including interactions with state agencies (e.g., social insurance agency, employment agency), healthcare, and other support services available in their current setting. Shadowing is a research technique in which participants are observed as they go through real-life situations, the goal being to gain first-hand experiences of their interactions within certain settings. Semi-structured interviews were also carried out with key individuals in the participants' lives, such as a parent, teacher, partner, or sibling.

[1] Participation for all? School and post-school pathways of young people with functional disabilities.

[2] Regional Ethical Review Authority in Linköping Reg. no. 2016/470–31.

The sample includes nine young people, between the ages of 18 and 31, six of them women and three men. Seven informants had been diagnosed after the age of 18 and two during their compulsory school years.

Being Inclusive of Diverse Voices: Navigating Access

Looking back, I see that I adopted a fieldwork approach that attempted to be inclusive of diverse voices—sometimes deliberately but most often not. This was done, for instance, by consciously deciding to use a variety of contact routes to increase the probability of greater heterogeneity among the study participants. In hindsight, however, it was even more evident in how I, in researching another neurotype, continuously worked on developing a broad communication competence (Stones, 2023) to enable different voices to be heard, before and after participants gave their consent.

Establishing Contact and Consent

The first stage of my fieldwork was to establish contact with young adults with ADHD and collect data, as laid out in the broader study. The aim was to include people with different educational journeys. However, keeping in mind the dominance of middle-class and socioeconomically well-to-do families in existing research on autism and ADHD (Cascio, Weiss & Racine, 2020; Taneja Johansson, 2023), I was keen to reach young adults who had lived in different spaces and places. To accomplish this, the strategy I adopted was to identify settings in which young people with ADHD (18–26 years of age) were likely to be present. This included a variety of educational settings, employment agency programmes targeting this group, healthcare settings, independent organizations likely to have contact with people with ADHD (e.g., *Attention*, *KRIS*) and social media ADHD groups. Within these spaces, I identified gatekeepers who might consider sharing information about the study with potential participants. This itself was a painstakingly slow process, leading to numerous emails, phone calls, and meetings. Nonetheless, these interactions with gatekeepers also provided important contextual insights that further helped me to refine my contact approach for including diverse voices. As a result of these conversations, and after learning that many individuals in Sweden are still completing their compulsory education at 26, the upper age limit in this study was extended to 32 years. Finally, I targeted a few public spaces by, for instance, placing posters in public libraries and sharing my research through word of mouth to university students and colleagues.

These different recruitment methods resulted in expressions of interest in the study, either directly from an individual or via an intermediary such as a parent, teacher, or another person in their professional role. One dilemma at this stage was to ascertain the individual's own wish to participate in cases where there was an intermediary coordinating the contact. For example, one parent expressed interest in her daughter participating in the study. I was informed that the daughter had given her consent for me to contact her and schedule a time for an introductory meeting. We decided on a place, date, and time to meet, which she later cancelled. We re-scheduled, but this too fell through. I picked up on her hesitation and sent a text message telling her that she did not have to feel compelled to meet me or participate in the study, adding that she could contact me at a later stage if she wanted to know more. This was my humble way of helping her retract her statement of interest, which might have been her mother's more than hers. In another case, while the situation was different and involved a student who had communicated to a teacher that he was interested, the events rolled out in a similar fashion. It was at these times important to listen beyond what was being said to the cues that their actions were expressing.

A key stage after expression of interest in the study was to talk—in person, online, or on the phone—with potential participants to share more details about the research and obtain their consent. All individuals with whom I met to further discuss the study subsequently consented to participate (either at the meeting or soon thereafter), and all but one continued to be part of the study for approximately two years. I have often wondered what happened at these meetings. Why did everyone agree to participate? This question takes on another meaning in the context of what I was actually asking for when it came to participation in this study. This was not just a one-off interview, but engagement over a *long period* of time, with multiple meetings, allowing me to accompany them in different settings, accessing past and current documentation and permitting me to speak to their significant others. This is probably too much to digest for anyone, let alone individuals whose life journeys have often been very rocky rather than rosy.

Did I, in any way, coerce or pressure participants into agreeing? The diverse backgrounds of the people I met, the different ways in which they received the initial information regarding the study, and the settings of our first meeting indicate that this was highly unlikely. The key to the positive outcome was probably my being receptive and listening to their spoken and unspoken communication at these first meetings (Stones, 2023). In reviewing my fieldnotes, I see that I was focused on listening to the person in front of me, the hesitations they communicated, and what might be embedded in their silences. Based on this, I adjusted both what study information was delivered and how, the duration of the meeting, and any additional questions I asked.

The long duration of the study and multiple forms of engagement meant that, for many, I needed to be careful regarding how I pitched their involvement. Asking participants to sign up for the study would rule out the option of them being able to participate to the extent they desired. Providing options for participation is also recommended by Anderson (2021), drawing on findings from her study with autistic librarians. My own decision to do so was driven by two main arguments. First, the fact that participants had shown an interest in the study and were here with me meant that this study was something they wanted to contribute to, and as a researcher I felt responsible for making it possible. Second, the expectation that they might consent to being 'shadowed' in private life situations seemed to me too much to ask for at a first meeting.

Seemingly embedded in many participants' communication was not disinterest in the study as such, but primarily a hesitation about being shadowed. These feelings were not usually stated explicitly (as in *"I do not want to be shadowed"*), but often implicitly in comments such as *"my life is of no interest to shadow"* or *"I don't think you will get anything from shadowing me"*. Without being certain, I do think these statements expressed a fear of and/or discomfort with being shadowed by a stranger, or for some it was even hard to imagine how they would feel about shadowing in the actual situation.

With all participants, there was an agreement that, to start with, we would meet a few times to talk about their experiences—past and present, and that they did not at this point have to commit to more than that. If after these meetings they felt comfortable with continuing, we would see what parts of the desired data collection could be pursued—it was nothing they needed to decide on right now. For many, being offered a gradual start was important, and I would argue an example of ethical and caring research practice. This approach took into consideration not only the individual's comfort level with the methods adopted in the study, but even more so to what *they* gave importance when sharing information about *their* life journeys.

Another factor potentially influencing participants' consent was my explicit positioning as a neurotypical researcher (though not using those words), who was seeking to understand and learn from their experiences—they were the experts on their own lives. This is very similar to the approach adopted by Goodall (2019) in his study with autistic teenage boys in Northern Ireland, where he consciously positioned himself as a knowing adult but keen to learn from the boys' experiences. It also partially relates to how Crane et al. (2021), in her evaluation study, approached participants in an autistic-led post-identification programme by being open about not identifying herself as autistic. In addition, she was advised by the autistic programme facilitator to spend time with potential participants, talk about her background and interest in autism research and other research interests. She found this to be a fruitful approach to increasing comfort levels and trust among programme attendees who were considering giving their consent to be interviewed.

My approach significantly deviated from this. While I did share information about my background (including being born and brought up in India) and the work I do, I did not overly engage with my interest in ADHD research per se. I never said to any participant at the first meeting, or later, that I was interested in their ADHD life, but emphasized time and again that my focus was on understanding their lives. This is an important distinction, because over the coming years I realized that, for most of them, the diagnosis, and/or ADHD category label, was peripheral in their accounts of who they were and their described lives. Moreover, my participants came from diverse socioeconomic and educational backgrounds, which was another reason to not delve into my research interest. For many of them, the word research probably carried limited meaning. I worried that talking too much about it would minimize their own 'expert' status. Downplaying my 'expert' status was my attempt to address this power imbalance.

Developing Trust and Respect

While participants had agreed to take part in the study, gaining access to different spaces of their experience was continuously navigated, and the realm of the private further extended. This was largely driven by my conscious effort to build a relationship that made participants comfortable enough to let me into other spaces in their life, both planned formal meetings (e.g., mentor, psychiatrist, social services, etc.) and other private spaces inhabited by them (home, family, etc.). The importance of relationship building is well established in ethnographical studies, where access to lives and places has been shown to be determined by the time invested in developing trust; as Tavory (2019, p. 727) describes: "Ethnography is made of relationships". The importance of a relationship built on trust has been argued to be even greater when researching a neurotype other than one's own (Crane et al., 2021).

It was during the first few meetings after obtaining consent that I laid the foundation for a respectful and trusting relationship. The usual outline for these meetings was looking back on the participant's school years and broadly mapping what had occurred after school, with an aim to re-create a timeline of the important phases and incidents in their lives. The atmosphere at these initial meetings was a bit 'formal', and I was keen to find ways to make participants feel comfortable. The approach I adopted was finding shared connections with each participant. A genuine interaction based on topics of mutual interest has been identified to be a key driver for social interaction, helping to create a sense of belonging among autistic adults (Bertilsdotter Rosqvist, 2019). These shared interests with my participants were not always obvious, but like a detective I looked for cues in the conversation on which I could build a connection. These varied from sharing some mundane everyday experiences of having school-age children, to having lived in the same foreign country, having hosted travellers as part of an online travel group, studying at university, having roots in another country, moving into a new flat, buying a bicycle, watching Netflix series, religion, disliking gyms, etc. It was never a matter of saying *"my life is like yours"*, but an acknowledgement of some shared experiences that often became topics of discussion when we met, and points on which I actively shared my views and experiences in the conversation.

Integral to developing the relationship with participants was respect, and not just for the ethical rights they have when participating in research (Swedish Research Council, 2017). It was my respect for their feelings and individual preferences—both stated and unstated—that made them feel comfortable. Allowing options for participation and engaging participants to the extent they wanted to be involved is emphasized as good practice by Anderson

(2021) in her study with autistic adults. Where did participants prefer to meet? What meeting duration was suitable? Did they prefer to do something else while talking (e.g., walking, eating)? Did they prefer direct questions to express themselves or talking freely? Did they prefer meetings booked well in advance, or the same week? At what time of day did they feel most up to meeting? How could I plan our meeting, so that it fitted into that particular phase of their lives? And when we met, respect involved reading the signals they communicated for when it was time to stop or to switch the topic of discussion.

It is important to note that, while some of these preferences on the part of participants might have been consistent over the years, in most cases they varied depending on how our relationship developed and how their own lives changed over the course of the research. For instance, one participant clearly stated that she wanted to meet me at the adult education centre, which seemed to be a safe space for her. I could have assumed this as a given for all our meetings, but I continued to ask her each time where she would prefer to meet. A year into the study she suggested we could meet elsewhere, such as a café, and that was the start of a new phase of interaction. Another participant preferred to meet in public spaces, such as libraries and cafes, but a few months into the fieldwork I was invited home on her initiative, which was one of the many times we met there. For someone else, the preference was to sit or walk outdoors. These cues were important to listen for, as they allowed me to continue building on a trusting relationship—trust that was connected to the emotional and physical experience of our interaction.

A relationship always involves two people, and here it is important to reflect on my own positionality. As an academic by profession and a researcher, I was perceived by participants as someone who had successfully navigated their educational journey—often a stark contrast to their own lived lives. With the focus of this research being on participants' educational pathways, the power imbalance between us could not have been greater. This could have become a hurdle in developing a trusting relationship and ultimately hindering my access to informants' lives. The key to addressing this challenge was being humble about recognizing the complexity of what it means to be knowledgeable and about what. In recordings from our meetings, I find myself regularly acknowledging participants as experts on *their* lives, who lived them every day, which is something I had little knowledge about and was the very reason I was with them.

More significantly, this was expressed in my actions through the non-judgmental interest I showed in their specific areas of interest and expertise. This is evident in the recordings from our later meetings, where I ask curious follow-up questions regarding topics I had no knowledge of and listen attentively to the informants, who are enlightening me. I learnt about large-scale gaming events and gaming in general, how to care for animals (e.g., horses and dogs), how gangs work, effectiveness of different diets, drugs available on the market—risks, benefits and possible ways to obtain them, the art of vaping, and much more. These conversations were not relevant to the study focus, nor could they be attributed to rapport building, because they took place at a later stage in the fieldwork and could be argued to be partly in contradiction to the stated ethical guidelines of good research practice (Swedish Research Council, 2017). However, I believe, in line with Mulinari (1995), that these conversations played an important role in reducing the power imbalance and contributing to a trusting and respectful relationship. Mulinari (1995), in her interactions with women in Nicaragua during her fieldwork, describes how she moved away from the interview script, to opening the room to let participants steer the conversation towards their own interests and what they wanted to talk about, something I will discuss further in the next theme.

An integral part of gaining participants' trust was being respectful of their time, what they shared with me, and their privacy. Essential here was showing respect for the individual's private space and only pushing to enter this space when they communicated their readiness. For some, entry to these spaces went faster than with others, and there were some spaces I was never allowed to enter. Meeting the important people in their lives was in many cases an acknowledgement of this trust. This included not only parents but also siblings, partners (both old and more recent relationships), and in one case a psychiatrist.

Building mutual respect through empathy and honesty is discussed by Anderson (2021) in the context of autistic adults. Drawing on her fieldwork experience, Nordstedt (2021) notes that establishing and maintaining relationships are not only accomplished through curiosity and time, but also require sharing on the part of the researcher. A relationship involves two people, and researchers sometimes need to move away from their neutral and objective role, towards displaying emotions and humanity, which I feel in many ways I did. For instance, I openly shared my experiences of moving to a small town in Sweden, my feelings of not fitting in, not being understood, and being perceived as different.

Being Mindful of Blurred Boundaries: Navigating Moral Dilemmas and Conflicts

Reading my scattered notes from my fieldwork and reflecting on my experiences, I realize that I was constantly questioning the definable boundaries that are the epitome of a high-quality study (Swedish Research Council, 2017). What are the boundaries to knowledge generation, what knowledge counts, when is something a risk, and what does it really mean to inform participants that the fieldwork has ended?

Caring—Knowledge

As the study progressed, a moral dilemma that frequently surfaced was the unclear boundary between caring for participants' well-being and generating data for the study. I started questioning whether these two things were mutually exclusive.

What probably started as relationship building to navigate access transformed over time into caring for the life and well-being of the participants, not in the long term but in the 'here and now'. I am acutely aware that many participants initially showed interest in the study because they were in a good place at that point in time, but in two years a lot can happen and change. My interactions with participants thus took shape based on what they were currently experiencing in their lives. Their lives were not just their lives, but were interconnected with others (Hutchinson, 2005). How partners, children, and parents were doing was closely related to their own experience in the here and now. At our meetings, participants took up fights with partners, breakups, frustrations with significant others, abusive partner relationships, children's school difficulties, economic hardships, parents' illnesses, and addictions, all of which I listened to.

Then there was also another side to listening, the happy moments—a new partner, a baby shower to organize, family members' birthday celebrations, moving to a new location, a special Christmas, etc. The approach I adopted through the study was listening to the participants at *all* times—listening to what they wanted to talk about, not just to what I found relevant to the study aim. Being interested in their lives, and not the 'research focus' life, seemed to be a small but sensitive act of caring in the context of a study that asked so much of participants.

Another example of blurred boundaries is when a participant reached out for assistance. This did happen, although it was quite rare and not in all cases. For instance, when they had misplaced a document that they knew I had scanned. One participant needed a copy of her diagnosis at the last minute to be able to take medicine with her when travelling to a foreign country. In another case, a participant couldn't locate her school certificates, which she needed to apply for a job. While these were small acts of caring, it was not always easy to resolve the dilemma in other cases, as in the account below.

One day I received a phone call from a participant who was facing a problem with The Swedish Board of Student Finance (CSN) and urgently needed to obtain written accounts describing her early participation in upper secondary school programmes. She asked me to accompany her to these places. While I did think for a moment about the limited relevance to the study, I could hear that, in her vulnerable state, she really needed someone with her. I spent over six hours going from one educational setting to the other, talking to strangers, looking for any teacher who could have taught her. I also helped her identify which people to contact and what might be good to write in the e-mail to them. At the end of this day, I knew deep down I had done this just because I cared about the participant, but was it wrong to do so? Getting too close to participants is often a point of contention in ethnography (Hammersley & Atkinson, 2019). Nordstedt (2020), based on her own ethnographic work on the everyday lives of people injecting narcotics, describes how caring for research participants is seen as a 'taboo' within academic circles and is rarely discussed.

Nonetheless, in hindsight I see that the dilemma was not just about the struggle between caring and generating data for the study, but something even greater. It was a conflict between what knowledge is perceived as valuable knowledge—the original focus outlined for the research study—or what participants want to share. This is also discussed by Peuravaara (2021), who considers how we as researchers often enter the field to ask certain questions, so focused on the goal we forget to be sensitive to participants' own subjective interests in what they want to discuss and what they desire to do. While Peuravaara's study focused on young students diagnosed with autism, the foundational question is the same.

I realize now that I was struggling with this question from the very initial stages of my fieldwork. Early into my fieldwork, I was frequently instructed by the project leader to solely focus my energies on that which was in 'our' interest—shadowing participants in meetings and other formal and informal activities as well as obtaining formal documentation. I was not to waste unnecessary time on meeting and engaging with participants. At these points, I felt I was being indirectly admonished for spending too much time with the participants, wasting project money, and overstepping ethical boundaries by focusing on data that were not needed. I tried to justify my approach by saying that doing the kind of fieldwork required by the project with a neurodivergent group requires one to continually obtain consent for access (which contrasted with the other group in focus in the larger project—people who are deaf). I still strongly believe this is true, but I think that, even though I was not able to articulate them at that point, the questions I was struggling with were: What is research and for whom? What knowledge is perceived as valuable to the study, the limited focus as originally outlined in the research proposal or what participants who are at the centre of the study want to share?

Reflecting on her own fieldwork experiences with women in Nicaragua, Mulinari (1995) eloquently highlights the importance of keeping the interactive space open to other topics than those the researcher is interested in, which might not always correspond to what the informant finds interesting. She described how moving away from the more research-driven conversation aided in opening up a space for a different focus (see also Jones, 2022, Chapters 7, 8 10,12, this volume). In many ways I did the same. After the first few more directed meetings, I always started each meeting by saying 'give me an update, what's been happening since we met last'. This gave participants a chance to direct the conversation to what they found important or thought was interesting, right from the beginning of our meeting.

Risk—Emancipatory

This study required participants to be engaged over a long period of time with me—a researcher who not only entered their personal sphere but was also privy to sensitive information about their past. One question I often reflected on throughout the study is the possible reasons why, despite this intrusion into their personal space, many of them wanted to continue to contribute. Some participants carried on for over two years, others disappeared for periods, some longer than others, only to return again.

The information they provided was not only sensitive, but also often emotionally charged. Part of the study looked retrospectively at the years behind them—both while at school and afterwards. For all these young people, school was a difficult phase, even though the intensity of their negative experiences differed. It was also a period rife with difficult and aggressive relationships with teachers and parents (see Taneja Johansson, 2021a). For some, thinking about their school years brought back painful memories of an absent parent, or the general feeling of never being good enough. The years following school were often worse, with accounts of abuse, drugs, juvenile homes, working in strip clubs, shop lifting, difficult relationships, and for some a long, slow, and troubled journey to complete their compulsory school courses (see Taneja Johansson, 2021b).

Reading my fieldwork notes, I see how, after some of these emotionally charged conversations, I was often concerned about having adequately informed participants of the possible risks of going back in time, that is, the risks of both generating new emotions and rekindling old ones. Even more importantly, I wonder if I could have in anyway helped them foresee their (unpredictable) reactions to their own and other people's actions in hindsight. In narrating different episodes of their life, they sometimes expressed deeply charged emotions regarding their own feelings of failure or, at the other extreme, intense anger with the system that had let it happen.

There was, however, one incident that made me acutely aware of the long-term risks. Social services sent me their documentation on one participant digitally, but also sent, unsolicited, a paper copy of it to the participant. The participant mentioned this to me, expressed no interest in reading it, and stated that she was going to discard it. Unfortunately, at a later stage she did read it, and at that point I had no contact with her. It was her parents who mentioned this to me when I interviewed them. They felt that, even if she knew about her past, reading it in print and with all the details had been too much for her. This along with other events in her current situation had led to a breakdown. While she had had a major life breakdown every few years, the cause of this one so close to completion of her education made me question the role I may have played in it.

Having discussed many of the implicit risks of participation that might have been hard to anticipate for participants at the start, I want to return to the question I started with: Why, despite many of the negatives associated with involvement in the study, did participants continue their engagement? Initially I was unsure, but towards the end of my fieldwork and working with the data, I wondered whether they continued interacting with me or chose to have me around in their lives because they got something in return from our interaction.

Participants had diverse reasons for initially choosing to participate in the study. Some were acutely aware of sharing their story, others were convinced when I explained the purpose of the study, and then there were those who were sceptical but considered it an interesting experience. However, as the study progressed, I think participation took on another meaning. I would go so far as to say that, in many ways, it was an emancipatory journey, though differently for each individual. Along with emotions of frustration, anger, and sadness, our conversations seemed to have a cathartic function. Many of the participants were still facing, in their present life, the consequences of the dysfunctional forms of support they had received in the past—from school, family, healthcare, or other care systems. Comments, like *"I have never had a chance to think about it"* or *"it all feels clearer now when I speak about it"*, were frequently heard in the meeting recordings. At a distance, participants seemed to see their life from a different perspective. This was at times painful, but also meaningful. My notes from after a conversation with Anna expand on this:

> Anna and I were walking back to her place after I had shadowed her during a lesson. In the recent weeks we had had long conversations about her chaotic years after school and I was concerned about how it was impacting her. I decided to take this opportunity when we were walking on this nice sunny day to ask if our recent conversations were troubling her and if it was painful to remember. She said it is intensive and tiring but she feels it gives her more understanding about herself and it helped her see things in another light at times. Then she was silent for a bit and said 'it's not always easy to remember'.

What I might have perceived as a risk seemed at times to be a different process in action. It was not just about the past, but looking back at the messy life situations the participants had been in also brought self-awareness of how far they had actually come in their lives. In moments including these realizations, participants expressed strong belief in the future ahead, a sense of hope.

Moreover, the conversations with me were free from judgements, opinions, and views. I was not a parent, sibling, partner, teacher, employer, government authority, or medical professional, but someone who was simply there to listen to what they had to say. In many ways, I was an outsider in the short term in their often-complicated lives, and this I think was an important reason why they continued.

Gradual Withdrawal—Abrupt Exit

I began my self-reflexive fieldwork discussion with how I initiated and gradually developed a short-term relationship with neurodivergent participants over approximately a 2-year period. This building of trust and respect, while integral to the research, also came with a risk. The prolonged duration of this study resulted in the boundaries of a researcher, friend, advisor, therapist, etc., blurring over time, especially for those with whom I met more regularly. For some, I had become a 'friend' and 'confidant', a person with whom they openly shared what they were struggling with now. In a strange way, I started looking forward to these regular meetings, engaged in what was happening in their lives—Did you apply for that job? Did you change flats? Is the assignment over? How is your mother doing now? I hope you've broken up with your boyfriend. The list is endless. With some participants, the depth of relationship I had been able to form went beyond a researcher-participant interaction, such that a couple of them were just as curious about what was happening in my life.

In the recordings of meetings towards the end of the study, I noticed that I often directly or indirectly mentioned my role as a researcher and how the study was reaching its end. While I continued to listen to whatever participants wanted to talk about, showed an interest in what had happened in their lives since we last met, and shared 'news' from my side, I also found it necessary to more frequently draw these boundaries. The moral dilemma was not in the here and now, but an acute awareness that these meetings would one day come to an end. How does one exit the field? And even more importantly: How does one withdraw from the relationships one has so carefully shaped together? While ethnographic studies often contain numerous accounts of the initial phases of fieldwork—from rapport building to the forming of relationships with participants—less frequently discussed is how one ends these relationships (Smith & Delamont, 2023, see Chapter 20, this volume). And this was something playing on my mind, as well as being visible in my fieldnotes:

> Today when I met Malin, she gave me a big hug before we went into the café and she said – 'I look forward to our meetings so much'. But I don't really need to continue meeting her, time has anyway run out and I have plenty of data too. I did tell her later that the study has come to an end. It was easier with Sandra who herself also withdrew over time, maybe for her the new phase in her life with different routines made it easier.

Nordstedt's (2020) reflection of the difficulties of exiting the field in an ethical and respectful way is related to my own fears when I was trying to exit. I had been in touch with some participants for over three years, which was much longer than required by the project. While it was easier with those who lived farther away, or with whom I, in some way, had less regular contact, the withdrawal was much slower and longer for the participants I met with frequently. The question I struggled with for a long time was whether a clean break signalling the end was really the most caring way of ending the study for the participants. The approach I ended up adopting was gradual withdrawal. I prolonged the intervals between our contacts and then over time I stopped, timing this to periods when participants were transitioning into a different life phase (e.g., had completed their education, started a new job) or after the long summer break. This, I hoped, was a less severe and a more natural end, slowly stepping back from people who had generously shared their thoughts, lives, and experiences—as opposed to making a decisive and abrupt break on a fixed date.

I received the following e-mail from a participant more than six months after we had last met.

> Hope life is kicking! I don't have your number in my phone and just wanted to say hi and that I'm alive because I just walked by your office and remembered the study. Haha! Do u need me for anything? Just let me know.

I replied and asked her to come up. We chatted, and she had a great deal to share. I told her I was working on all the data from the study as well as about another project I was involved in. She gave me a hug when she left. That was the end.

Conclusion

The importance of the ADHD voice is gaining traction, but research in this field continues to be strongly shaped by an ADHD deficit discourse, as evident in the limiting focus often adopted in the research. Voice has no meaning or value if little effort is made by the receiver to 'truly listen'. Through a self-reflexive account of ethnographical fieldwork with young adults with ADHD, the chapter has provided insights into what ethical and caring research practices could be like when researching a neurotype other than one's own, central to which is listening effectively and sensitively (Goidsenhoven & Schauwer, 2020; Peuravaara, 2021; see also Chapters 15, 20, this volume). It is through deeper reflection on the challenges, dilemmas, and tensions faced during fieldwork that I now better understand the underlying reasons for many of my own choices and actions. As researchers we are often urged to create distance, maintain objectivity, and avoid emotional engagement, neglecting who we are and the central role we play in engaging and creating these sustaining relationships with our participants (Brannelly & Barnes, 2022). Ethical and caring fieldwork practices are integral to quality in longitudinal studies, particularly in relation to neurodiversity.

References

Anderson, A. M. (2021). Exploring the workforce experiences of autistic librarians through accessible and participatory approaches. *Library and Information Science Research*, *43*(2). https://doi.org/10.1016/j.lisr.2021.101088.

Bertilsdotter Rosqvist, H. (2019). Doing things together: Exploring meanings of different forms of sociality among autistic people in an autistic work space. *Alter*, *13*(3), 168–178. https://doi.org/10.1016/j.alter.2019.03.003

Brannelly, T., & Barnes, M. (2022). *Researching with care*. Policy Press.

Cascio, M., Weiss, J., & Racine, E. (2021). Making autism research inclusive by attending to intersectionality: A review of the research ethics literature. *Review Journal of Autism and Developmental Disorders, 8*(1), 22–36. https://doi.org/10.1007/s40489-020-00204-z

Crane, L., Hearst, C., Ashworth, M., Davies, J., & Hill, E. L. (2021). Supporting newly identified or diagnosed autistic adults: An initial evaluation of an Autistic-Led Programme. *Journal of Autism and Developmental Disorders, 51*(3), 892–905. https://doi.org/10.1007/s10803-020-04486-4

Czarniawska, B. (2007). Shadowing and other techniques for doing fieldwork in modern societies. Liber.

Eccleston, L., Williams, J., Knowles, S., & Soulsby, L. (2019). Adolescent experiences of living with a diagnosis of ADHD: A systematic review and thematic synthesis. *Emotional and Behavioural Difficulties, 24*(2), 119–135. https://doi.org/10.1080/13632752.2019.1582762

Fricker, M. (2007). *Epistemic injustice.* Oxford University Press.

Goidsenhoven, L. V., Schauwer, E.D. (2020). Listening beyond words: Swinging together. *Scandinavian Journal of Disability Research, 22*(1), 330–339. https://doi.org/10.16993/sjdr.756

Goodall, C. (2019). 'There is more flexibility to meet my needs': Educational experiences of autistic young people in mainstream and alternative education provision. *Support for Learning, 34*(1), 4–33. https://doi.org/10.1111/1467-9604.12236

Hammersley, M., & Atkinson, P. (2019). *Ethnography: Principles in practice.* Routledge.

Hutchison, E. D. (2005). The life course perspective: A promising approach for bridging the micro and macro worlds for social workers. *Families in Society, 86*(1), 143–152. https://doi.org/10.1606/1044-3894.188

Jones, S. C. (2022). Measuring the wrong thing the right way? Time to rethink autism research tools. *Autism in Adulthood: Challenges and Management, 4*(2), 104–109. https://doi.org/10.1089/aut.2021.0050

Martin, N. (2021). Perspectives on UK university employment from autistic researchers and lecturers. *Disability and Society, 36*(9), 1510–1531. https://doi.org/10.1080/09687599.2020.1802579

Merriam, S. B. (2016). *Qualitative research: A guide to design and implementation.* Jossey-Bass.

Mulinari, D. (1995). Motherwork and politics in revolutionary Nicaragua. Bokbox Publications.

Nordstedt, M. (2020). Att skapa (och avsluta) tidsbegränsade nära relationer. In K. Gustafsson & S. Gruber (Eds.), *Etnografisk socialt arbete. Fält, analys och text* (p. 52–63). Linnéuniversitetet.

Pellicano, E., & den Houting, J. (2022). Shifting from 'normal science' to neurodiversity in autism science. *Journal of Child Psychology and Psychiatry, 63*(4), 381–396. https://doi.org/10.1111/jcpp.13534

Peuravaara, K. (2021). Emotional nuances: Critical reflections on emoji methods. *Scandinavian Journal of Disability Research, 23*(1), 305–316. https://doi.org/10.16993/sjdr.822

Ringer, N. (2020). Living with ADHD: A meta-synthesis review of qualitative research on children's experiences and understanding of their ADHD. *International Journal of Disability, Development and Education, 67*(2), 208–224. https://doi.org/10.1080/1034912X.2019.1596226

Stones, E. (2023). Cross-Neurotype communication competence. In M.S. Jeffress, J. M. Cypher, J. Ferris, & J. A. Scott-Pollock (Eds.), *The Palgrave handbook of disability and communication*. Palgrave Macmillan. https://doi.org/10.1007/978-3-031-14447-9_4

Swedish Research Council. (2017). *Good research practice*. Vetenskapsrådet.

Taneja Johansson, S. (2021). Looking back on compulsory school: Narratives of young adults with ADHD in Sweden. *Emotional and Behavioural Difficulties, 26*(2), 163–175. https://doi.org/10.1080/13632752.2021.1930904

Taneja Johansson, S. (2023). Whose voices are being heard? A scoping review of research on school experiences among persons with autism and attention deficit/hyperactivity disorder. *Emotional and Behavioural Difficulties, 28*(1), 32–51. https://doi.org/10.1080/13632752.2023.2202441

Taneja-Johansson, S. (2021b). Turning points in the educational pathways of young people with attention-deficit/hyperactivity disorder in Sweden. *International Journal of Inclusive Education*. https://doi.org/10.1080/13603116.2021.1965806

Tavory, I. (2019). Beyond the calculus of power and position: Relationships and theorizing in ethnography. *Sociological Methods and Research, 48*(4), 727–738. https://doi.org/10.1177/004912411987596

CHAPTER 12

Doing Participatory ADHD-led Research

Lill Hultman and Maya Hultman

Introduction

Within academia, there is a dual process of transmitting and producing knowledge, a process that often involves both conveying our own perspectives and making sense of our social experiences. But what happens when certain voices are not provided with spaces or places to be heard? This could be seen as epistemic injustice (Fricker (2007), which occurs when a person is wronged in their capacity as a "knower," i.e., a person's opinions are downgraded and not considered as legitimate knowledge due to both credibility deficit and lack of hermeneutical resources to make sense of their experiences.

In this chapter, our aim is to reflect upon the collaborative research and writing process between two ADHD:ers in which we will utilize our own autoethnographic research process to both show how this type of process can unfold and to describe and reflect upon challenges and strengths that are related to ADHD functionality and how we have tried to overcome challenges and utilize our ADHD functionality as a strength, or to "work with it."

L. Hultman (✉)
School of Social Sciences, Södertörn University, Huddinge, Sweden
e-mail: lill.hultman@sh.se

Department of Neurobiology, Care Sciences and Society, Karolinska Institute, Huddinge, Sweden

M. Hultman
Nacka, Sweden

H. Bertilsdotter Rosqvist and D. Jackson-Perry (eds.), *The Palgrave Handbook of Research Methods and Ethics in Neurodiversity Studies*,
https://doi.org/10.1007/978-3-031-66127-3_12

Challenges related to collaboration can involve collective, individual, and sometimes conflicting needs regarding time management, balancing energy levels, pacing and the ability to hyperfocus (HF). We contextualize our experiences in the framework of neurodivergence, which contributes to de-medicalize ADHD and understand it as neurological difference instead of deficit (see also Chapters 3, 23, in this volume). Our dual roles as neurodivergent researchers and participants emphasize the importance of reflexivity. We are bonded by our common experience of being mother and daughter as well as recipients of disability support. One of us is a middle-aged, Swedish, female academic researcher within the field of social work and disability studies. The other is a young woman with lived experience of having personal assistance in her everyday life. Our ADHD functionality is both similar and different, and it entails fluctuating energy, high intensity compared to neurotypical pacing, variable attention, and interest-based motivation. Although we have different needs and pacing, we believe that our intertwined experiences make it easier for us to relate to each other's challenges and we can understand each other's experiences due to each of us having ADHD functionality. In our text we will also describe how an ADHD writing process can differ from a neurotypical writing process (see also Chapter 17, in this volume). For example, in terms of handling the constant flow of new ideas, within a collective writing process, there is a need for re-thinking the definition and value of co-writing, and utilizing the group members' collective strengths to create individualized working processes within a collective writing process.

The Interest-driven Research Process

For us, interest-driven writing was a prerequisite for starting a joint research project. In our case, having an interest-based motivation means engaging with a task that we think is important and to some degree challenging. Dodson (2022) underlines the association between ADHD and an "interest-based nervous system," a nervous system which does not implicate attention deficit, but inconsistent attention. From this perspective, ADHD is not a defective nervous system. It is a nervous system that works well using its own set of rules which is an interest-driven nervous system. In line with this Hallowell and Ratey (2021) suggest that ADHDers have variable attention where the attention is depending on interest. When a deep interest is discovered, the attention becomes focused and is transformed into hyperfocus. Closely associated with interest-based motivation and attention are the challenges of boredom, which Hallowell and Ratey (2021) have referred to as the "kryptonite" for the ADHD brain. Interest-based motivation consists of four overlapping aspects: novelty, interest, urgency, and challenge (Fig. 12.1).

In our joint research process, we started by identifying a topic we wanted to write about. Both researchers had a strong commitment to the research topic of mental health illness. For Maya, the involvement had its origin in deeply rooted feelings and affects connected with specific situations that she found herself in while she was admitted to an inpatient psychiatric unit. Her experience of powerlessness created a desire to change and make visible the

Fig. 12.1 Aspects of the interest-driven motivation

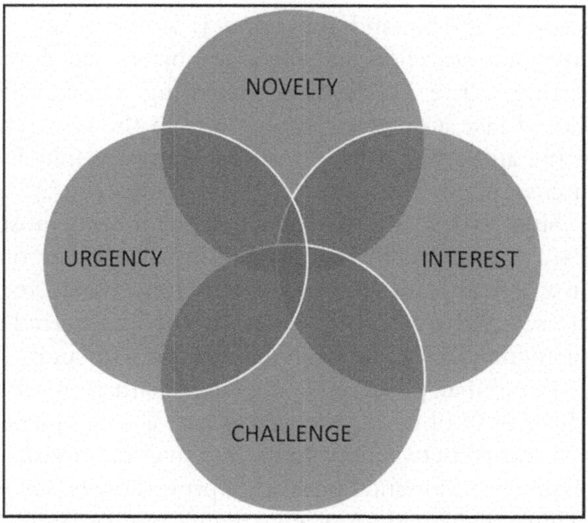

embedded power structures. For Lill, Maya's commitment was contagious. At first, she was hesitant to write about private experiences due to the sensitive nature of the topic. On the other hand, being able to speak up about issues of power, particularly in relation to being in extremely vulnerable situations, convinced her of the importance of "giving an insider perspective" of being exposed to those types of experiences.

Sometimes a call for a certain kind of text or a topic can facilitate the start-up of a project. Lill had been told that *Frontiers in Psychiatry* would have a thematic issue on mental illness and epistemic injustice. Since we already were interested in exploring mental illness and power, we thought that it could provide us with an opportunity to make our voices heard. Lill thought it would be a good fit to use Miranda Fricker's (2007) concept of epistemic injustice to demonstrate the embedded power structures that we had experienced in contact with adult in-psychiatric care facilities. Being part of a larger context (a theme issue on mental illness and epistemic injustice) also meant that we did not have to put effort and energy into identifying a suitable journal that would be interested in the topic we wanted to explore. In this chapter, we describe and reflect upon the process of co-writing the article for *Frontiers in Psychiatry* (Hultman & Hultman, 2023).

Mutual Exploration Processes

After we made the decision to explore our experiences, we wanted to find a suitable research method where we could use our experience-based knowledge about inpatient psychiatric care. We discussed our respective roles, whether both should be authors or whether Maya should be an informant. Since we wanted to get away from the power asymmetry that arises when one of us is an informant and the other is a researcher, we decided that we would both be researchers and research objects, which resulted in autoethnographic writing. Since Lill is used to writing academic texts, we decided that she should take a greater responsibility in the research process. However, as well as the autoethnographic writing, Maya was fully involved in both analysis and discussion.

Since Maya was the one who had been most dedicated to writing this text, we decided that she would initially be the one to identify key situations on which the autoethnographic writing would focus. The key situations that she selected were critical incidents that occurred while she was admitted to inpatient care at an adult psychiatric care unit. The most important thing in this initial phase was to identify situations that illustrated circumstances where both of us experienced negative consequences of the power asymmetry that exists between patients and mental healthcare professionals or family members and mental healthcare professionals. We wanted the critical incidents to illustrate situations when patients and/or family members end up at a disadvantage either because their experiential knowledge is disqualified (testimonial injustice) (cf. Fricker, 2007) or because it has been difficult to find the appropriate words to describe the experience since established medical concepts do not take into account the lived experience of having mental health issues (hermeneutical injustice) (cf. Fricker, 2007). (c.f. Chapter 5, this volume).

Hermeneutical injustice often goes hand in hand with testimonial injustice. What is considered as credible knowledge is filtered through medical staffs' conceptions of what will benefit the patient. This often results in patients' experiences being dismissed as "behavior related to the patient's diagnosis" or "not knowing what is best for themselves." A power asymmetry where the distance between the patients' lived experience and medical terminology reinforce the hermeneutical injustice. This prevents healthcare staff from integrating patients' unique experiences, which contributes to making individual experiences invisible.

Establishing Co-creation, Flow, and Meaning-Making

According to Schneid and Raz (2020), participatory research can move us beyond neurotypical frames of analysis by looking from different perspectives at the same phenomenon. Key components in our research process were flexibility and an individualized design, adapted to our variable needs (see Chapters 7,8, in this volume), which included different processing time between us, and different ability to stay in hyperfocus. A common denominator was our interest-driven motivation and the commitment to co-write a text. When we started to work with the article, we did not have a clear idea about study design. Thus, the *planning phase* of the project developed organically, and we utilized a mind map to help us figure out appropriate study design and method. After considering various methods, we decided to use what Chapman (2004) refers to as Critical Personal Narratives (CPN), also known as critical autoethnography. CPN address issues of power and knowledge in practice. It is personal because it is based on autoethnographic accounts, and it focuses on a series of incidents that seek to transform the personal experience so that it can be useful politically. We have used CPN to identify incidents that have been experienced as unfavorable in relation to situations that have occurred in relation to treatment at mental health inpatient-care units.

After Maya had written the initial narratives, we moved on to *getting entangled with data*, where we started to co-write and reflect upon our experiences. To avoid overload, we needed to take a break from the data. Due to the sensitive topic, we had to take a long rest from the writing process. If we had not been able to adapt the pace to our needs, we might not have been able to finish the article. In that sense it was important to have enough time to move between phases 2 and 3, since writing is not always a "straight forward process." Sometimes you have to go through phase 2 and phase 3 twice, which should not be considered as a failure but rather as an individualized adaptation)(Fig. 12.2).

By writing together, a continuous dialogue was created, which increased our individual motivation to write and immerse ourselves in the collective creation of meaning-making. The advantage of mixing- "brainstorming" with Lill trying to clarify and translate text into English was the ability to further develop the text when Maya's interest in the text was limited. In our writing process, translation and intertextuality was a bit challenging, since translating Swedish sentences and words from English to Swedish, or vice versa, also entailed understanding theoretical concepts. Since we had already co-written

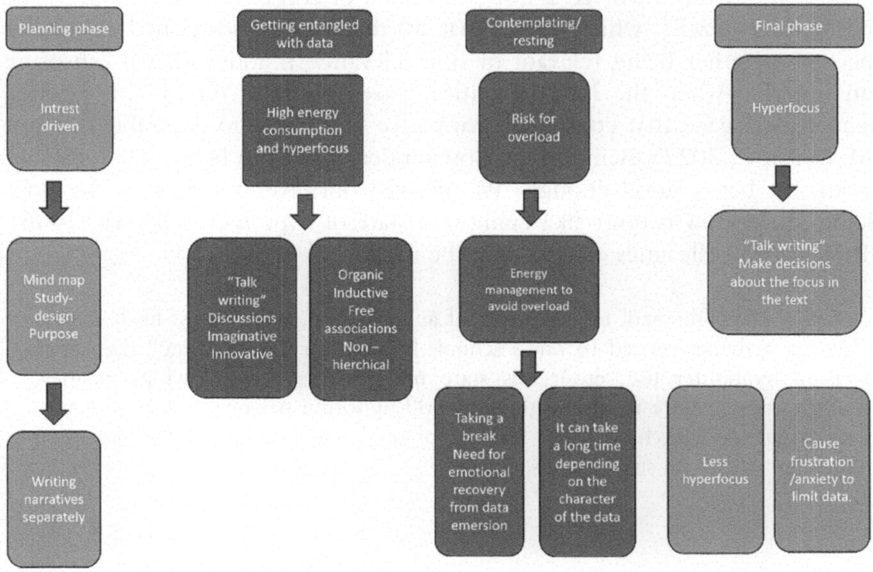

Fig. 12.2 Our interest-driven research process

the article to *Frontiers in Psychiatry*, Fricker's (2007) concept of epistemic injustice was familiar to both of us. Nevertheless, hermeneutical injustice became pertinent in our own writing process regarding language translation as well as writing academic language. However, the process of translation also functioned as a catalyst for new ideas and understanding, since trying to convey meaning encourages reflection and questioning one's understanding that is informed by the situated knowledge and one's standpoint (c.f. discussion about unknowing in Chapter 22). Translation demands devotion. The act of translation is the attempt to "becoming-other that makes the constant movement towards the Other and that remains attentive to the silences and the in-between where no guarantees are provided" (Bertilsdotter Rosqvist, Hjort & Nygren, p. 410). Partnership and reciprocity generate equitable relationships that have the capacity to generate imaginative co-engagement (Aaltola, 2014) while the critical gaze of peers or people in powerful positions can be associated with negative emotions such as fear and shame (Dolezal, 2017). Drawing on Foucault, (2002) power can be understood as a productive force, that operates relationally and reinforces itself and the object/subject it works upon. Through such a conception, power is materialized in various ways. Furthermore, power and knowledge are intertwined in that there is no power relation without the correlative constitution of a field of knowledge, nor any knowledge that does not presuppose and constitute at the same time power relations. Thus, when the critical gaze is directed inwards it re-enforces a sense of internalized hermeneutical injustice, i.e., one is so used to being questioned in how knowledge is contextualized that it becomes easy to question oneself.

Initially, writing involved a flow—a stream of consciousness that overrode the "critical gaze"[2], which meant that we deliberately postponed evaluating the text as either being relevant or non-relevant. In addition, self-censoring can implode when the internal critical gaze is reinforced by the external neurotypical gaze that constructs normative ideas of who is disabled or not (McDermott, 2022). Being in the flow hindered us from being self-conscious about not being good enough. We allowed ourselves to go with the flow. Being in the flow is quite like being in a state of hyperfocus (Brown, 2006). Hupfeld and colleagues (2019) describe it as:

> A state of heightened, intense focus of any duration, which most likely occurs during activities related to one's school, hobbies, or "screen time" (i.e., television, computer use, etc.); this state may include the following qualities: timelessness, failure to attend to the world, ignoring personal needs, difficulty stopping and switching tasks, feelings of total engrossment in the task, and feeling "stuck" on small details (p.119).

The feeling of being in the flow allowed us to stay in our comfort zone of interest-driven writing that made us forget about time and place while interacting with the text. Being in the flow implies having no impairments since the flow allows us to stay within our interest-based comfort zone.

Being in the flow liberated our creativity and gave rise to many openings and possibilities. Since we both are ADHD:ers, largely motivated by interest-driven motivation, it was important to maintain the joyful and playful flow of writing as long as possible, since we both know that parts of the writing process involve a higher degree of structure where we need to adapt the text into a readable, structured form.

We started to write in Swedish, sometimes just as bullet points, sometimes drawing mind maps on paper: the important thing at this stage was to capture ideas and thoughts when they arose. Sometimes it was only a short sentence scribbled down on the back of a receipt, or a short notation on the cellular phone's notepad. Most of the time it took a while for both of us to process the text to fill in the blanks where parts of the sentences were still inside our mind. In the "in-between-time," there was time for us to process our unfinished thoughts or fragments that could either be developed to more coherent text or rejected as not relevant. For Lill, the process of consolidating fragments often occurs before falling asleep or when stacking dirty dishes in the dishwasher, folding clothes in the closet, or doing some other household chores.

Since Maya found it difficult to write, think, and articulate her thoughts at the same time, we decided that Lill should write down what she said without asking clarifying questions or making any comments about the content. Only when Maya felt that she had nothing more to say was Lill allowed to ask clarifying questions or contribute with her reflections.

Throughout the entire research process, we tried to modify our working methods to suit our varying and sometimes conflicting needs. By having a bottom-up perspective in all parts of the research process, we tried to keep up the motivation and enhanced the opportunity for co-creation. When we felt confident that the text was underway, Lill started to translate the text into English and from that point we continued to write the text both in Swedish and English.

Challenges of Co-creation with Another ADHD Researcher

One of the biggest challenges when we wrote our joint paper for *Frontiers* was to stop writing, knowing when to finalize the research process and "close the text." In that situation, the journal's word limitation became helpful since it created a structure for what was possible to write in terms of length and content. The difficulty lay in being able to accept that the text was good enough, since there are always little things to tinker with. In addition, the text can't cover every interesting aspect of a phenomenon, even though there is more to say about the subject matter, which was a huge challenge for our

ADHD minds, that wanted to include more aspects, which in the end could have made the text shallow and less coherent. Another related challenge lay in not getting stuck in interesting but irrelevant details, since it could derail you from the focus of the article and in the worst case stall the entire writing process. Likewise, closing the scope of the article can also be challenging since it demands an adaption to a fixed structure that may including "killing your darlings" (cutting out possible subthemes) to ensure readability for a broader audience.

Co-creation can also trigger issues of self-doubt in us (which can be seen as a form of internalized epistemic injustice) and at times, turn the joint venture into performance anxiety and "writer's cramp." When one of us started to question the "bottom-up-process" or tried to pick up the pace, it triggered thoughts about the outcome, for example, is the text interesting to anybody else? Will the article be accepted for publication? In addition, one of us feared that it would be the last chance in a long time for her to write a research article and thus it would be sad to finish it too soon. Apart from hindering you from finishing the mutual writing process, interest-driven writing can get you stuck in the initial flow writing phase. Uneven motivation and interest-driven motivation can both be an advantage and an obstacle since it is very hard to produce any text when there is a lack of motivation or a lack of hyperfocus. Sometimes, when one of us needs to take a pause, it can become difficult if the other wants to pursue the writing and finish the article. Thus, different writing paces can make co-creation challenging. To prevent this from happening, it was important for us to create our own deadlines, and not worry about the final product (the finished chapter or the article). Committing to agree upon deadlines was important to avoid procrastination and to maintain a professional relationship which was particularly challenging since we are family members and live in the same household.

To mitigate this, it became important to have some ground rules, and committing to deadlines was one of them. We decided to set goals for every writing session, although both had vetoes to pause the process if they became too tired or felt like they were on the verge of having an overload. However, this was challenging since motivation and inspiration can fluctuate a lot, particularly for Maya who was not as used to writing and producing academic texts as Lill. Sometimes it was difficult to agree on writing time since we had different pacing and energy levels at different times of the day. Sometimes Lill had to remind Maya that we needed to pursue the writing, which she felt was a tedious chore, particularly since it reminded her of the roles of mother/daughter. One way to solve this was to allow Lill to proceed with the text without having to involve Maya in the writing process. The uneven work pace was challenging, particularly since both of us find it difficult to calculate how long it takes to write a text, cook a meal, walk the family dog, and so on. Both found it difficult to stop writing once we got started. The difficulty in "measuring time" also affects goal setting, i.e., what is reasonable in terms of production? How much text should you write in a day?

Focus shifting is extremely difficult for both of us. We are often at risk that we forget to eat, drink water, and take breaks which can result in one of us having an overload at the end of the day. Reluctance to shift focus from the text makes it difficult to work when there are a lot of competing things or chores to perform. Often, we chose to clear the calendar and devote an entire day to talking about, reflecting upon, and writing on the text, even though we sometimes ended up with four hours of co-creation and four hours of start-up time. Start-up time implies overcoming the resistance to recapitulating what we did the last time and slowly work ourselves back into the text which also includes adapting our different energy levels and pace to each other's needs and internal time. For us it is equally difficult to start and stop writing. Too many start-and-stops were draining for our energy levels.

However, voluntary stops can be necessary to avoid overload. Voluntary stops for refilling energy or processing data are different from external interruptions such as checking the mailbox, meetings, telephone calls, chit chatting with colleagues. When this happens, it interrupts the flow in the writing process. Nevertheless, we believe that our collaborative strengths overcome our individual difficulties.

Performance Anxiety and Variable Attention

It can be difficult to write within given time limits. Challenges with time management combined with performance anxiety can lead to writer's block. It is easy to get stuck in the text, when there are too many options and once you get stuck, there can be a tendency to get stuck in the same train of thought or float off into new paths that are not directly relevant. It can also be difficult to start writing after a long break, since it can feel like starting all over again. In the worst-case scenario, a "short break" from the text can imply "leaving it" for 2 months and then you return to a text with many "loose ends" and then you cannot understand the meaning of disjointed sentences or supporting words. The challenge of having challenges with working memory means that a lot of pausing from the text could be experienced as starting all over with the text each time you have a long break from writing it. As many ADHDers know, executive functioning can be exhausting since it requires frequent mental pauses and ceaseless self-regulation. Having a challenging working memory influences the amount of effort and type of actions required to modify what our brains ideally would do automatically. The stronger your working memory, the less effort to start-up or take on new tasks and challenges (Fig. 12.3).

Since our writing process was interest-driven, parts of the research process that were perceived as boring, for example writing references, were not done continuously or systematically. This resulted in a lot of work with the reference list before submitting the manuscript. Our writing process had an irregular pace in which different pacing among us did not allow for either a regular or speedy writing process. The start-up phase was easy because everything was

Fig. 12.3 Difficulty with poor working memory

new and exciting. In particular the autoethnographic writing that involved narratives that were based on one's own experiences felt rewarding to write down and share with each other. The finalizing stage of transforming the text into a coherent entity, writing up the paper, was experienced as more challenging than the start-up phase. By the less experienced researcher (Maya) the finalizing writing stage was perceived as more mechanical, since it allowed less freedom, and required a higher degree of stringency, coherence, and structure. The final phase of the writing process, which mostly consists of revising the manuscript, was perceived as more challenging in relation to the power balance of co-writing, since negotiating meaning entailed the use of shared hermeneutical resources, such as making sure that both of us understood and agreed upon why certain aspects of the text needed to be further developed or omitted, or why certain concepts did not fit in with others, etc. This part of the process was also difficult in relation to time management where different pacing and levels of fatigue was experienced as challenging. Challenges occurred when one of the writers wanted to continue when the other began to lose focus and experience sensory overload in terms of noticing the dishwasher turning, for example. We also had to figure out when it was best for us to write together, often we had to compromise and started around 10 am, although we never managed to find a fixed time for writing since one of the us's energy levels fluctuated daily.

To reduce the risk of getting stuck, we realized that it was good to work with different parts of the text simultaneously, i.e., writing a little under each heading and not initially worrying too much about the parts not fitting together. In this sense, we utilized our variable attention to fuel the interest-based writing process. Worrying about progress, or having an idea of perfect language, and coherent writing (c.f. (Freer et al., 2011), only contributed to writer's cramp, so we concentrated on content and flow, instead of language and form. Knowing each other made it easier to fill in gaps or ask for clarification. However, knowing each other well can also contribute to "taking things for granted" which can result in missing an opportunity to go into depth. Nevertheless, the benefits of feeling safe outweigh the challenges. Lack of energy makes it impossible to force the process, and since the text was much more personal than usual, it was emotionally challenging to write it, which some days made it much harder to proceed with analysis. Thus, when we planned this project, we knew we had to respect each other's need to process the text emotionally. This can become a problem when it comes to meeting deadlines, which means that identifying effective strategies to complete the text, being aware of difficulties with time management, and knowing one's weaknesses and strengths, can be helpful when planning the writing process.

Possibilities with ADHD-led Research

First and foremost, we believe it is important to identify a mutual interest (see Chapters 7, 11, in this volume), to pin down and decide what to explore, and to allow this first step to take time, "to try out" different possibilities without having to decide immediately and to be able to change the aim once the research process is underway. This is a way to take advantage of "the wandering mind," the playfulness of "what if…."

In relation to writing together, we put considerable effort into discussing and negotiating meaning, to consider what is needed to contextualize the points we try to make, or whether we clarify some parts of the text and delete others. In this phase, and in relation to any possible power asymmetry, it is important not to be hierarchical, to listen to each other's thoughts instead of getting stuck in your own frames of analysis. An advantage of collaborating with another person who is interest-driven is the lack of hierarchical thinking since position becomes less important than being able to engage with new ideas or the task at hand. An advantage of both of us knowing each other well and having similar types of ADHD is that we have clear and direct communication. If we want to add or change what the other person has written, we do not have to worry about offending the other. Nevertheless, it can be important to not be too quick in dismissing each other's ideas when they do not "fit in" with the current text or with a preconceived idea or concept, but to allow them to "rest" for a while, since they might be relevant in a later stage. It is equally important to be able to delete redundant text, which we believe is one of the most challenging aspects of ADHD, since "the wandering" mind likes to incorporate new aspects of the studied phenomena (Jackson-Perry et al., 2020). In this sense, it becomes important to value the experience-based knowledge and to take advantage of the ADHD functionality, which allows the wandering mind to discover what arouses an interest-based attention. Living with ADHD can entail a sense of being used to feeling like having one's thoughts standing out, coming from another direction. In a safe space it is easier to be open with one's needs and ask clarifying questions.

Going with the Flow

In our writing process, a lot of inspiration came from "thinking together" and "speaking out loud." Since we already knew each other, we did not have to put effort into creating a safe space (see Chapter 7, in this volume). This enabled us to focus on the task at hand. Initially we tried our ideas on each other, without the requirement to have a clear concept or thoughtful ideas. When we collaborate, we often sit in the same room, we think when we write, i.e., talk-writing (talking as we write down our thoughts in a joint document). By having the writer speak out loud, the monologue can transform into a dialogue whereby the person who is not producing any text can ask questions or associate freely (Fig. 12.4).

Communicating one's thoughts can also compensate for having a short memory span, since both can support each other to remember what has

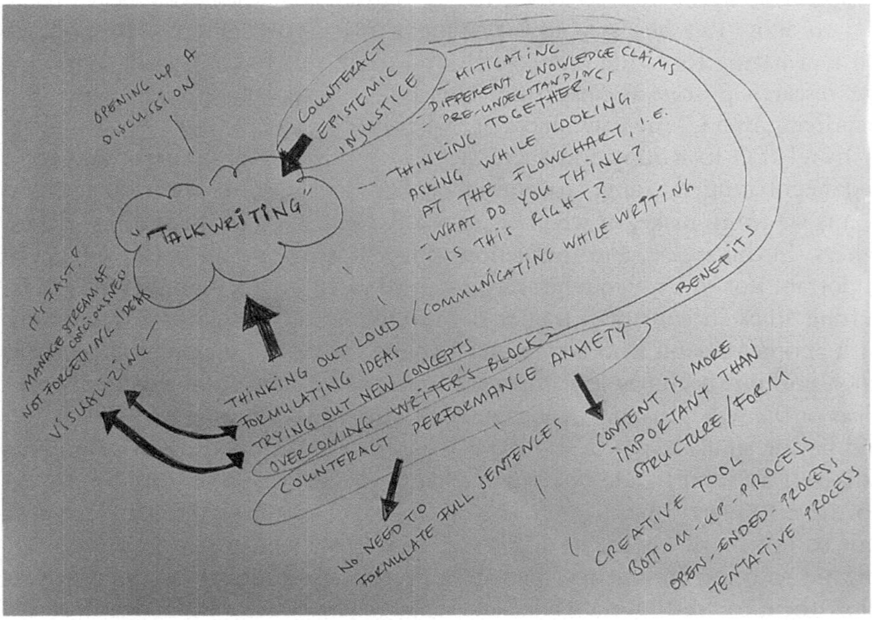

Fig. 12.4 Mindmap created at the start of our collective writing process

been said. Turn-taking allows for reflexivity but can be more demanding than writing by yourself in silence. Initially, in the brainstorming/start-up phase, "speaking out loud" can make the process more creative and the text more vivid, but toward the end, it was easier to work alone and take turns working with the text.

Working together with a person with whom you share everyday life can make it difficult to set boundaries between work and leisure. This may particularly be the case if you have an interest-driven motivation which reinforces the tendency to be occupied with the text around the clock. This can be both nice and exhausting. When you are in hyperfocus, it is difficult to stop working with the text. It can both be difficult to know when you are done (have done your share/workload) or stop because it is fun (as part of one's spare time). This can lead to exhaustion since it results in less time to do other things; household chores, shopping, exercising, walking the dog, eating dinner, seeing friends, etc. There is a risk for imbalance between doing research and tending to other chores. If something is interesting, it's easy to keep working, even though you do not have the energy for it. When the body signals that it is time to stop, and you keep on going, there is a risk of being low on energy the following day. Thus, it can be helpful to structure time in Pomodoro, i.e., work for 25 minutes and then take a 5-minute rest, even if it feels unnecessary at the time.

However, without commitment our joint research process would not have been possible to do. We both find it difficult to write about phenomena that we do not find interesting. Neither of us would write about a topic that does not engage us. It's about authenticity, we want to understand something because it interests us and because we think it is important to share this knowledge with other people. We also believe that there is an ethical dimension to having lived experience, since it is more likely to diminish the processes of "othering." Our work process being bottom-up has allowed us more creativity than a top-down process and has worked well in relation to our interest-driven motivation.

A Word of Advice

When we embarked on this mutual research journey, we did not have any experience of working together. We did not have a plan beforehand, but early on we created a common roadmap that considered our ADHD functionality. Based on that we made some ground rules. Every research process is unique. That means you must make your own "roadmap" and rules. However, we think this checklist may be helpful when creating a roadmap:

Checklist

- To do research based on a mutual interest.
- To allow yourself to be creative, not to worry about structure in the start-up phase.
- To jot down ideas before they disappear.
- To utilize visual aids such as flowcharts and mindmaps to get started when you are stuck.
- To allow for the research process to take time.
- To not consider several start-ups as a failure since it is part of the research process.
- To not postpone the "boring parts" (for us, filling in references while working with the text) until the end.

References

Aaltola, E. (2014). Varieties of empathy and moral agency. *Topoi, 33*, 243–253. https://doi.org/10.1007/s11245-013-9205-8

Alexander, S. *ADDitude*. (9 February 2024). Your brain's GPS Is glitchy: Why working memory fails and how to bolster It. https://www.additudemag.com/working-memory-powers-executive-function.

Bertilsdotter Rosqvist, H., Hjorth, E., & Nygren, A. (2023). Meeting up in broken word/times: communication, temporality and pace in neuromixed writing. *Medical Humanities*, *49*, 407–415. https://doi.1136/medhum-2022-012384

Brown, A. (2023). The univocity of attention: ADHD and the Case for Re-newed Self-Advocacy, [PhD Diss. York University].

Chapman, V. L. (2004). Using critical personal narratives: A poststructural perspective on practice. New Dir Adult Cont Educ, 95–103. https://doi.org/10.1002/ace.142.

Dodson, W. *ADDitude*. (6 November 2023). Secrets of your ADHD Brain. https://www.additudemag.com/secrets-of-the-adhd-brain/

Dolezal, L. (2017). Shame, vulnerability and belonging: Reconsidering Satre's account of shame. *Human Studies, 40*(3), 421–438. https://doi.org/10.1007/s10746-017 9427-7

Flanagan, J. (1954). The critical incident technique. *Psychological Bulletin, 51*, 327–335.

Foucault, M. (2002). *Sexualitetens historia. Band 1: Viljan att veta*. Daidalos.

Fricker, M. (2007). *Epistemic injustice*. Oxford University Press.

Hallowell, E. M., & Ratey, J. J. (2021). *ADHD 2.0: New science and essential strategies for thriving with distraction-from childhood through adulthood*. Ballantine Books.

Hultman, L., Hultman, M. (2023). Believe me, only I know how I feel. An autoethnographic account of experiences of epistemic injustice in mental health care. *Frontiers in Psychiatry*, 14. https://doi.org/10.3389/fpsy.2023.1058422

Hupfeld, K. E., Abagis, T. R., & Shah, P. (2019). Living "in the zone": Hyperfocus in adult ADHD. *Atten Def Hyp Disord*, *11*, 191–208. https://doi.org/10.1007/s12402-018-0272-y

Jackson-Perry, D., Bertilsdotter Rosqvist, H., Kourti, M., & Layton Annable, J. (2020). Sansory strangers: Travels in normate sensory worlds. In Bertilsdotter Rosqvist, H., Chown, N., & Stenning, A. (Eds.), *Neurodiversity studies. A new critical paradigm*. Routledge.

McDermott, C. (2022). Theorising the neurotypical gaze: Autistic love and relationships in The Bridge (Bron/Broen 2011–2018). *Medical Humanities, 48*(1), 51–62.

Schneid, I., & Raz, A. E. (2020). The mask of autism: Social camouflaging and impression management as coping/normalization from the perspectives of autistic adults. *Social Science and Medicine* (1982), *248*, 112826–112829, 10.1016j.socscimed.2020.112826

CHAPTER 13

An Inclusive Methodology for Research with Multiply Neurodivergent Participants

Romy Hume

Introduction

We can actively resist epistemic injustice, create more trust in researchers, and tailor research to community priorities by engaging neurodivergent people authentically and respectfully in research. However, studies show that autistic populations, in particular, are considered hard to reach and that researchers tend to perceive great difficulty in face-to-face communication with autistic participants (Haas et al., 2016). This chapter illustrates how an adapted version of Interpretive Interactionism (Denzin, 2001) has facilitated a neurodiversity-affirming interview study with 17 autistic adults living in Aotearoa New Zealand. The study was about therapeutic relationship-building. Rather than following the traditional blueprint in autism research, which aims for interventions that help autistic people adapt to neurotypical relational norms, I took a different approach. I sought to learn from autistic people's experiences in therapeutic relationships. Based on their input, I suggested 'interventions' for non-autistic health and support professionals instead.

All except one autistic participant were, like myself, multiply neurodivergent, representing both innate and acquired neurodivergence (autism, dyslexia, dyspraxia, and ADHD; depression, anxiety disorders, OCD, and PTSD). Finding flexible, individualised, and interdependent ways of allowing everyone to feel 'at home' in the research was therefore paramount. To create a respectful, inclusive research methodology that would satisfy this

R. Hume (✉)
Auckland, New Zealand

H. Bertilsdotter Rosqvist and D. Jackson-Perry (eds.), *The Palgrave Handbook of Research Methods and Ethics in Neurodiversity Studies*,
https://doi.org/10.1007/978-3-031-66127-3_13

goal, I drew on strategies found in participatory and person-oriented, ethical autism research, focusing on community priorities, accessibility, trust, and respectful language (Cascio et al., 2020; Chown et al., 2017; Nicolaidis et al., 2019; Raymaker, 2018). I then turned to Interpretive Interactionism (Denzin, 2001), a qualitative approach for insider researchers. My insider status functioned at two levels: the first was my professional experience supporting autistic clients[1]; the second was my lived experience as an autistic adult with suspected ADHD, which allowed me to inscribe myself in the study via personal reflections. Reflections of this kind can infuse richness in interpretive research, adding a layer that transforms the narrative from being a story *about* to a dialogue *with*. By inscribing myself in the study, I also aimed to balance power relations between me as the researcher and participants as the researched. If autism is indeed a stigmatised or "spoiled identity" (Goffman, 1991), I wanted to communicate to participants that I share their stigma.

As Raymaker (2017) noted, being an insider researcher may help build trust with participants and reduce the risk of faulty assumptions or bias. However, this position also has drawbacks, such as the potential for being disadvantaged in one's career (Mogendorff, 2013), and the discomfort that comes with public self-identification (Rinaldi, 2013). My strong emotional investment in my research question was also a challenge, particularly during the first few months. My research journal during this time devolved into a litany of complaints like the following:

> Attended a public adult autism seminar organised by Minds for Minds and the Centre for Brain Research [...] Opening speech of the director a disaster. The only mitigating factor was his own admission that he didn't know much about autism. (Oh to have the confidence of an old white cis male! To be aware of own ignorance but keynote anyway.) Still, the *level* of ignorance seemed inexcusable for someone in his position: he called autism "a terrible disease." I'm surprised none of the autistic people in the audience threw a shoe at him.

Later on in the process, gut-wrenching empathy often had me reeling during and after participant disclosures of the injustices they faced. I attempted to sit with these tensions, allowing them to fuel my passion for the project.

[1] I was employed by Autism NZ, the biggest organisation supporting autistic people and their families in Aotearoa New Zealand. They assist in finding and accessing services, navigating 'the system', and self-advocacy. Autism NZ educators deliver programmes that increase public understanding of autism and upskill professionals involved with autistic people (e.g. early childhood educators, physicians and clinicians, therapists, schools teachers, parents, and employers). It was my task to create an employment support programme that integrated career's counselling, vocational support, and employer education. I then implemented this programme, supporting autistic jobseekers to find and thrive in a career of their choosing.

In addition to Interpretive Interactionism, I also drew on the idea of 'cripping'[2] methodology (Kafer, 2013; Price & Kerschbaum, 2016; Price & Siebers, 2011), on access intimacy (Mingus, 2011), and on strategies found in participatory and person-oriented, ethical autism research (Cascio et al., 2020; Chown et al., 2017; den Houting et al., 2020; Fletcher-Watson et al., 2019; Nicolaidis et al., 2019; Raymaker, 2018). However, I could not integrate all the strategies suggested for ethical autism research. I could not pay participants for their time or offer capacitation opportunities,[3] as I did not have the budget or the personal power to do so. Most importantly, I could not ensure an equal sharing of power and responsibility, as I followed the traditional approach for PhD research. In other words, I made all key research decisions before engaging with participants. Although I made every effort to reduce power dynamics between participants and me in other ways, obvious inequalities remained. Many participants inadvertently highlighted these when they thanked me at the end of our meetings for *giving them the opportunity* to make their voices heard. Each time they thanked me, I became deeply uncomfortable, hearing an unspoken postscript that referenced the decades of research where autistic voices had been at best ignored, and at worst misrepresented or erased.

In the following, I outline how I synthesised these influences, inspirations, and tensions into a 'cripped interactionism'. I highlight errors made and lessons learned.

Interpretive Interactionism

Beyond its call for insider researchers, Interpretive Interactionism has two other major advantages that recommend it for engagement with neurodivergent populations. First, it is "an approach that involves minimal theory, seeks to show or perform rather than tell, and is based on a belief that less is more" (Denzin, 2001, p. 2). As respectful research should be accessible to the community with which it engages, its methodology should allow the researcher to present findings in a way that allows the community easily to connect with its essence. Interpretive Interactionism's *show; don't tell* is therefore well-suited. Second, Denzin (2001) urged that Interpretive Interactionist research must include a call to action with the purpose of effecting social change, allowing service providers to assist service users more effectively, as well as bringing "new value and meaning to identities and texts that are marginalized and stigmatized by the larger culture" (pp. 6–7). As one of the

[2] The verb 'to crip' derives from Crip Theory (McRuer, 2006), which challenges the idea that to be non-disabled is to be 'normal'. Crip theorists subvert normative logics, assumptions, and practices, drawing attention to non-normativity as an equally desirable way of living. To crip a practice involves re-envisaging it for non-normative bodyminds.

[3] For example, via internships or paid research assistant roles.

top research priorities of the autistic community is the improvement of well-being in day-to-day life (Roche et al., 2020), more effective support services and the dismantling of stigma are crucial.

Interpretive Interactionist researchers proceed through five stages: deconstruction, capture, bracketing, construction, and contextualisation. During deconstruction, the researcher is required specifically to discuss truth claims that contradict their own thinking, contextualising them historically and morally, and demonstrating "how these standpoints disadvantage and disempower members of a specific group, how particular texts directly and indirectly misrepresent persons and reproduce prejudice and stereotypes" (Denzin, 2001, p. 6). The resulting deep dive into ableist and dehumanising truth claims on autistic lack and deficit and brokenness was confronting. Deconstruction may also include analysis of popular cultural texts. In the case of my research, these texts were blogs and memoirs by autistic adults, mobilised to counterpoint deficit-focused truth claims.

During the second stage, the researcher captures multiple experiences of the phenomenon they want to investigate. Unlike the group capturing of stories in Denzin's (2001) research with Alcoholics Anonymous, the disparate access needs of my participants and myself led me to focus on individual stories, captured in semi-structured, in-depth, private interviews. Bracketing then isolates key phrases in the data, interpreting them and reflecting on meanings. During construction, the researcher orders and reassembles the key features isolated during bracketing. Finally, during contextualisation, the researcher "locates the phenomenon back in the worlds of lived experience" (Denzin, 2001, p. 51), reconstituting findings into a coherent whole to reflect participants' life worlds. Contextualisation is achieved by creative analytic practice to bring the research alive and engage with the reader on a deeper level. Details on cripping capture and contextualisation are discussed later in this chapter.

Access Intimacy: Access as a Relational Act

The meaningful, enabling, and respectful inclusion of neurodivergent research participants and the mitigation of power differentials were priorities of my project. A concept from the disability justice movement offered itself as a guiding light: that of access intimacy (Mingus, 2011, 2017). The concept defies attempts at definition. Mingus (2011), instead, gave a description: "Access intimacy is that elusive, hard to describe feeling when someone else 'gets' your access needs" (para. 4). It means that the disabled person feels safe to state their access needs, always. It is more than ensuring equal access—it "recognizes and understands the relational and human quality of access, while simultaneously deepening the relationships involved. It moves the work of access out of the realm of only logistics and into the realm of relationships" (Mingus, 2017, para. 21). In short, access intimacy has the power to bridge the emotional distance between two people, transforming both of them rather than merely adding a ramp one of them can climb while the other takes the stairs.

This emphasis on the relational aspect of access also highlighted the notion of interdependence between my participants and me. We navigated our sometimes divergent access needs together (c.f. Chapters 8, 12, this volume), with me attempting to prioritise participants' comfort over my own (c.f. Chapters 20, 21, this volume). This was not always successful, as with one participant who felt most at ease in cold environments. I was unable to hide my difficulty hearing him over the drone of the air-conditioner and, being most at ease in hot environments myself, must have looked pitiful trying to suppress my shivering. He suggested turning the air-conditioner off, which I gratefully accepted. However, even mishaps like this appeared to enhance the relationship between us by way of putting me, the researcher, on a more equal footing with my participants. I was an imperfect human being who sometimes needed accommodations, just like them, not an interchangeable agent of the academe. This humanising of the researcher, and the consequent deepening of the relationship between them and the participant may lead to richer conversations, as it did in my research.

Access intimacy can never be 'provided' or even promised, depending, as it does, not on logistics but on individual perceptions, dynamics of ease and safety, and in-the-moment readiness to adjust course. Therefore, I can only say that I *strove to* create access intimacy with my participants. Reflecting one of the fundamentals of access intimacy, I wanted to give participants the reassurance that I am a researcher who 'gets' their access needs. My overarching goal, therefore, was to anticipate needs as much as possible and prepare for them. This was important to me based on my own struggle with internalised ableism, which nearly always leads to shame when admitting access needs, as well as a sense of guilt about inconveniencing the other person with requests they may not have prepared for. So, instead of waiting for participants to request accommodations and thus reinforcing uneven power dynamics, I wanted to communicate that their bodyminds were taken into account from the outset. I did this primarily by being proactive in my use of adaptations. I took the following steps to create an enabling environment.

In the lead-up to the meeting:

- I asked participants to choose their preferred meeting time and location, either online via video call, or a public place within the Auckland region, or at the University of Auckland, Epsom Campus. If they elected to meet on Campus, I booked a room with natural light, minimal auditory and visual noise, and a kitchenette. This flexibility and attention to sensory aspects of the meeting space reflect known recommendations for sensory accessibility (Hallett & Kerr, 2020).
- For those who wanted to meet on Epsom Campus, I provided an adapted map of the location with irrelevant information blanked out for visual

clarity. This map showed sites of potential triggers for those with sensory sensitivities, and the location of bus stops and types of available parking.

- I also provided a 'photo journey' of the walk from possible arrival points to the interview location, including inside the building. Both the map and the photo journey reflect lessons learned from my previous interactions with neurodivergent clients. Following the principles of access intimacy, I did not ask participants if they needed this visual info but provided it proactively. One responded, "Thanks for those great directions; if only everyone could be so thoughtful!"
- I offered to meet participants at the car park or bus stop and walk in with them.
- I encouraged the use of noise-cancelling headphones.
- I suggested that participants may bring a support person or animal to sit in on our conversation (Cascio et al., 2020; Nicolaidis et al., 2019; Raymaker, 2018).
- I provided a brief bio of myself, including disclosure that I am autistic and what my enthusiasms are. This decision reflects the benefits of self-disclosure for equalising power and being congruent, one of the 'big three' factors of effective relationships (Kramer, 2013). My word choice in the bio appeared to be important. For example, one participant commented on my use of the word 'enthusiasms' instead of 'special interests'. Smiling, she exclaimed, "I noticed that you were saying that; I was like, Oh, that's *cool*!"
- I provided all written information materials on a light blue background to reduce contrast, accommodating those with visual stress. Visual stress has been linked to some types of neurodivergence, particularly dyslexia, ADHD, and autism (Imhof, 2004; Ludlow et al., 2006; Singleton & Trotter, 2005). Again, reaching for access intimacy, I was proactive with this accommodation for everyone instead of first asking who needed it.
- I explained expectations regarding time investment and procedure for communicating before and after interviews (Cascio et al., 2020). I offered participants the opportunity to change any elements with which they were uncomfortable.
- I offered to email participants my interview questions in advance to reduce any anxiety they may have had around unexpected or intrusive questions and allow time to contemplate responses (Chown et al., 2017; Nicolaidis et al., 2019). I explicitly offered the opportunity to strike any questions, again attempting to equalise power. One participant used the interview guide to prepare extensively for our meeting, arriving with several pages of hand-written notes. Her preparations made our conversation one of the most in-depth, full of richly detailed, vivid recollections. Some declined, explaining that they did not want to risk overthinking their answers. All appreciated being given the option.
- I encouraged the use of augmentative and alternative communication devices (Lebenhagen, 2020).

- I asked how participants self-identified to enable me to individualise the language I used in our conversations (Cascio et al., 2020).

Immediately before the start of the meeting:

- I was proactive in asking if any environmental factors were present that made the participant uncomfortable. I invited them to make me aware of any disturbances that may arise later to enable me to address these factors as far as possible.
- I made clear that participants may choose not to answer questions and suggested that they either say "pass" or shake their head to indicate their desire to move on to the next question. In offering these concrete options for refusing a question, I aimed to decrease any uncertainty participants may have felt around protocol.
- I made clear that participants may leave the interview at any point without having to explain themselves, again to reduce power differentials (Cascio et al., 2020; Nicolaidis et al., 2019).

During the meeting:

- I provided stim objects and a weighted blanket to cope with any noise or distractions that could not be avoided. I based these adaptations on my own experience of using weighted blankets and fidgets for sensory reintegration.
- I refrained from wearing any scents or clothing with high-contrast colours and patterns (Pellicano et al., 2017).
- I refrained from touching participants in any way. Beyond avoiding handshakes, this included metaphorical 'touching' with the eyes if they showed signs of discomfort with being looked at. I drew on my professional experience for this adaptation, remembering one former client in particular. They often took time to ponder a question before answering. The amount of time they took appeared to shorten after I had learned not to look at them while they were thinking.
- I gave ample time to process each question, rephrased, and clarified (Cascio et al., 2020; Nicolaidis et al., 2019). I was also prepared to visualise questions in the form of hand-drawn mind maps, a method that had been useful with former clients at Autism New Zealand; however, this was not necessary.
- I offered a break if participants showed any indication of becoming tired or stressed, as frequently as necessary. I offered a break after one hour even if participants did not show any indication of tiring (Cascio et al., 2020).

- I brought my own herbal teas to Campus to be able to offer a hot drink. Offering a drink, even if only a glass of water, had proved an easy way to create a welcoming, relaxed atmosphere with my former clients.

After the meeting:

- I offered continued synchronous and asynchronous communication options should participants think of anything they wanted to add or correct (Cascio et al., 2020; Nicolaidis et al., 2019). Anyone who wished to qualify their initial answers after reviewing their transcripts had the opportunity for further conversations, until they felt that their viewpoints were accurately represented (Chown et al., 2017). Eleven participants used this option, some extensively.

All participants appeared comfortable during our conversations, sharing freely and openly. Fifteen specifically remarked afterwards how much they enjoyed the conversation, and how different the experience was compared to what they expected a research interview to be like. Nine spontaneously indicated their readiness to engage further should I think of more questions or need any other help.

Errors Made

- One participant was confused by the photo journey through the inside of the building on Epsom Campus. My error was not being specific enough in my second-to-last photo description. I had written, "through the doors by the lockers, then immediately turn left", with the following photo showing the meeting room after the left turn. It was not obvious that this last photo showed the back of the 'doors by the lockers'. The participant therefore continued along the corridor to the next left turn and got lost. I should have been more precise and written, "through the doors by the lockers; then the meeting room is on your left, within 1.5 m of the doors you just went through". However, the experience created an excellent opportunity for trust-building: I assumed full responsibility for the unclear directions and apologised right away, acknowledging the potentially high anxiety my error may have caused. This readiness to apologise arose from my own experiences of confusion by unclear instructions, where I was usually blamed for not reading the instructions properly. By not re-enacting this pattern, I could reassure the participant that I would always respect and centre *their* experience.
- Another participant had requested a meeting at City Campus, where I was not familiar with the buildings. Although I attempted to provide directions, my error was relying on publicly available maps, photos, and site plans, and not visiting the location before the meeting. We both

became confused by ambiguous signage inside the building, got lost, and spent some time searching for our meeting room on the wrong floor. However, this experience led to instant rapport between us, as we laughed and bonded over our shared inclination to the same perceptual quirks.

- My original demographic survey included a space for participants to indicate their 'preferred pronouns'. When a transgender participant made me aware of the bias inherent in this question (i.e. cis people have 'pronouns', not 'preferred pronouns', so the same should apply for transgender people), I was mortified. I apologised, thanked the participant for letting me know, and deleted the 'preferred'.
- My original interview guide included a question about whether participants had heard of the term 'camouflaging', and whether they had ever camouflaged with their therapists. After the first time a participant seemed confused by the term and asked if I meant 'masking', a more common term in the community, I recognised my error in assuming that the academic term 'camouflaging' was widely known. I referred to 'masking' instead from then on.

Cripping Methodology

For the second stage of Interpretive Interactionism, capture, I conducted semi-structured interviews. Designed to be used flexibly, they are ideally suited to collecting data from people with atypical social communication (Smith, 1996). Semi-structured interviews allowed me to engage with responses and enabled participants to raise any additional issues they perceived as important. Throughout the process, my goal was for interviews to become dialogues, reflecting a deep and genuine reciprocity. This goal was supported by my efforts towards access intimacy, my self-disclosure, my openness about my sensory sensitivities when organising mutually agreeable meeting spaces, and generally not referring to meetings as 'interviews' but as 'conversations'. Although I anticipated some success with this approach, I was surprised at how organically conversations developed with most participants. Their naturally told recollections contained the answers to most of my questions before I could even ask them, leaving my interview guide mostly forgotten and unused on the table. A check at the end of the conversation to see what areas I'd missed rarely resulted in more than one further question.

Smith (1996) asks the researcher to avoid prompts and overly explicit or leading questions unless the participant has difficulty understanding the question. I anticipated that this could be an issue in my project, as some autistic people may find answering open-ended or vague questions confusing (Attwood, 2007). As I had learned in my professional experience, some may be pitched into cognitive overload by a single vague question, with cascading effects for the remainder of the conversation. In a research setting, cognitive overload may lead participants to withdraw early from an interview (Griffith et al., 2012). Therefore, following Price and Kerschbaum's (2016) observation that disability "crips methodology" (p. 20), I cripped the interview. I phrased my questions as unambiguously as possible from the start, being more explicit and concrete than may be typical in a qualitative interview. For example, instead of asking, "What made you feel like you had a good relationship with your counsellor?", I asked, "Can you describe the things your counsellor did to try to make you feel comfortable and understood?" If I noticed any signs of confusion in participants, I supplemented prompts phrased as closed questions or offered examples of common scenarios. These prompts allowed participants to elaborate on whether or how the scenario played out in their lives. The resulting frequent prompts for some participants raised the spectre of leading questions, risking mere confirmation of my ideas rather than opening a door to unexpected or disconfirming findings. To address this possibility, I followed up questions on specific relationship-building methods with checks such as: "Was this important for you?".

I was also concerned about participants agreeing with my interpretations to avoid discord, as this had been an issue in my professional experience. Therefore, I was extremely cautious in how I articulated my thoughts during conversations. For example, I prefaced any interpretations with phrases such as, "just wondering if I understood this correctly". I also relied on the more equitable power dynamics I tried to create throughout my research relationships, which stood in contrast with my professional experience. In my research, I consulted *participants* as the experts, whereas in my previous role, my clients had consulted *me* as the expert. This approach appeared successful, with most participants correcting my tentative interpretations at least once.

I planned for each conversation to take up to two hours. This extended time frame acknowledged the concept of 'crip time' (Kafer, 2013; Price & Kerschbaum, 2016; Price & Siebers, 2011). Flowing with crip time in our conversations allowed for rest periods, moments of silence to process questions and formulate an answer without pressure, and idiosyncratic communication patterns. Participants appreciated the unpressured environment as it gave them the scope to talk about their experiences in a way that suited them. With some, this involved letting their answer hide behind other stories for a while, approaching it from different angles, and allowing it to come out on its own terms. These circumambulatory conversations seemed to be a welcome departure from participants' expectations of a conversation with a researcher. As one reflected,

> I had sort of expected being interrupted. I'm used to going off course and people want to bring me back to the straight and narrow, and you let me have free reign. But then I'm also conscious that you brought me back but still kept me going around those areas. It's useful for me as well, to get a sense of where all those connected things happen. And so that was very positive.

My conversation with this participant lasted 2 hours 37 minutes, taking us well beyond my planned time limit. Having set the meeting for the end of the day, however, we did not need to rush off and enjoyed the slightly eerie feeling of walking out together through the university's empty, crepuscular stairwell, echoing only with our own footsteps.

Other participants appreciated how being on crip time allowed me genuinely to connect with them and to listen to their thoughts around certain issues attentively, regardless of whether they were related to my questions. One participant contrasted our conversation with a recent experience she'd had with a psychologist who did not make her feel as if they were interested in her as a person.

> Romy: "How would you have ideally had that meeting [with the psychologist] go? Like, what would have been a good attempt to connect with you from her side?
>
> Participant: "I would have liked her to talk to me about why I felt that way. And probably *this* conversation is probably *exactly* what I would have expected."

Participants had full control over their data at all stages of the research process. I offered them a copy of their recording and transcript to enable them to review their data and ensure their views were represented accurately. As per the guidance of the Ethical Review Board, the consent form specified that they had one month to make any changes or request another meeting. However, I did not enforce this limit, aware that some participants may have needed extra time to process and reflect on their conversations. Most participants took around two months to review their transcripts.

Errors Made

The first participant to review his transcript alerted me to a potential pitfall in offering this option. I had described the process as an opportunity to make any changes he saw fit, redacting statements he was not happy with and adding additional information if anything sprang to mind. As the participant was not happy with his spoken communication style, he proceeded with a time-consuming in-depth edit of his transcript. Recognising my error, I revised the way I communicated the purpose of the transcript review process. For all others, I specified that revisions were only required for content, not style, and that I would tidy up any quotes used in the final study (deleting stammers and filler words such as 'like' or 'um') unless they requested that quotes be kept exactly as spoken. I opted to make these slight stylistic edits of quotes the default, both in response to the first participant's obvious discomfort with seeing his speech transcribed, and to prevent any discomfort others may have felt around the same issue. This decision aligns with the latest best practice principles in ethical autism research, which suggest that the researcher should be proactive in offering options to participants, including around "how results are written" (Cascio et al., 2020, p. 8).

Member Checks

To contribute to the trustworthiness of the research, I engaged in four forms of member checks[4] (Chown et al., 2017; see Chapter 10, this volume). First, four participants had indicated an interest in a behind-the-scenes look at how I would interpret their answers. I extracted each of their codebooks from NVivo12, showing bracketed phrases and assigned codes, and then I emailed them their codebook with an invitation to check if they agreed with my interpretations. No corrections were necessary as a result of these checks. Second, opportunities for member checks arose organically during conversations, as participants asked about my findings from previous conversations. I gave a brief summary of major themes at the end of each conversation to avoid influencing answers, and then elicited participants' opinions on if and how they thought those themes may relate to their perceptions. Third, nine participants had indicated that they wanted to contribute more to the project if I had any follow-up questions at a later stage. I contacted these participants via email to invite feedback on emerging themes that had not formed part of their conversations with me, describing the theme and including specific examples drawn from other transcripts. Fourth, I shared the easy-read version of my study with all participants three months before finalising my discussion of findings, including the question: "If there is anything in the document that you'd like to comment on, those three months would give me enough time to incorporate your feedback in my final study. I would love to hear from you!" The second, third, and fourth types of member checks led to valuable additional examples, further discussion and refinement of my interpretations, and a correction to the easy-read version.

Accessible Dissemination and Creative Analytic Practice

Accessible dissemination, crucial in ethical autism research, is not a novel idea associated only with disability studies. Indeed, sociologist Richardson (1988) suggested in the late 1980s that research should be, first and foremost, "for the people" (p. 206). Her goal, and mine, was to "meet a different standard of science and truth: accessibility to lay audiences" (p. 206). The alternative, producing research that makes a difference only to the author's career, seemed "foolish at best, and narcissistic and wholly self-absorbed at worst"

[4] Members checks are used in qualitative research to improve study authenticity and validity. They involve checking in with research participants at several stages of the project to get their feedback on the researcher's interpretations of their data.

(Richardson & St. Pierre, 2005, p. 960). However, accessibility to lay audiences has not traditionally been a forte of academic writing. As Tsao (2011) pointed out, scholars write for other scholars, often couching their ideas in registers that are incomprehensible to all but the initiated (c.f. Chapters 8, 10, this volume). I therefore attempted to keep jargon to a minimum from the outset, choosing clear and non-academic language not only in publications intended for dissemination, but within my thesis itself. This was a priority as most participants had indicated an interest in reading the thesis.

I attended to accessibility in several other ways. First, I created a comprehensive easy-read version of my study, which I shared with all participants, and a brief lay summary for Autism NZ to share in their newsletter, which reaches 12,000 community members. This summary included my university email address with an invitation for community members to request the easy-read version of my study. Second, I co-authored an article with a participant (Hume & Burgess, 2021), which included a personal narrative to bring life to an aspect of my research that could not be included in the final thesis. And third, I presented my findings via creative analytic practice (Hume, 2022). Creative analytic practice is intended to make research accessible to a wider readership beyond academia, offering

> one way to break down the academic–public divide, and perhaps to more profoundly influence public understandings and perceptions of issues, peoples and places than the traditional factual texts that we are encouraged, if not pressured, to produce. (Bruce, 2019, p. 70)

Here, Bruce (2019) alludes to the 'pressure' academics may feel to uphold standards of writing for their own sake, sacrificing understandability in exchange for intellectual clout. Poetic enquiry scholar Katie Fitzpatrick describes the act of resisting this pressure as "an act of rebellion in my usual academic world; a challenge to prose and to neoliberalism" (Fitzpatrick & Fitzpatrick, 2021, p. 6). I felt the same pressure, and the same sense of rebellion, when working on my first publications: the constant push–pull between wanting to be taken seriously and wanting my research to be 'for the people'.

Denzin (2001) suggested that creative analytic practice can be deployed in many ways; for example, poetry, collective stories, autoethnography, or drama. For my project, I chose 'found poetry', a form of research poetry that rearranges "words, phrases, and sometimes whole passages that are taken from other sources and reframed as poetry by changes in spacing and/or lines (and consequently meaning), or by altering the text by additions and/or deletions" (Butler-Kisber, 2021, p. 23). Found poetry thus "offers social researchers an opportunity to write about, or with, people in ways that honor their speech styles, words, rhythms, and syntax" (Richardson, 2001, p. 880). In doing so, it can "permit silenced voices/stories to be heard" (Butler-Kisber, 2021, p. 22).

Poetry is a "*practical* and *powerful* means for reconstitution of worlds" (Richardson, 1993, p. 705, emphasis in original), particularly when those worlds elude normative interpretations. It does not narrow the interpretive space to suggest only one meaning, as may be the case with prose. Instead, it can speak to "the multiple meanings and the possible meanings that we create together" (Glesne, 1997, p. 218). In other words, as autistic poet Nathan Spoon pointed out, poetry in itself is "a kind of neuroqueer language" (Spoon & Stone, 2021, para. 9). Poetry also has the power to captivate our senses even if our mind rejects it, opening up "our hearts and ears to different ways of seeing and knowing" (Butler-Kisber, 2021, p. 21). It is this opening up to difference that makes poetry invaluable in the context of autism. In allowing readers to *feel* my research findings, I hoped that it may become easier for them to suspend old beliefs about the nature of autism than if they engaged only on an intellectual level.

Nevertheless, poetry may seem an incongruent means of communicating the life worlds of autistic adults. After all, academic texts have historically portrayed autistics as 'arhetorical' (Yergeau, 2018), unable to understand, much less deploy, metaphor or figurative language (Morsanyi et al., 2020). Similarly, popular representations of autistics in literature, film, and television tend to show unemotional, hyper-logical savants fascinated by science and facts, not literature (Freeman Loftis, 2015; Murray, 2008; Prochnow, 2014). It is unsurprising, then, that although poetic inquiry is relatively common in research with people from other marginalised groups (Schoone, 2020), it has been absent from research with autistic people. My vision of participants' utterances *as* poetry aimed to dislodge the idea of the computeresque autistic brain. In doing so, I followed the scores of autistic poets who have published their poems in online collections, on blogs, in poetry journals, or in books. My participants' poetic voices resonate with those of autistic poets the world over, joining them in challenging prejudice and stigma.

Summary

The neurodiversity-affirming methodology described in this chapter draws on Interpretive Interactionism and access intimacy, supported by a readiness to crip methodology as and where needed.

The main nuances that distinguish Interpretive Interactionism from other qualitative methodologies can be summarised as:

- Interpretive Interactionism is a methodology for insider researchers. It therefore represents an insider perspective at all levels.
- Participants' voices are not only privileged, but also influence how the research is presented and disseminated.
- Deconstruction is not restricted to academic sources, and its focus is on demonstrating how the examined texts disempower members of the social group to which the researcher and the participants belong. This focus adds a critical perspective.

- The discussion of findings deploys creative analytic practice, which allows for more accessible dissemination.

Striving towards access intimacy led to:

- Proactive use of adaptations.
- Minimisation of power dynamics.
- Enhanced trust and comfort in research relationships.

Cripping research methodology involved:

- Flexibility in timing, location, structure, and modality of interviews.
- Engaging in circumambulatory conversations rather than pushing for immediate, direct answers.
- Flexibility in the consent and review process.
- Changes in how participant quotes were represented.

References

Attwood, T. (2007). *The complete guide to Asperger's syndrome.* Jessica Kingsley.

Bruce, T. (2019). The case for faction as a potent method for integrating fact and fiction in research. In S. Farquhar & E. Fitzpatrick (Eds.), *Innovations in narrative and metaphor: Methodologies and practices* (pp. 57–72). Spinger.

Butler-Kisber, L. (2021). Poetic inquiry. In E. Fitzpatrick & K. Fitzpatrick (Eds.), *Poetry, method and education research: Doing critical, decolonising and political inquiry* (pp. 21–40). Routledge.

Cascio, M. A., Weiss, J. A., & Racine, E. (2020). Person-oriented ethics for autism research: Creating best practices through engagement with autism and autistic communities. *Autism, 24*(7), 1676–1690. https://doi.org/10.1177/1362361320918763

Chown, N., Robinson, J., Beardon, L., Downing, J., Hughes, L., Leatherland, J., Fox, K., Hickman, L., & Macgregor, D. (2017). Improving research about us, with us: A draft framework for inclusive autism research. *Disability and Society, 32*(5), 720–734. https://doi.org/10.1080/09687599.2017.1320273

Denzin, N. K. (2001). *Interpretive interactionism* (2nd ed.). Sage.

Fitzpatrick, E., & Fitzpatrick, K. (2021). What poetry does for us in education and research. In E. Fitzpatrick & K. Fitzpatrick (Eds.), *Poetry, method and education research: Doing critical, decolonising and political inquiry* (pp. 1–18). Routledge.

Fletcher-Watson, S., Adams, J., Brook, K., Charman, T., Crane, L., Cusack, J., Leekam, S., Milton, D., Parr, J. R., & Pellicano, E. (2019). Making the future together: Shaping autism research through meaningful participation. *Autism, 23*(4), 943–953. https://doi.org/10.1177/1362361318786721

Freeman Loftis, S. (2015). *Imagining autism: Fiction and stereotypes on the spectrum.* Indiana University Press. http://www.jstor.org/stable/j.ctt17kmw55

Glesne, C. (1997). That rare feeling: Re-presenting research through poetic transcription. *Qualitative Inquiry, 3*(2), 202–221.
Goffman, E. (1991). Stigma: Notes on the management of spoiled identity. Simon & Schuster.
Griffith, G. M., Totsika, V., Nash, S., & Hastings, R. P. (2012). I just don't fit anywhere: Support experiences and future support needs of individuals with Asperger syndrome in middle adulthood. *Autism, 16*(5), 532–546. https://doi.org/10.1177/1362361311405223
Haas, K., Costley, D., Falkmer, M., Richdale, A., Sofronoff, K., & Falkmer, T. (2016). Factors influencing the research participation of adults with autism spectrum disorders. *Journal of Autism and Developmental Disorders, 46*(5), 1793–1805. https://doi.org/10.1007/s10803-016-2708-6
Hallett, S., & Kerr, C. (2020). You need support, validation, good coping skills. You need and deserve acceptance: Autistic adult experiences of counselling. *Autistic Mental Health and Autistic Mutual Aid Society Edinburgh (AMASE).* www.autisticmentalhealth.uk/counsellingreport
den Houting, J., Higgins, J., Isaacs, K., Mahony, J., & Pellicano, E. (2020). 'I'm not just a guinea pig': Academic and community perceptions of participatory autism research. *Autism.* https://doi.org/10.1177/1362361320951696
Hume, R. (2022). Show me the real you: Enhanced expression of Rogerian conditions in therapeutic relationship building with autistic adults. *Autism in Adulthood, 4*(2), 151–163. https://doi.org/10.1089/aut.2021.0065
Hume, R., & Burgess, H. (2021). "I'm human after all": Autism, trauma, and affective empathy. *Autism in Adulthood, 3*(3), 221–229. https://doi.org/10.1089/aut.2020.0013
Kafer, A. (2013). *Feminist, queer, crip.* Indiana University.
Kramer, C. H. (2013). Revealing ourselves. In M. Baldwin (Ed.), *The use of self in therapy* (3rd ed., pp. 36–63). Routledge.
Lebenhagen, C. (2020). Including speaking and nonspeaking autistic voice in research. *Autism in Adulthood, 2*(2), 128–131. https://doi.org/10.1089/aut.2019.0002
McRuer, R. (2006). Crip theory: Cultural signs of queerness and disability. New York University.
Mingus, M. (2011). Access intimacy: The missing link. *Leaving Evidence.* https://leavingevidence.wordpress.com/2011/05/05/access-intimacy-the-missing-link/
Mingus, M. (2017). Access intimacy, interdependence and disability justice. *Leaving Evidence.* https://leavingevidence.wordpress.com/2017/04/12/access-intimacy-interdependence-and-disability-justice/
Mogendorff, K. (2013). The blurring of boundaries between research and everyday life: Dilemmas of employing one's own experiential knowledge in disability research. *Disability Studies Quarterly, 33*(2). https://doi.org/10.18061/dsq.v33i2.3713
Morsanyi, K., Stamenković, D., & Holyoak, K. J. (2020). Metaphor processing in autism: A systematic review and meta-analysis. *Developmental Review, 57*, 1–18. https://doi.org/10.1016/j.dr.2020.100925
Murray, S. (2008). *Representing autism: Culture, narrative, fascination* (1 ed.). Liverpool University Press. http://www.jstor.org/stable/j.ctt5vjmwc
Nicolaidis, C., Raymaker, D., Kapp, S. K., Baggs, A., Ashkenazy, E., McDonald, K., Weiner, M., Maslak, J., Hunter, M., & Joyce, A. (2019). The AASPIRE practice-based guidelines for the inclusion of autistic adults in research as co-researchers and

study participants. *Autism, 23*(8), 2007–2019. https://doi.org/10.1177/1362361319830523

Price, M., & Siebers, T. (2011). *Mad at school: Rhetorics of mental disability and academic life*. University of Michigan.

Price, M., & Kerschbaum, S. (2016). Stories of methodology: Interviewing sideways, crooked and crip. *Canadian Journal of Disability Studies, 5*(3), 18–56. https://doi.org/10.15353/cjds.v5i3.295

Prochnow, A. (2014). An analysis of autism through media representation. *ETC: A Review of General Semantics, 71*(2), 133–149.

Raymaker, D. (2018). *Authentic inclusion in autism research*. https://researchautism.org/authentic-inclusion-in-autism-research/

Raymaker, D. M. (2017). Reflections of a community-based participatory researcher from the intersection of disability advocacy, engineering, and the academy. *Action Research, 15*(3), 258–275. https://doi.org/10.1177/1476750316636669

Richardson, L. (2001). Poetic representation of interviews. In J. F. Gubrium & J. A. Holstein (Eds.), *Handbook of interview research* (pp. 876–891). SAGE Publications, Inc. https://doi.org/10.4135/9781412973588

Richardson, L., & St. Pierre, E. (2005). Writing: A method of inquiry. In N. K. Denzin & I. S. Lincoln (Eds.), *The Sage handbook of qualitative research* (3rd ed., pp. 959–978). Sage.

Richardson, L. (1988). The collective story: Postmodernism and the writing of sociology. *Sociological Focus, 21*(3), 199–208. https://doi.org/10.1080/00380237.1988.10570978

Rinaldi, J. (2013). Reflexivity in research: Disability between the lines. *Disability Studies Quarterly, 33*(2). https://doi.org/10.18061/dsq.v33i2.3711

Schoone, A. (2020). *Constellations of alternative education tutors a poetic inquiry*. Springer.

Smith, J. A. (1996). Semi structured interviewing and qualitative analysis. In J. A. Smith, R. Harré, & L. v. Langenhove (Eds.), *Rethinking methods in psychology* (pp. 9–26). Sage.

Spoon, N., & Stone, J. (2021). Poetry and autistic advocacy: An interview with Nathan Spoon. *Little Lobbyists*. https://littlelobbyists.org/blog/2021/3/11/poetry-amp-autistic-advocacy-an-interview-with-nathan-spoon

Tsao, E. (2011). Walking the walk: On the epistemological merits of literary ethnography. *Anthropology and Humanism, 36*(2), 178–192. https://doi.org/10.1111/j.1548-1409.2011.01091.x

Yergeau, M. R. (2018). Authoring autism: On rhetoric and neurological queerness. Duke University.

CHAPTER 14

Developing a Phenomenology of Autism

Nicola Shaughnessy, Jonathan Green, Emma Williams, Jonathan A. Smith, Bastian Greshake Tzovaras, Georgia Aitkenhead, Damian Milton, Emily Jones, and Karen Leneh Buckle

Introduction: The Need for a Phenomenology of Autism

Most mental health conditions and diagnoses are defined by a combination of key subjective experiences and outward behaviours. For example, Obsessive Compulsive Disorder (OCD) is defined by the experience of intrusive, obsessive *thoughts* and other mental experiences alongside compulsive *behaviours;* Schizophrenia similarly. Whilst initial understanding of 'autism' in the early twentieth century was similar, modern developments of the autism concept

N. Shaughnessy (✉)
School of Arts, University of Kent, Canterbury, UK
e-mail: n.Shaughnessy@kent.ac.uk

J. Green
The University of Manchester, Manchester, UK

Health Innovation Manchester and Manchester Academic Health Sciences Centre, Manchester, UK

Royal Manchester Children's Hospital, Manchester, UK

E. Williams
School of Psychological Sciences, University of Surrey, Guildford, UK

J. A. Smith
School of Psychological Sciences, Birkbeck, University of London, London, UK

H. Bertilsdotter Rosqvist and D. Jackson-Perry (eds.), *The Palgrave Handbook of Research Methods and Ethics in Neurodiversity Studies*,
https://doi.org/10.1007/978-3-031-66127-3_14

since the 1960s have focused on an agreed set of solely *behavioural* observations by others (clinicians, researchers or parents) without including the quality of subjective autistic experience itself, its 'phenomenology' (Green & Shaughnessy, 2023). The behavioural observations have focused on 'deficits in communication and social interaction', as well as particular qualities of sensory processing, and repetitive or restrictive behaviours (APA, 2013). This essentially behavioural definition has also been mirrored in an evolving clinical and diagnostic language (for instance codified within ICD and DSM), and through this into a diagnostic and clinical usage in everyday practice, defining a clinical and research concept of autism, and influencing a wider cultural concept of 'what autism is'. Whilst this approach has had aspects of real value for clinical and developmental science work, facilitating a reliable shared global consensus in identifying 'autism' and enabling scientific enquiry and clinical communication, autistic advocates and academics have for many years been robustly critiquing its serious limitations and partiality (Dinishak, 2016; Sinclair, 1993; Yergeau, 2018).

Advocacy from the autistic community as to their experience, as well as theoretical work from scholars working in the area of neurodiversity studies and more recently clinical scientists and others, has led to a revolution in thinking, including a taking more seriously of the corpus of autistic experience literature that already exists, as well as stimulating more (Chown et al., 2017; Fletcher-Watson et al., 2019; Gillespie-Lynch et al., 2017; Green & Shaughnessy, 2023; Milton, 2014). This has generated new information and ideas about the nature of autistic life and experience, necessitating a radical reconceptualization, or augmentation, of the current diagnostic behavioural autism concept through integrating the subjective and experienced in the form of phenomenology into cultural, scientific and clinical usage. And yet, there is

B. G. Tzovaras
The Alan Turing Institute, London, UK

G. Aitkenhead
The Alan Turing Institute, London, UK

D. Milton
Tizard Centre, University of Kent, Canterbury, UK

E. Jones
Centre for Brain & Cognitive Development, Birkbeck, University of London, London, UK

K. L. Buckle
The University of Manchester, Manchester, UK

to date no established systematic phenomenology of autism with which to characterize the subjectivity of the condition.

We consider that the time is right now to begin developing such a systematic and representative phenomenology. In this chapter, we outline an approach to developing such a phenomenology empirically, and through co-production within a neurodiverse team. This includes neurodivergent and neurotypical scholars. For clarity and as a neuromixed team (autistic and non-autistic researchers) we use the terms phenomenology of autism and autistic phenomenology throughout this chapter to indicate our positions and approach. We envision contributing to a new conceptualization of autism that moves beyond external and behavioural observation (as in the diagnostic criteria) to refer to interior experiences and suggest a method that can be used by a diversity of researchers (neuromixed) as a phenomenology of autism. Concurrent with this, we are questioning the dichotomies between neurotypicality and autism to suggest a more nuanced configuration (a human spectrum, Murray et al., 2023), sensitive to context and intersectionalities. At the same time our approach encompasses and acknowledges autistic phenomenology as a way of being in the world, a process of natural enquiry (supporting survival) which is about understanding the self and surroundings (a form of autoethnography). This is reflected in the work about autistic experiential states (e.g., empathy, attention, flow states, masking) informed by the lived experience of autistic scholars (McDonnell & Milton, 2014; Milton, 2012; Murray et al., 2005; Pearson & Rose, 2023).

Methods Towards a Phenomenology of Autism

New methods are required to establish a phenomenology that is capable of distinguishing autistic experience from a range of other experiences, a requisite for subjective experience to be included within a future nosology of autistic diagnosis. Recent systematic studies of specific groups and particular experiences of autistic people have begun to generate a body of phenomenological literature, but within relatively small and selected samples with their own partialities. We propose a more inclusive, systematic and empirical approach to developing a phenomenology of autism. This will involve the use of online digital technology platforms employing a citizen science approach to collect experiential data at scale without sacrificing the deep and intimate subjective qualities of traditional phenomenology (see Chapter 15, this volume). Citizen science describes a growing set of transdisciplinary practices that engage members of the public in creating scientific knowledge to enable the collection of phenomenological data at scale.

The key constituents of our proposed method, combining depth with breadth and reach, are briefly outlined below, followed by a more detailed description of each in subsequent sections of the chapter.

- **DEPTH. Co-construction and application of an in-depth phenomenological method** (Interpretative Phenomenological Analysis) **to work with autistic accounts of lived experience.** We will build on the existing experience and work of our neuromixed group of researchers to co-construct a trusted engagement with the autistic community, using in-depth Interpretative Phenomenological Analysis (IPA) to explore autistic (and comparative non-autistic) subjectivity in specific areas of lived experience (e.g. senses, inertia, attention). Purposive sampling will be employed across various intersectionalities, (e.g. age, dis/ability, gender, and sexuality).
- **BREADTH. Development of online methods at scale to capture the breadth of autistic subjectivities:** The in-depth IPA data will feed into a broader approach to data collection offered by online digital technology platforms employing a citizen science approach. This will test the idiographic phenomenological data collected in the first stage of our research at scale across the range of autistic experience.
- **REACH. Innovative methods for representing communication differences:** A key methodological challenge is that of representativeness across ability, particularly within intellectual disability and non-speaking participants. New methods of developmental science will be combined with ethnographic participant observation to enable us to infer, beyond mere verbal communication, core elements of autistic people's lived realities, including the social-spatial–temporal fields in and through which experiences arise.
- **TRANSFORMATION. Synthesis of methods and findings towards a systematic phenomenology of autism**: The information collected will be iteratively integrated, working towards the possibility of a distilled but representative phenomenology of autism, with the aim of characterizing key aspects of autistic experience. A key question is whether we will be able to separate out early aspects of autistic difference from the developmental outcome of a cascade of self-environment interactions over time.

DEPTH. Co-construction and Application of Interpretative Phenomenological Analysis (IPA) within Autism

In this section, we outline an approach enabling an in-depth examination of the lived experience of core features of autism. Our approach is based on a well-established qualitative methodology, interpretative phenomenological analysis (IPA) (Smith et al., 2022) explicitly developed to explore individual personal experience but we are presenting adaptations of the methodology to optimize its use in autism research.

The IPA Method

IPA was first developed in the 1980s and 1990s by chapter author Jonathan A. Smith. It has three key theoretical foundations:

- *Phenomenology*: This is the philosophically grounded concern with attempting to provide an account of lived experience which is, as far as possible, conceptualized in its own terms.
- *Hermeneutics*: IPA sees phenomenology as being an interpretative endeavour whereby researchers are attempting to make sense of participants who are trying to make sense of events, objects and interactions they encounter.
- *Idiography:* This approach steadfastly values the examination of each case in great detail, only after which is there a move to look for patterns of convergence and divergence across case.

We think these key underpinnings point to what makes IPA an especially valuable methodology for investigating the phenomenology features of autistic experience. There has been a recent upsurge in the number of IPA papers examining the sensory experiences of autistic people, either as a primary focus (Kyriacou et al., 2023; Taels et al., 2023) or as a key aspect of another experience, such as being a mother (Dugdale, Thompson, Leedham, Beail & Freeth, 2021) or listening to music (Venter et al., 2023). Prior to this, while the corpus of qualitative work in the autism field has grown exponentially, studies have predominantly been externally focused, concentrating on examining the secondary impacts of being autistic such as the challenges faced in navigating everyday contexts (e.g., care services, education, the workplace). Although such research is valuable, we see a pressing need for studies drilling down more deeply into the core, potentially foundational, elements of the lived experience of people in this community, including how these intersect with the negotiation of everyday settings. Such elements represent more likely candidates for differentiating autistic from non-autistic phenomenology, making it particularly relevant to our eventual goal of developing an enhanced descriptive nosology in which key subjective lived experiences discriminatory for autism can be identified alongside current behavioural and developmental descriptors.

Sequence of IPA Investigation

The research question for an IPA study should reflect the theoretical underpinnings outlined above and the particular interests of the researchers. Box 1 presents the research questions from recent studies using IPA to find out more about the core sensory experiences of autistic people and so illustrates this process.

Orientating the Interviews
Most IPA studies collect data through in-depth one-to-one interviews with participants and we would consider this the starting point for research with autistic people (although group and ecological methods have been used in related research; Boldsen, 2022).

Box 1: The research questions formulated in some recent IPA studies examining core autistic subjective experiences

1. How do autistic individuals describe and make sense of their experiences of sensory hypersensitivity, and how do they contextualize these experiences in day-to-day activities and interactions? (Taels et al., 2023)
2. How do participants make sense of and interpret their lived music listening experiences? (Ventner et al., 2023)

Researchers construct an interview guide to facilitate the interview, but this is used very flexibly in the interview itself. Participants are encouraged to talk in detail about the elements they consider most important, and the interviewer attempts to facilitate this process, offering prompts and probes to enable delving deeper. Box 2 provides examples of the type of questions asked in an IPA interview designed to find out more about autistic subjective experiences.

Box 2: Examples of the type of questions that might be asked in an IPA interview study on autism

Q1 Could you describe the key aspects of your own autistic experience

Q2 How do these aspects of your autistic experience affect your everyday life?

Q3 How have these experiences evolved over time?

Interview Analysis
IPA analysis involves a slow step-by-step, inductive process of engagement with the transcript of each interview. Limited space permits only a brief overview of these steps here. For further details see either Smith et al. (2022) or Smith and Nizza (2022). The transcript is prepared for analysis by constructing a Word document with three columns and the interview material is copied into the central column. As one reads through the transcript, one margin is used

for making exploratory notes on what is arising—this can be comments on the language employed, notes on especially interesting content, pointers to elaborations or ambiguities arising in the narrative and so forth. Having completed this for the whole transcript, one then returns to the beginning of the interview and records experiential statements in the other margin. These push the analysis to a next level of conceptualization, attempting to provide a short concise and pithy summary of what we are learning about the participant's experience from this section of text, grounded in but moving beyond their verbatim report. See box 3 which illustrates exploratory notes and experiential statements for a small piece of text.

The next stage in analysis involves looking for patterns within the experiential statements and clustering them accordingly. Clusters are labelled as personal experiential themes (PETS) because they represent the core experiential features for that individual emerging from the analysis.

Box 4 shows the beginning of a patterning for the material in the extract in Box 3. In an actual study, experiential statements from the whole interview are considered in a similar way. Having completed a table of PETS for the first case, one starts from scratch again with the second case, repeating each of these steps described above. After having conducted this process with each case and having a table of PETS for each case, one then looks for patterns of similarity and difference across case—this culminating with a table of group experiential themes (GETS). The table of GETS provides the skeleton structure for the write up, but the narrative needs to be enlarged to provide a detailed account of the participants' experience accessible to a reader who has played no part in the study. IPA, like much qualitative research, is a writerly activity and much of the power of the report or paper depends on the way in which it is written. Good practice in IPA involves taking each GET in turn, introducing it and then constructing a narrative whereby extracts from several participants are drawn on to illustrate the range of ways in which different respondents make manifest the theme in question. Each extract is followed by interpretive commentary from the researchers highlighting the experiential features being demonstrated. See Box 5 for a short illustrative example and Nizza et al (2021) for a more detailed account of constructing high-quality IPA write up.

Box 3: Exploratory notes and experiential statements from study by Murray et al (2023).

Experiential statements	*Original transcript*	*Exploratory notes*
Heightened awareness of specific sensory stimuli Forced to deal with difficult sensory stimuli at school Felt understood by mother, who recognised and accommodated sensory need Feels physically sick and overwhelmed when encounters specific smells Some tactile stimuli feel intrusive; they cannot be shut out Feels sick just thinking about some kinds of tactile stimuli Avoids joining a social situation when involves encountering specific sensory stimuli Learned to cope in adulthood with some previously intolerable sensory stimuli	I *hated* wearing shoes for the first 5 years or so of my life and always took them off when possible and continued a barefoot preference throughout my teens, although I did cope with shoes throughout my schooling. I have always cut labels from all my clothes, nothing even slightly scratchy—my mother accommodated this as a real need. I have a lifelong difficulty with fake flowery smells. I have often entered a shop and immediately left it because some ghastly perfume is being pumped out and it makes me feel sick; petrol smell does too; laurels in bloom ditto. The seeds in figs, the fuzz on peaches, the peaches, the slime of avocadoes and the different slime of okra, cannot be ignored! The very thought of them can make my gorge rise. Ughhh! The slimy and unpredictable nature of washing-up-water helped crystallise all my reasons for not wanting to be a Brownie—not worth the slime for sure, argh!! However, I discovered in my late 60 s that I can eat very small, thin slices of fresh figs in a good mixed salad without throwing up…so breakthroughs can occur and open-up new possibilities!	*hated*. Strong emphasis: suggests wearing shoes was close to being intolerable in early years Wearing shoes avoided; chooses to be barefoot whenever possible Coped with wearing shoes at school: suggests wearing shoes was a problem she learned to deal with at school when had to Tactile intolerance: cuts labels from clothes due to their 'scratchiness'. Suggests heightened awareness of some stimuli Mother accommodated need to cut out labels. Highlights importance of how others react (can validate/invalidate subjective experience) Specific smells intolerable/ disruptive. Encountering them instigates a flight response. Ghastly: suggests experience of horror at smell Pumped out. Sense of being physically overpowered by fumes leading to feeling physically sick Attuned to fine differences in kinds of slime: Does this indicate heightened awareness of stimuli experienced as problematic?

(continued)

(continued)

Experiential statements	*Original transcript*	*Exploratory notes*
		Certain tactile (e.g. fuzz/ slime) stimuli intrusive/cannot be ignored Even the thought of certain stimuli results in feeling physically sick. Ughhh!: Language conveys a physical shudder in response to washing up water. Unpredictable nature of stimuli also an issue Sensory intolerance given as a reason for not joining Brownies [youth organization for girls]: *Intersection between sensory and social experience.* Argh!!: Again, language conveys physical shudder in response to slime Learned to cope with sensory experiences previously intolerable. Suggests some change over time is possible if worked on

Box 4: Personal experiential themes (PETS)

Theme 1: Subjective sensory experiences (sensorium) central to sense-of-self

Heightened awareness of specific sensory stimuli
Feels physically sick and overwhelmed when encounters specific smells
Some tactile stimuli feel intrusive; they cannot be shut out
Feels sick just thinking about some kinds of tactile sensory stimuli

Theme 2: Intersection of social and sensory experience

Avoids joining a social situation when involves encountering specific sensory stimuli

Felt understood by mother who recognized and accommodated sensory need

Forced to deal with difficult sensory stimuli at school

Theme 3: Experience of sensory stimuli can change over time

Learned to cope in adulthood with some previously intolerable sensory stimuli

Box 5: Short extract from a phenomenology paper to show how it should be written up. Murray et al., 2023.

Subjective sensory experiences (sensorium)

We developed the use of the term 'sensorium' as a mutually agreed way to cover the totality of our subjective sensory experience and processing of the world, both interpersonal and physical. We found that this was a domain in which the differences across our 'spectrum' [encompassing autistic and non-autistic experience] were among the most defined and early emerging. They have much to do with one's sense of developing an implicit awareness of one's own being-in-the-world:

> "I *hated* wearing shoes for the first 5 years or so of my life and always took them off when possible and continued a barefoot preference throughout my teens. I have always cut labels from all clothes, nothing even slightly scratchy – my mother also accommodated this as a real need. (…) Fake flowery smells – I have often entered a shop and immediately left it because some ghastly perfume is being pumped out and it makes me feel sick; petrol smell does too; laurels in bloom, ditto. The seeds in figs, the fuzz on peaches, the slime of avocados and the different slime of okra, cannot be ignored! The very thought of them can make my gorge rise…

We talked about how these acute sensory feelings are directly pertinent to what we termed above the 'setting conditions' for trust. The autistic experience was rife with abrupt discontinuities that often feel incompatible with a sense of trust and which are experienced as direct threats to a sense of self in the world.

Particular Considerations in Using IPA for Autism Research

The phenomenological method is grounded in an effort of imaginative empathy across experiential difference. In an autistic phenomenology, differences across and within 'neurotypes' have been associated with the 'double empathy problem' (Milton, 2012). Whilst this is often represented as differences between neurotypical (NT) and neurodivergent (ND), these problems of cross-neurotype interaction occur between differently neurodivergent people, with the impact perhaps being increased with greater 'distance' between the neurotypes in question. Specific features of IPA, including its commitment to an equality of voice between participant and researcher as well as researcher reflexivity, mean it may have a particularly valuable role to play in making steps towards bridging this 'double empathy' gap (Howard, Katsos & Gibson, 2019).

Equality of voice

IPA subscribes to a model of the participant as an intrinsically sense-making agent (Taylor, 1985). This gives rise to a 'double hermeneutic' in which the researcher is trying to make sense of the participant trying to make sense of what is happening to them (Smith, 2007). By placing the informant at the centre of the research as experts in their own experience and facilitating their meaningful participation in the sense-making process, IPA necessitates a respect for the participant and their perspective (MacLeod, Allan, Lewis & Robertson, 2017).

Reflexivity

Another important step towards bridging the 'double empathy' divide is IPA's requirement for researchers to consider how their own experiences, pre-conceptions and positionality (location in relation to various social identities, e.g. being neurodivergent or neurotypical, gender, ethnicity, age) might shape both interpretation of participants' experiences and choices made about research design and procedures. Such reflexivity acknowledges and foregrounds the potential for misunderstandings to arise between a participant's words and researchers' interpretation of them due to divergent perspectives (Howard et al, 2019). The presence of neurodivergent interviewers may thus also help to narrow this 'distance'.

Co-construction

The neurodiversity paradigm calls for a need to engage autistic people more centrally in co-producing meaningful research that identifies the priorities of the autistic community and ensures methods are respectful, consensual and sensitive to the potential power and experiential differentials involved in such engagement. Whilst IPA positions the participant at the centre of research and requires researcher reflexivity, it is possible to go further in terms of bridging the 'double empathy' gap through taking a more participatory approach. IPA lends itself well to a co-construction paradigm shift.

Methodological Considerations

Sampling

The depth of work required in IPA means that sample sizes are small compared with those of most other methods. However purposive sampling and a stratified cell design afford considerable analytic leverage. For example, one may wish to consider the impact of age and gender on the phenomenon under investigation. Through stratified sampling one creates a series of 4 cells each containing 4 participants and a total sample of 16 (4 younger women, 4 older women, 4 younger men, 4 older men). This then enables multiple levels of analysis, for example comparing 8 males v 8 females and then probing further into the impact of age on gender differences. This is a simple version to give an idea of what is possible. It is of course for the researchers to decide which and how many variables to include in a design. We see 3 or 4 as a useful minimum cell size to allow meaningful group comparisons.

Comparing Autistic and Non-autistic Experiences

In order to identify a phenomenology characteristic of autism we need first to positively identify the qualities of autistic experience, and then establish meaningful differences with non-autistic experience by comparison. Such a 'figure-ground' technique will be valuable in understanding what is specific to autism; and necessary in work towards an updated nosological description of autism which includes the subjective.

Offering Flexible Formats for Data-gathering

Autistic people vary in terms of how they express themselves verbally and their preferred means of communication. Therefore, researchers need to offer a range of possible formats to facilitate inclusion (e.g. offering alternative ways of doing interviews, for example via high-quality web-conferencing, over the phone, via e-mail or face-to-face interviews; see Pellicano et al., 2020). Recent years have seen IPA researchers themselves include a variety of visual methods for data collection, such as drawings and photographs, which serve as a springboard for the conversation between interviewer and interviewee (Shinebourne & Smith, 2011; Kirkham et al., 2015; King, Williams & Gleeson, 2017).

Quality

There is now a very large corpus of IPA research but, as with any methodology, the quality is mixed. There are two useful papers pointing to how to conduct and write up high-quality IPA. Smith (2011) indicates the importance of combining creativity with rigour, so that a strong IPA write-up presents insightful analysis of rich data from participants but also ensures that the methodology is presented clearly and transparently, and that care is taken to sample the corpus reasonably proportionately in the report. Nizza et al. (2021) focus on the hallmarks of high-quality IPA writing and similarly stress the importance of a detailed in-depth analysis of extracts from participants. They also point to the need for the construction of a compelling narrative which presents the evidence and interpretation in a sequence which, therefore, provides a clear and accessible exploration of the important experiential themes. This narrative should also illustrate the different ways in which participants speak to those experiential themes, thus offering a nuanced examination of both convergence and divergence in participants' accounts.

BREADTH: Development of Online Methods at Scale to Capture the Breadth of Autistic Experience

This section outlines general principles of citizen science and examines its capacity to broaden and upscale the IPA method described above for understanding the phenomenology of autism.

Introduction to Citizen Science

As noted above, citizen science describes a growing set of transdisciplinary practices that engage the public in creating scientific knowledge (see Chapters 4, 9, this volume). It is highly variegated, with numerous labels and associated practices, including crowdsourcing, participatory action research, community science and more that can be classified according to degrees of participation from non-researchers. A commonly used model distinguishes between "contributory" citizen science: primarily designed by researchers with some public contributions, e.g. crowdsourcing, and "co-created" citizen science, with researchers and citizens working together on most aspects (Shirk et al., 2012; see Box 6). This creates a continuum, from least to most level of decision-making potential of patients/members of the public. Key benefits of co-created citizen science include engaging participants more fully in research outcomes (Ottinger, 2017) and providing opportunities for the public to contribute to knowledge production, leading to more active scientific citizenship (Irwin, 1995). These benefits are relevant to the challenges of "undone science" (c.f discussion of unknowing in Chapter 22, this volume)—science which, despite being deemed important or necessary, is unfunded or ignored by traditional academic/industrial research (Frickel et al., 2010; see also Chapter 10, this volume). Similarly, community-driven research offers pathways for the inclusion of populations typically underrepresented or missing from data collection, such as indigenous communities, racialized minorities, or people with mental health conditions (The Ada Lovelace Institute, 2021). As community-led or co-created citizen science frequently centres the lived experiences of public contributors or communities, these approaches can help address challenges around 'undone science' in health research, either by formulating research agendas that address shortcomings, or by actively contributing to knowledge generation (Greshake Tzovaras et al., 2021).

Key considerations when planning and conducting citizen science include: Who should I engage with? At which stage? *How* should I engage? Trade-offs vary depending on research topics and methods, with no one-size-fits-all answers. Deep continuous contributor engagement might require smaller-scale research than shallower engagement—e.g. for crowdsourcing data. Furthermore, multiple modes/levels of engagement can be adopted at different stages. For generating a large-scale phenomenology of autism, the 'DEPTH' method can be understood as an in-depth, smaller-scale approach to citizen science. The resulting findings will support further citizen science to generate a wider dataset.

Box 6: Overview of research steps involving public engagement, based on Shirk's (2012) model, annotated by no. using participation method (Hecker et al., 2018). Figure by Kloppenborg, (CCB-Y 4.0).

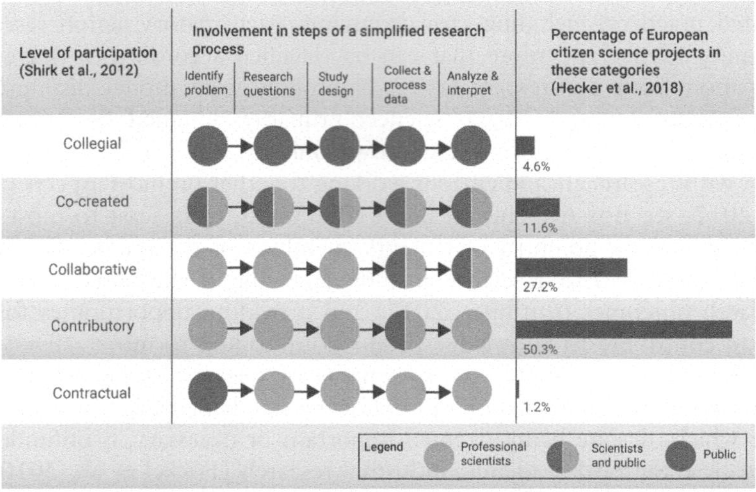

Citizen Science Methods

Taking AutSPACEs as a case study (a citizen science project investigating sensory processing in autism, which employs community engagement throughout; see Box 7), we will highlight how varying methods can be used at different stages to equitably work with research communities. We will further show how this approach can be adapted for a phenomenology of autism.

Research Agenda-Setting: Surveying at Scale

Key decisions occur before research commences—a major one being agenda-setting. As 'undone science' highlights, mismatches between public/societal needs and research agendas can be problematic, leading to failures to target real needs, or even to disrespectful or pejorative studies. Autism research by non-autistic scientists can in this way raise the risk of outputs that increase stigma and discrimination (Fletcher-Watson & Bird, 2020). Community agenda-setting and co-design can mitigate such risks.

Box 7: AutSPACEs' research stages

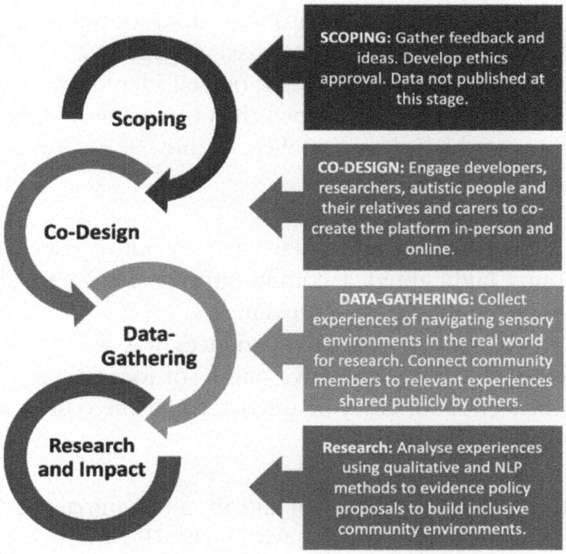

AutSPACEs' research agenda was directly informed by the collaborative prioritization work of Autistica and the James Lind Alliance (JLA) (Autistica, 2016).[1]Core to the JLA's approach are "priority-setting partnerships", in which service-users, patients, carers & clinicians jointly prioritize research areas. The JLA typically collects viewpoints through online & paper surveys, using open-ended questions to encourage in-depth results, though face-to-face meetings are also used (Nygaard et al., 2019). Larger-scale surveys can sample a wider range of opinions, allowing for more representative demographics, however, unidirectional, text-based responses do not allow for deepening or clarifying dialogue. Hence, focus groups and workshops can be useful. Focus groups produce unique data thanks to group interactions/dynamics and a more natural, social context (Wilkinson, 1999). Typically, they last 1–2 hours and include 6–10 participants, who engage in an organized, flexible conversation, facilitated by a moderator who either records the discussion or takes notes (Wilson, 2012).

[1] Autistica is a UK-based research charity that engages in campaigning and funding of research around autism and neurodiversity more broadly. The James Lind Alliance is a UK-based non-profit initiative that is supported by the UK's National Institute for Health and Care Research (NIHR). The JLAs goal is to bring together patients, carers and clinicians to identify and prioritise research efforts.

Operationalizing Priorities and Refining Research Questions: Focus Groups and IPA

For AutSPACEs, in-person focus groups were used for initial scoping (Aitkenhead et al., 2023). Recruitment prioritized engaging a wide demographic across age, gender, whether diagnosed or self-identified autistic, level of support need, etc.). Focus groups shaped the project methodology and participation process (e.g. which data to collect & how?), resulting in two major adjustments:

1. Initially, only sensory experiences from contributors were to be collected. Focus groups highlighted a community preference to also optionally share recommendations for improvements
2. Moderation of publicly shared experiences was identified as a priority. The more open-ended, interactive nature of focus groups means uncertain or ambivalent issues may be uncovered, at the cost of shrinking how many participants can take part.

For research on the phenomenology of autism, much initial operationalization and question-refinement will be done within the IPA framework, outlined in the section on DEPTH above. Subsequent learning can inform adaptions to an existing citizen science platform, such as AutSPACEs, expanding to additional topics identified from in-depth accounts. Focus groups are valuable to bridge the gap between the narrow but deep focus of the IPA, and the broad but comparatively shallow approach of crowdsourced citizen science, to further prioritize and refine research topics.

Co-creation of Research Implementations

Co-creation can use multiple methods, including regular (virtual) community meetings or co-working sessions and distributed collaboration software methods, such as collaborative online text editors, software development tools (e.g. version control), and more. AutSPACEs used multiple methods in concert to develop, implement and iterate research tools. A similar approach can be used to adapt IPA into a broader survey.

Online Community Meetings

Larger groups can be engaged with via recurring synchronous online community meetings using video-conferencing, supporting community members to discuss research topics or make decisions on direction and implementation. For example, it can be used to gather feedback on research outcomes/methods and software design to ensure they are fit for purpose. The agenda can be co-created by researchers and participants, and frequency can vary depending on their needs and interests.

Online Co-working

Whilst community-meetups can be used for decision-making and discussion, online co-working sessions can support collaborative research. Co-working sessions were used in AutSPACEs to co-create moderation policies, culminating in a jointly written academic paper by autistic volunteers and core research team members, describing moderation dilemmas and resulting strategies (Aitkenhead et al., 2023).

Bringing it Together

As discussed, citizen science takes multiple forms. The best approach depends on features such as research, stage, type and other contextual factors. Projects need not limit themselves to a single mode of engagement: methodologies can be adapted and iterated. After initial survey-based priority-setting, AutSPACEs transitioned to small focus groups, before engaging simultaneously in highly open participation—via. community meet-ups and open-source development—and smaller, focused co-working and user testing. All aimed to deploy a citizen science platform for large-scale data collection. Regardless of method, participatory approaches have the potential to improve research outcomes and ensure research topics align with the needs and priorities of affected populations. These methods will coalesce to co-create a research tool to collect phenomenologically informed data at scale. The adapted AutSPACEs platform, informed by in-depth IPA findings, can test idiographic phenomenology, as generated from a small in-depth dataset. Comparisons of experiences across a range of autistic perspectives are possible, reflecting greater diversity. This approach can also uncover sub-groupings of autistic experiences across different strata, such as age, gender, ethnicity, level of support need, co-occurring conditions or other intersectionalities.

REACH: Innovative Methods for Representing Communication Diversities and Maximizing Autistic People's Participation

Developing an autistic phenomenology that represents the diversity of autistic experience requires inclusive methods to maximize participation and include a diversity of voices (Harrington et al., 2014). This would include people with learning disabilities (LD) and those with communication and complex needs (CCN). This is a complex task in view of the challenges presented by the double empathy problem and the potential impact on how 'different' the researcher and participant are to each other. Much depends on the researcher's ability to manage that difference, prioritizing the participant's needs and preferences rather than their own (see Chapters 13, 20, 21, this volume). Having autistic researchers as part of the team (and actively engaged in data collection) is helpful to an inclusive process attuned to autistic diversity but researchers need to stay alert to their perspectives and sensitive to those of others as they negotiate the complexities of communication differences between and within neurotypes. Research approaches similarly need to be attuned to the individual and their expressive capacities and preferences (e.g. physical, embodied communication, Van Goidsenhoven & De Schauer, 2020). Creative research methods are gaining traction in these contexts, using participatory arts to engage with the experiences of hard-to-reach groups who might otherwise be excluded from research. Such methods can be neuroaffirmative ways of exploring autistic interests and communication. This might involve the use of creative media (drawing, visual art, music, drama, movement, puppetry/objects) and sensory materials to engage with these populations (Ockelford, 2016; Trimingham & Shaughnessy, 2016). Media technologies are also effective in engaging autistic people with CCN, facilitating communication and phenomenological enquiry through interactions with microphones and motion capture (Frauenberger et al., 2011, Keay-Bright, 2023).

Studying autistic phenomenology requires not only approaches that can encompass the viewpoints and internal experiences of autistic people with CCN, but also those that are applicable to early development. Many autistic children begin to show externally recognized behavioural differences in toddlerhood, making it likely that experiential differences are also present from a young age. Early in development, language skills are not only still nascent, but young children have limited skills in recognizing and expressing their thoughts and emotions. If an autistic phenomenology is to be applied to the earliest stages at which autistic children may diverge in their experiences, we require innovative ways to capture this (Bulgarelli & Jones, 2023).

New methods may help us more directly measure a child's engagement and interactions with their environment, and the effects of this on their subjective experiences. Recording a child's actual experienced environment at home or school with wearable technologies could provide rich insights into how their experience is likely to be shaped by external influences on what they encounter, and internal influences on what they choose to engage with. Combining creative practices with neuroscience (e.g. measuring a child's arousal in a play-based sensory environment) offers new interdisciplinary ways of investigating and representing a more diverse range of autistic voices in phenomenological research. We have suggested a practical context within which these aspects of information could be used to enable or infer subjective experience in autistic people (and indeed others) who have CCN. Such methods will be central to a more inclusive understanding of the range of autistic being-in-the-world, sensitive to cognitive and sensory processing and diverse communication preferences.

Transformation: Synthesis of Methods and Findings Towards a Systematic Phenomenology of Autism

The final stage of the inquiry will be a process of integration from these diverse perspectives and varieties of information, into a narrative inference about autistic phenomenology. This inference or set of inferences will be co-constructed within the diverse team involved and also with iterative consultation with other groups. We will use structured group discussion processes, creative facilitation exercises and interdisciplinary analytic models to support and justify the inferences as it is crucial that the best inferences are arrived at that are transparent in their processing and convincing to the wider autistic, clinical and scientific community.

To What End?

To what end do we understand this process of developing a phenomenology of autism? To be sure at one level there is a total benefit *in itself* which is to construct accounts of autistic experiences which give coherence and voice to experienced autistic identities. We do not of course expect a completely simple narrative. This is likely to be complex and heterogeneous with (as elsewhere in the field) unclear boundaries between the autistic and non-autistic states. But at the very least we would hope that the comparative method we advocate would help mutual understanding across this notional divide—and towards the ability to exercise moral imagination and empathy towards each other across it (Bervoets, 2022).

Secondly, we would expect that in identifying key dimensions of autistic experiences in this way we would provide a radically different perspective on which to base scientific investigation into the nature and origins of autistic development; being optimistic, to help unlock some of the difficulties and recursive cul-de-sacs in current scientific inquiry—thus for instance to provide insights into where to direct attention in neuroscientific investigation.

Thirdly, we might expect this inquiry would be able to have a profound impact on clinical understanding and the diagnostic criteria for autistic states (Chapman, 2019).[2] We would not expect the phenomenological aspect to replace the current behavioural criteria but rather to complement and illuminate them, just as it does in other classifications within mental life. Indeed, we have argued elsewhere (Green & Shaughnessy, 2023, Murray et al. 2022) that phenomenology is likely to help us reconsider behavioural criteria as to how much certain behavioural features are core features of autistic states and how much secondary manifestations of lived autism in certain environments. For instance, the phenomenological emphasis on differences in sensory experiences (the sensorium) as a primary experiential feature distinguishing

[2] This refers to Robert Chapman's concept of 'states of being' in *Autism as a Form of Life* (2019).

autistic and non-autistic states, can help illuminate how 'behavioural' aspects of social avoidance, social understanding or social behavioural difference are downstream consequences of living with sensorial differences in complex, or neurotypically orientated, environments.

Development Sequencing and Inference

This implication of developmental sequencing raises other issues for our inquiry. Since a central issue in interpreting subjective statements about autistic phenomenology will be its developmental nature. Does what one is tapping into constitute a primary aspect of autistic difference or the developmental outcome of a cascade of self-environment interactions over time? This is going to be a complex issue to work out, but one that epidemiology in other areas has addressed methodologically. One beginning will be the careful reconstruction of developmental experience in a sequenced fashion, for instance using large datasets classified by age, and this will begin to help the discrimination of early experiences from later secondary ones. An example of the potential power of this approach is suggested by evidence from reported phenomenology in early development (Murray et al., 2023), plus results from early intervention science (Green, 2022), and supported by some theoretical literature. These suggest that some at least of the social interaction differences and impacts described currently within the autistic behavioural phenotype are in fact *secondarily emergent* rather than primary, resulting from transactions between neurodivergent individuals and their environment in the early years. A developmental approach of this kind can be applied to any neurotype. Here it may be highly illuminating for how we understand the development of autistic lives and indeed to how we describe autism itself.

Conclusion

Our group's combination of leading expertise in, and experience across, neurodivergence and neurotypicality, clinical and research science, creative practices, medical humanities, sociology, education and computer science develops the interdisciplinary route towards our overarching aim: a representative empirical phenomenology of 'autism'.

The diversity associated with the autistic population has created challenges for nosology, theory and the development of structures for support and care. This chapter has identified the need for new resources for the characterization and conceptualization of the phenomenology of autism. But a similar approach could be extended to investigate other experiential differences within and across neurodivergencies more generally (for instance to establish a phenomenology of ADHD or AuDHD). This will increase interdisciplinary understanding of how autism is experienced and can ultimately be used to inform training and approaches to education, health and social care as well as enhancing awareness and understanding that could be transformative for the autistic community.

Integrating phenomenology and lived experience into a new conceptualization of autism in this way promises major benefits. It would (i) contribute to an increased mutual empathy and common language between NT professionals and the autistic community; (ii) provide crucial new foci for research through aspects of autistic experiences previously neglected, and more detailed and deeper information to guide scientific investigation of the neural system underpinnings of these individual differences; (iii) add richness and utility to clinical phenotyping and diagnostic definitions; and (iv) help identify new care pathways and supportive innovations for healthcare and education. To integrate autistic experiences into the autism concept in a systematic and representative way is novel and ambitious, responding to wider calls for a paradigm shift to rethink the way in which autism science is framed and conducted (Milton, 2014; Pellicano & den Houting, 2022; Green, 2022).

References

Aitkenhead, G., Fantoni, S., Scott, J., Batchelor, S., Duncan, H., Llewllyn-Jones, D., Greshake Tzovaras, B., Taylor, R., Stoffel, M., Smith, O., & Whitaker, K. (2023). How to co-create content moderation policies: The case of the autspaces project. *SocArXiv:* https://doi.org/10.31235/osf.io/c2xe7

American Psychiatric Association, DSM-5 Task Force. (2013). *Diagnostic and statistical manual of mental disorders: DSM-5*™(5th ed.). American Psychiatric Publishing. https://doi.org/10.1176/appi.books.9780890425596

Autistica. (2016). *Your questions: Shaping future autism research.* https://www.autistica.org.uk/downloads/files/Autism-Top-10-Your-Priorities-for-Autism-Research.pdf

Beadle-Brown, J., Wilkinson, D., Richardson, L., Shaughnessy, N., Trimingham, M., Leigh, J., Whelton, B., & Himmerich, J. (2017). Imagining Autism: Feasibility of a drama-based intervention on the social, communicative and imaginative behaviour of children with autism. *Autism, 22*(8), 915–927.

Bervoets, J. (2022). Neurogradualism: Neurodiversity without categorical difference, a case study of autism. *InterCultural Philosophy*, (1), 65–76. https://doi.org/10.11588/icp.2022.1.90132

Boldsen, S. (2022). Autism and the sensory disruption of social experience. *Frontiers in Psychology, 13*, 874268. https://doi.org/10.3389/fpsyg.2022.874268

Buckle, K. L., Leadbitter, K., Poliakoff, E., & Gowen, E. (2021). No way out except from external intervention: First-hand accounts of autistic inertia. *Frontiers in Psychology, 13*(12), 631596.

Bulgarelli, C., & Jones, E. J. (2023). The typical and atypical development of empathy: How big is the gap from lab to field? *JCPP Advances, 3*(1), e12136.

Chapman, R. (2019). Autism as a form of life: Wittgenstein and the psychological coherence of autism. *Metaphilosophy, 50*(4), 421–440.

Chown, N., Robinson, J., Beardon, L., Downing, J., Hughes, L., Leatherland, J., & MacGregor, D. (2017). Improving research about us, with us: A draft framework for inclusive autism research. *Disability and Society, 32*(5), 720–734.

Dinishak, J. (2016). The deficit view and its critics. *Disability Studies Quarterly, 36*(4). https://doi.org/10.18061/dsq.v36i4.5236.

Dugdale, A. S., Thompson, A. R., Leedham, A., Beail, N., & Freeth, M. (2021). Intense connection and love: The experiences of autistic mothers. *Autism: the international Journal of Research and Practice, 25*(7), 1973–1984. https://doi.org/10.1177/13623613211005987

Fletcher-Watson, S., & Bird, G. (2020). Autism and empathy: What are the real links? *Autism, 24*(1), 3–6. https://doi.org/10.1177/1362361319883506

Fletcher-Watson, S., Adams, J., Brook, K., Charman, T., Crane, L., Cusack, J., Leekam, S., Milton, D., Parr, J. R., & Pellicano, E. (2019). Making the future together: Shaping autism research through meaningful participation. *Autism, 23*(4), 943–953.

Frickel, S., Gibbon, S., Howard, J., Kempner, J., Ottinger, G., & Hess, D. J. (2010). Undone science: Charting social movement and civil society challenges to research agenda setting. *Science, Technology, and Human Values, 35*(4), 444–473. JSTOR.

Gillespie-Lynch, K., Kapp, S. K., Brooks, P. J., Pickens, J., & Schwartzman, B. (2017). Whose expertise is it? Evidence for autistic adults as critical autism experts. *Frontiers in Psychology, 8*, 438.

Green, J. (2022). Autism as emergent and transactional. *Front. Psych.* 13. https://doi.org/10.3389/fpsyt.2022.988755

Green, J. & Shaughnessy, N. (2023). Autistic Phenomenology past, present, and potential future**.** *Front. Psychol.Sec. Psychopathology, 14.* https://doi.org/10.3389/fpsyg.2023.1287209

Greshake Tzovaras, B., Rera, M., Wintermute, E. H., Kloppenborg, K., Ferry-Danini, J., Aidelberg, G., Aronoff, R., Lindner, A., & Misevic, D. (2021). Empowering grassroots innovation to accelerate biomedical research. *PLOS Biology, 19*(8), e3001349. https://doi.org/10.1371/journal.pbio.3001349

Harrington, C., Foster, M., Rodger, S., & Ashburner, J. (2014). Engaging young people with autism spectrum disorder in research interviews. *British Journal of Learning Disabilities, 42*, 153–161. https://doi.org/10.1111/bld.12037

Hecker, S., Hacklay, M., Bowser, A., Makuch, Z., Vogel, J., Bonne, A., & Moedos, C. (2018). *Citizen Science: Innovation in Open Science, Society and Policy*. https://www.uclpress.co.uk/products/107613

Howard, K., Katsos, N., & Gibson, J. (2019). Using interpretative phenomenological analysis in autism research. *Autism: The International Journal of Research and Practice, 23*(7), 1871–1876. https://doi.org/10.1177/1362361318823902

Irwin, A. (1995). *Citizen science a study of people, expertise and sustainable development* (1st ed.). Routledge.

Kanner,. (1943). Autistic disturbances of affective contact. *Nervous Child, 2*, 217–250.

Keay-Bright, W. (2023). Amplifying ability: Engaging young people with autism spectrum disorders through gesture, movement and sound technologies. In E. Brooks, & D. J. Brown (Eds.) *Virtual reality games for rehabilitation. Virtual Reality Technologies for Health and Clinical Applications*. Springer.

Kirkham, J. A., Smith, J. A., & Havsteen-Franklin, D. (2015). Painting pain: An interpretative phenomenological analysis of representations of living with chronic pain. *Health Psychology, 34*(4), 398–406. https://doi.org/10.1037/hea0000139

Kloppenborg, K., Ball, M. P., & Greshake Tzovaras, B. (2021). A Peer Production Model for Citizen Science: Comparative Analysis of Three Online Platforms. *SocArXiv*. https://doi.org/10.31235/osf.io/rw58y.

Kyriacou, C., Forrester-Jones, R., & Triantafyllopoulou, P. (2023). Clothes, sensory experiences and autism: Is wearing the right fabric important? *Journal of Autism and Developmental Disorders, 53*(4), 1495–1508. https://doi.org/10.1007/s10803-021-05140-3

MacLeod, A., Allan, J., Lewis, A., & Robertson, C. (2017). 'Here I come again': The cost of success for higher education students diagnosed with autism. *International Journal of Inclusive Education, 1–15*. Advance online publication. https://doi.org/10.1080/13603116.2017.1396502

Martinez, A. M. (2019). The promises and perils of automated facial action coding in studying children's emotions. *Developmental Psychology, 55*(9), 1965.

McDonnell, A., & Milton, D. (2014). Going with the flow: Reconsidering 'repetitive behaviour'through the concept of 'flow states'. In G. Jones, & E. Hurley (Eds.) GAP: *Good Autism Practice: Autism, Happiness and Wellbeing* (38–47). BILD publications.

Milton, D. (2012). On the ontological status of autism: The 'double empathy problem.' *Disability and Society, 27*(6), 883–887.

Milton, D. E. (2014). Autistic expertise: A critical reflection on the production of knowledge in autism studies. *Autism, 18*(7), 794–802. https://doi.org/10.1177/1362361314525281

Murray, D., Milton, D., Green, J., & Bervoets, J. (2023). The human spectrum: A phenomenological enquiry within neurodiversity. *Psychopathology, 56*(3), 220–230.

Murray D, Lesser M, Lawson W. (2005). Attention, monotropism and the diagnostic criteria for autism. *Autism*. 9(2):139–56. https:// https://doi.org/10.1177/1362361305051398.

Nizza, I. E., Farr, J., & Smith, J. A. (2021). Achieving excellence in interpretative phenomenological analysis (IPA): Four markers of high quality. *Qualitative Research in Psychology, 18*(3), 369–386. https://doi.org/10.1080/14780887.2020.1854404

Nygaard, A., Halvorsrud, L., Linnerud, S., Grov, E. K., & Bergland, A. (2019). The James Lind Alliance process approach: Scoping review. *British Medical Journal Open, 9*(8), e027473. https://doi.org/10.1136/bmjopen-2018-027473

Ockelford, A. (2016). The sounds of intent project: modelling musical development in children with learning difficulties, *Tizard Learning Disability Review, 20*(4), 179–194.

Ottinger, G. (2017). Reconstructing or reproducing? In *The Routledge handbook of the political economy of science* (1st ed.). Routledge.

Pearson, A., & Rose, K. (2023). *Autistic masking: Understanding identity management and the role of stigma*. Pavilion Publishing.

Pellicano, E., Lawson, W., Hall, G., Mahony, J., Lilley, R., Davis, C., Arnold, S., Trollor, J., Yudell, M. (2020). Documenting the untold histories of late-diagnosed autistic adults: A qualitative study protocol using oral history methodology. *BMJ Open, 10*(5). https://doi.org/10.1136/bmjopen-2020-037968

Pellicano, E., & den Houting, J. (2022). Annual research review: Shifting from 'normal science' to neurodiversity in autism science. *Journal of Child Psychology and Psychiatry, 63*, 381–396. https://doi.org/10.1111/jcpp.13534

Shinebourne, P., & Smith, J. A. (2011). Images of addiction and recovery: An interpretative phenomenological analysis of the experience of addiction and recovery as expressed in visual images. *Drugs: Education, Prevention and Policy*, 18 (5), 313–322 https://doi.org/10.3109/09687637.2010.514621

Shirk, J. L., Ballard, H. L., Wilderman, C. C., Phillips, T., Wiggins, A., Jordan, R., McCallie, E., Minarchek, M., Lewenstein, B. V., Krasny, M. E., & Bonney, R. (2012). Public Participation in scientific research. *Ecology and Society*, *17*(2). JSTOR. http://www.jstor.org/stable/26269051

Sinclair, J. (1993). Don't mourn for us. *Our Voice, 1*(3). Retrieved from http://www.autreat.com/dont_mourn.html.

Smith, J. A. (2011). Valuating the contribution of interpretative phenomenological analysis. *Health Psychology Review, 5*(1), 9–27. https://doi.org/10.1080/17437199.2010.510659

Smith, J. A., Flowers, P. & Larkin, M. (2022). Interpretative phenomenological analysis: Theory, *Method and Research.* Sage

Smith, J. A., & Nizza, I. E. (2022). Essentials of interpretative phenomenological analysis. *American Psychological Association.* https://doi.org/10.1037/0000259-000

Taylor, C. (1985). Self-interpreting animals. *W: Philosophical Papers 1: Human Agency and Language (s. 45–76).* Cambridge: Cambridge University Press.

Taels, L., Feyaerts, J., Lizon, M., De Smet, M., & Vanheule, S. (2023). I felt like my senses were under attack: An interpretative phenomenological analysis of experiences of hypersensitivity in autistic individuals. *Autism, 27*(8), 2269–2280. https://doi.org/10.1177/13623613231158182

The Ada Lovelace Institute. (2021). *Participatory data stewardship.* https://www.adalovelaceinstitute.org/report/participatory-data-stewardship/#:~:text=The%20report%20provides%20evidence%20that,can%20contribute%20towards%20increased%20public

Trimingham, M., & Shaughnessy, N. (2016). Material voices: Intermediality and autism. *Research in Drama Education: THe Journal of Applied Theatre and Performance, 21*(3), 293–308. https://doi.org/10.1080/13569783.2016.1195121

Van Goidsenhoven, L., & De Schauwer, E. (2020). Listening beyond words: Swinging together. *Scandinavian Journal of Disability Research*, *22*(1), 330–339.https://doi.org/10.16993/sjdr.756

Venter, F., Morelli, J., & Erasmus, E. (2023). Understanding the lived music listening experiences of adults on the autism spectrum. *Psychology of Music, 51*(3), 971–985. https://doi.org/10.1177/03057356221126201

Ward, J., Richardson, G. O., Hunter, K. Sensing interpersonal synchrony between actors and autistic children in theatre using wrist-worn Accelerometers. In Proceedings of the 2018 ACM international symposium on wearable computers (ISWC '18). *Association for Computing Machinery* (pp. 148–155). https://doi.org/10.1145/3267242.3267263

Wilkinson, S. (1999). Focus groups. *Psychology of Women Quarterly, 23*(2), 221–244. https://doi.org/10.1111/j.1471-6402.1999.tb00355.x

Wilson, V. (2012). Research methods: Focus groups. *Evidence Based Library and Information Practice*, *7*(1), 129–131. https://doi.org/10.18438/B8K32T

Woods, R., Milton, D., Arnold, L., & Graby, S. (2018). Redefining critical autism Studies: A more inclusive interpretation. *Disability and Society, 33*(6), 974–979. https://doi.org/10.1080/09687599.2018.1454380

Yergeau, M. (2018). *Authoring autism: On rhetoric and neurological queerness*. Duke University Press.

CHAPTER 15

A Phenomenological Methodology for Inclusively Researching the Views and Experiences of Autistic People with Profound Learning Disabilities

Ned Redmore

Introduction

Autistic people with profound learning disabilities have historically been situated as 'voiceless subjects' (Mietola et al., 2017, p. 264), understood as incapable of contributing their own perspectives and widely excluded from representing themselves within research or their related academic fields (Redmore, 2023c). The tools with which to conduct research with people from this group are underdeveloped (Simmons & Watson, 2014, p. 148) and the ways in which they can be ethically and meaningfully included in research continues to be an imperative question within inclusive research fields (Gjermestad et al., 2022).

'Autistic people with profound learning disabilities' is a term I use to refer to people with a diagnosis of autism and severe/profound learning disability, autism and complex or high support needs, or severe/profound learning disability (but not autism). As I have argued (Redmore, 2023c), people with these diagnoses share characteristics, needs, ways of being and life circumstances, and separating them in research according to diagnostic labels is unjustified, exacerbating their marginalisation from research. I adopt identity-first language in reference to autism as it is contended that it reduces stigma (Bottema-Beutel et al., 2020), though, in contrast, use person-first language

N. Redmore (✉)
Faculty of Wellbeing, Education and Language Studies, The Open University, Milton Keynes, UK
e-mail: ned.redmore@open.ac.uk

H. Bertilsdotter Rosqvist and D. Jackson-Perry (eds.), *The Palgrave Handbook of Research Methods and Ethics in Neurodiversity Studies*,
https://doi.org/10.1007/978-3-031-66127-3_15

in relation to learning disabilities, as it is the term most commonly adopted in research conducted by, or in partnership with, people with learning disabilities (e.g., Brownlee-Chapman et al., 2018).

Although autistic people with profound learning disabilities are often defined through deficiencies and incapacities, they are a heterogeneous group with diverse competencies and communicative abilities (Vehmas & Mietola, 2021). The term 'non-verbal' pervades the way this group's communication is described (e.g., Koegel et al., 2020), yet their communication is complex, including vocalisations, body language, facial expressions, 'emotional and sensory expressions' (Gjermestad et al., 2022, p. 11), and limited use of Makaton[1] and picture/visual communication. To those close to them, family members, support staff, friends and peers, they regularly express their views, characters and experiences, which, over time, develops dialogues and shared understandings (Nind, 2018).

Exploring the perspectives and experiences of autistic people with profound learning disabilities requires interpretation, presenting risks and challenges relating to the authenticity of voice. This group have had their ways of being rebuked by researchers (mis)interpreting their psychologies, experiences and values (Vehmas & Curtis, 2017), though this does not remove the significant need to increase their participation in research (Redmore, 2023c). The epistemic risk of misinterpretation, argue Vehmas and Mietola (2021, p. 33), should not suppress the moral obligation to 'represent these voices'.

Research methods to involve autistic people with profound learning disabilities have thus developed, now including collage and narrative diaries (Ridout, 2016), talking mats (Stewart et al., 2018), multi-modal approaches (Doak, 2018), photo-voice (Cluley, 2017), close participatory observation (Gaudion, 2015; Goode, 1994; Mietola et al., 2017; Simmons & Watson, 2014) and life-story work (McCormack, 2017). Such methods have created dialogue and meaning-making opportunities between researchers, autistic people with profound learning disabilities and their wider communities, yet their representation remains marginal within the broader learning disability and autism fields (Palmer & Walmsley, 2020; Redmore, 2023c).

Paradoxically, this marginal position is highlighted by frameworks for inclusive autism research (Chown et al., 2017) and learning disability research (Walmsley & Johnson, 2003) which exclude autistic people with profound learning disabilities through requirements of them being leaders, owners or instigators of research (Nind, 2013). Outside of these specific frameworks, inclusive research is also used as an umbrella term that covers several approaches involving leading and active roles for community members, such as emancipatory or participatory research. However, people with learning disabilities have voiced a preference that *all* such approaches be termed inclusive as this has a clearer statement of intent (Seale et al., 2015, p. 488). Adopting the

[1] Makaton is form of sign language that uses straightforward and figurative signs to communicate key content or meaning.

term 'inclusive' over other terminologies, as I will do in this chapter, can thus be seen as an initial step in developing accessible research opportunities (c.f. maximising participation in chapters 5, 14, this volume) for autistic people with learning disabilities.

The exclusionary nature of the learning disability framework led Nind (2019) to turn towards a more expansive idea of 'doing research inclusively'. At the heart of doing research inclusively is the concept of ***dialogue***; the ethical need for researchers to develop shared understandings with autistic people with profound learning disabilities through the ways in which they communicate, express themselves and participate in the world. Others have added to this (e.g., de Haas et al., 2022; Gjermestad et al., 2022), and a number of elements for researchers seeking to do research inclusively now exist.

This chapter presents a phenomenological methodology which, it is argued, incorporates several elements of doing research inclusively (see also Chapter 14, this volume). The purpose of this methodology is to explore the experiences of autistic people with profound learning disabilities, so that issues or topics can be researched in ways relevant to their perspectives. The chapter is presented as follows:

1. An overview of elements of doing research inclusively with autistic people with profound learning disabilities.
2. A discussion of phenomenology and the framework of Lifeworld Fractions (Ashworth, 2016), a tool for analysing experiences.
3. An overview of thinking phenomenologically while doing phenomenology (Berndtsson et al., 2007), a process to introduce theories and methods relevant to participants' experiences.
4. A reflection on how I shared perspectives with Ben, an autistic adult with profound learning disabilities, about how I could research inclusively with him.
5. The methodology illustrated through a research study into service culture (Redmore, 2023a) involving Ben.
6. A discussion of how the methodology promotes the involvement of autistic people with profound learning disabilities, as well as its limitations.

Doing Research Inclusively

A key notion of inclusive research is meaningful participation for those people with whom the research is concerned: 'Nothing About Us Without Us' (Charlton, 2000, p. 14) (c.f. Chapters 4, 8, 10, 19, 21, this volume). For autistic people with profound learning disabilities, 'being-with' is central to meaningful research participation, referring to researchers embedding themselves in the day-to-day lives of people from this group to create dialogue, share experiences and develop understandings of their lives and contexts (de

Haas et al., 2022; Mietola et al., 2017; Morris, 2003). It is a process which, over time, can enable researchers to challenge the etic perspective, that of the clinical outsider, and gain the emic perspective, that of the attentive insider (Gillman et al., 1997; Goode, 1992).

Taking 'being-with' as a central starting point for doing research inclusively enables other key elements to follow. *Dialogue* and trusting relationships can begin to form (Nind, 2018) as researcher, participants and those around them begin to learn of and from one another. This requires an openness to all forms of communication, expression and engagement and the minutiae of the contexts in which they occur: the people, the spatial and temporal details, the activities, the sensory landscapes (Simmons & Watson, 2014, pp. 155–156). In these situations, voice and its meanings can be seen as mediated, dynamic and relational, suggest Skarsaune et al. (2021) and Gjermestad et al. (2022), both drawing from Teachman et al. (2018). Similarly, Van Goidsenhoven and De Schauwer (2020) refer to 'voice without subject', something that emerges through the 'intra-action' between researchers, autistic people with profound learning disabilities and those people embedded in their day-to-day lives. No matter the terminology, required is a commitment to developing meaning and understanding through an open, measured and time-committed process.

An openness to all forms of communication necessitates an openness to a broad array of methods and methodologies. This is an element of the inclusive autism and learning disability fields more broadly (Fletcher-Watson et al., 2019; Seale et al., 2015), a need for 'flexibility, responsivity and creativity' (Brownlee-Chapman et al., 2018, p. 891). As researchers learn how autistic people with profound learning disabilities express experiences and views within their day-to-day contexts, so researchers must adapt our approaches accordingly. For example, Simmons' and Watson (2014) drew from multiple methods, theories and methodologies to explore the lifeworld of Sam, a person with profound learning disabilities. Adapting and reimagining traditional research approaches can extend and enhance dialogue, encouraging participation that otherwise may be inaccessible (Nind & Vinha, 2014).

Working with autistic people with profound learning disabilities, researchers are at risk of elevating their own perspectives without consideration for their positionality and preconceptions. Some researchers seek to tackle this issue by consulting with those close to participants, family members, support staff and so on (e.g., Simmons & Watson, 2014; Van Goidsenhoven & De Schauwer, 2020), reflecting on interpretations of communication and experience with those who've developed understanding over many years. In such cases, the interpretation becomes shared and mediated across research processes: note-taking, fieldwork, analysis, dissemination and so on. Interrogating preconceptions and interpretations can also be methodological in nature, Andrews et al. (2019) for example utilise a framework to redirect interpreters to aspects of experience they may not have yet considered. Such processes can support those close to autistic people with profound learning disabilities, as well as researchers, to challenge embedded assumptions and bring about new interpretations.

It is argued that this chapter's methodology enables, to greater or lesser extents, the described ways of doing research inclusively. This methodology uses the phenomenological framework, lifeworld fractions (Ashworth, 2016) and draws from the approach 'thinking phenomenologically while doing phenomenology' (Berndtsson et al., 2007), both of which are described below.

Phenomenology and Lifeworld Fractions

Phenomenology is concerned with the study of experience, its structures and meanings. It emanated from philosophers such as Merleau-Ponty and Husserl who sought to challenge the positivist perspective that dominated science, arguing that positivism was dishonest (Merleau-Ponty, 1962, p. 11), taking for granted the 'subjective element' that guides 'everything in the world we call organic life' (Husserl, 1970, p. 6). A phenomenon is a thing or event, and a phenomenological approach aims to understand the meaning of a phenomenon through how we experience it.

A key phenomenological concept is that of the lifeworld, referring to 'a person's subjective construction of reality' (Kraus, 2015, p. 4). It is the taken-for-granted nature of our daily lives (Finlay, 2011) in which our experiences are embedded with meaning though rarely reflected upon (Ashworth, 2016). Across phenomenology, the lifeworld is considered the structure within which the world appears for any given person (Ashworth, 2006).

Drawing from a range of theory, Ashworth (2006, 2016) defined the lifeworld through eight constituent elements termed fractions. All eight fractions are felt in any experience, some stronger than others. Considering how a person experiences a phenomenon through the scope of the fractions can reveal significant and complex aspects of their lifeworld in a fully formed and expansive manner. To describe the fractions, the following refers to the phenomenon of speaking to a doctor about a minor ailment, a cough:

- Selfhood—How do we experience our identity? Empowered, or a self that lacks autonomy? Disclosing a cough to a Doctor, I may experience myself as a fraud. I am not ill enough, and I fear judgement.
- Sociality—How others affect our experiences, and how we experience relationships. I am inferior to the Doctor. They cast decisions on what I am. How they act and attend to me affects how I experience myself.
- Embodiment—The way we experience our bodies. Do we feel strong or incapable? We may experience aspects of gender, age, disability or ethnicity. My body feels out of place when disclosing a cough, it is not ill enough. It tenses, and a nervous energy grows inside.
- Spatiality—The meaning of space and place. Space is embedded with customs and practices; they feel welcoming or unfamiliar. A Doctor's

office is strange, exacerbated by the ambivalent way they face half towards me, half towards their computer.
- Temporality—How the passing of time is experienced. Our biographies may challenge us or provide a welcomed sense of nostalgia. We experience our histories and our futures. Sitting in a waiting room feels slow, while speaking to the Doctor flashes by. I am confronted by a moment from my childhood.
- Discourse—The discourses we draw from. The way our actions and expressions are embedded in social and cultural contexts. Talking to a Doctor, I perform and exaggerate my cough to conform to the conventions of illness.
- Moodedness (mood as atmosphere)—The mood of a situation. The experience and feeling of an atmosphere. I experience a cloud of dread when sitting in a doctor's waiting room, though the cloud lifts as I disclose my ailment.
- Project—The pursuit of what we care for. A situation may make us feel enabled or challenged, proud or angry. Speaking to a doctor, I want to see my cough treated. A caring response elicits comfort and reassurance, and I am not embarrassed for disclosing such an illness.

The fractions prompt people to look more closely at how they have previously interpreted experience, to consider experience in a complex and open way, and to reframe prior assumptions and bring about new interpretations. Andrews et al. (2019) demonstrated this when interviewing parents about the developing sexualities of their autistic children, using the fractions to support them to explore their children's experiences from fresh perspectives. The fractions thus prompt even those with years of experience of engaging with autistic people with profound learning disabilities to look more closely at how they have previously interpreted experience and so, in this sense, promote elements of doing research inclusively.

Thinking Phenomenologically While Doing Phenomenology

A limiting factor in the approach of Andrews et al. (2019) and others that have used fractions (e.g., Finlay & Molano-Fisher, 2008) is that they are presented as a tool to guide interviews and data analysis, which has limited use when researching with autistic people with profound learning disabilities. To value all forms of communication, expression and experience, and to be able to adapt the research approach accordingly, it is suggested that the fractions be incorporated into the process of 'Thinking phenomenologically while doing phenomenology' (Berndtsson et al., 2007, p. 256). This is a theory-led approach that asks researchers to be guided by how people experience a phenomenon so that they can explore it with relevant methods and theoretical perspectives. It involves longitudinal research with people in their everyday lives, adopting methods apposite to how they communicate their experiences, and introducing theory and adapting foci relevant to their experiences. Theory, argue Vehmas and Mietola (2021, p. 55), enables researchers to 'see beyond the obvious and push our understanding of the lives' of autistic people with profound learning disabilities, and so with consideration for its broader approach, this phenomenological-led methodology has potential for doing research inclusively with autistic people with profound learning disabilities.

The inclusive promise of thinking phenomenologically while doing phenomenology is, however, negated by its inaccessibility to researchers without an established phenomenological knowledge base, a challenge only exacerbated by the complex and difficult to access writing so often found in phenomenological works (Dahlberg & Dahlberg, 2019). Furthermore, the approach does not provide a framework to challenge a researcher's assumptions or direct them to alternative, unconsidered areas of experience and so risks elevating their understandings and perspectives. However, both of these challenges can be addressed by using fractions within the approach. As discussed, they help researchers to reconsider, reframe and open up interpretations of experience. The fractions are also a framework based in a broad range of theory including phenomenology, discourse theory (e.g., Harré & Gillett, 1994) and human geography (e.g., Seamon, 1979), and so, as Dahlberg and Dahlberg (2019, p. 5) note, represent a directory of theories relevant to the exploration of experience. In essence, the fractions can act as signposts to support researchers to explore theories, methodologies and approaches relevant to a person's lifeworld and day-to-day context. Utilised within the approach of Berndtsson et al. (2007), the fractions can thus enable researchers with different levels of academic experience to promote the elements of doing research inclusively when working with autistic people with profound learning disabilities. An example of this process is detailed below.

Ethics

The methodology is described below by drawing from a study in which I explored day-service cultures from the perspectives of autistic people with profound learning disabilities (Redmore, 2023b). The study was granted a favourable ethical opinion by The Coventry and Warwickshire Social Care Research Ethics Committee (18/WM/0322). All names are pseudonyms.

Sharing Perspectives Across Intellectual and Communicative Capacities

The study initially involved a key participant, Ben, a young autistic adult who attended a day-service three days a week. He primarily communicated through vocalisations and body language. His days were busy, pursuing activities and interests in the service and out in local areas. The service was also busy with between 20 and 30 members and staff accessing it each day.

Ben and the majority of his peers would be considered to have severe or profound learning disabilities, meaning that they had significantly limited capacity to understand my role or the project more broadly, no matter the way in which this was communicated. My perspective was that I wanted to enhance and facilitate participation for Ben wherever possible, to allow him to shape the research focus and direction, though it was not possible to explain this to him. Similarly, asking Ben for his perspective through methods normative to me, i.e. speaking or writing, was impossible not only due to his intellectual capacity but also because he was extremely distressed by direct communication or minute changes to his routine.

Hultman et al. (Chapter 8, this volume) describe the sharing of perspectives in a community and academic group of neuro-mixed researchers, including ADHD:ers and people with acquired brain injury. In this dynamic a disharmony of communicative preferences and abilities existed, including people that spoke at different speeds and used different complexities of language, requiring the development of a shared practice to facilitate meaningful participation. Developing a shared practice with Ben was in some ways very easy, as he had no interest in participating or expressing himself other than through the way he lived every day, involving the schedules and activities that defined his life. Across the eighteen days I spent with Ben, there was almost no direct communication between us other than a nod at the start and end of each day.

Ben did, however, understand that I was not a support worker, as he had specific ways of engaging with them that I was not part of, and over the course of the project, showed a keen interest in what I was there for, observing me regularly and at one point taking and looking through my notepad. In this sense, he could share that he was comfortable and accepting of me and what participation meant for him, and it was subsequently a task for me to ask how this form of participation could enable the exploration of his experiences. This process showed that it is important not to presume the limitations of autistic people with profound learning disabilities in their understanding of research and participation, but rather to learn how you understand one another's roles over the course of researching inclusively.

Illustrating the Approach

The purpose of the approach was for me to develop a dialogue with Ben in which he highlighted areas of experience important to him, as well as methods and theories relevant to his day-to-day circumstances and that of his peers. The approach required a multi-stage, cyclical process of fieldwork, analysis, fieldwork, analysis and so on. Analytic stages did not represent a final analysis and did not require an in-depth interpretation of the data, rather they prompted me to explore the relationship between Ben's experiences, the day-service context, and relevant theories and approaches. The intention was for Ben's views and experiences to guide the study's development and focus, thus promoting meaningful participation.

As described below, the research began with close participatory work with Ben and broader ethnographic observations of service life. A subsequent period of analysis with lifeworld fractions highlighted areas of experience important to Ben and the broader context of the service community, pointing to theories with which to explore these experiences. By returning to the service and once again being-with Ben in his day-to-day context, I reflected on the analysis and whether it still felt pertinent and plausible, to see if there were aspects of his lifeworld and service life to explore, and to consider the relevance of theoretical perspectives highlighted through the analysis. It was also an opportunity to consider whether the methods being used were appropriate to the way Ben and his peers were communicating their experiences. Over two stages of fieldwork and analysis, the process enabled me to learn about Ben and his peers' experiences and communicative preferences, leading me to understand how I could research service culture through a relevant theoretical perspective and method.

Examples of the analytic method are provided below through screenshots of Nvivo 11, a qualitative data analysis software.

Stage one

Fieldwork

The aim of the initial stage was to observe and learn about the milieu in which Ben, his peers and their support staff lived day-to-day and to develop an understanding of their social and cultural practices. This required me to develop trusting relationships, ensuring any engagement and research activity was sensitive to their etiquettes and circumstances. By coming to know and understand how they lived and responded to their everyday situations, and by sharing in these situations together, I was able to begin to collect evidence of the ways they were experiencing their worlds.

The initial stage involved nine days (9am–4 pm) of fieldwork over three weeks. Ben was very accepting of me accompanying him and his support worker on his day-to-day pursuits, though he found direct communication distressing. Developing a dialogue with Ben, I began to understand that he disliked direct engagement and instead valued a sociality in which he enjoyed other peoples' presence through observing, listening and sharing situations. I thus chose methods that reflected this preference and etiquette, and which helped to form a body of evidence and understanding over the course of the nine days, they were:

Participatory Observation

This method was drawn from Simmons and Watson (2014) and involved spending approximately 5-hours per-day with Ben, taking fieldnotes detailing the minutiae of his day-to-day life including his communication and expressions, the surrounding contexts, his activity, and how others engaged with him and vice versa.

Ethnographic 'hanging out' Observations
The day-service was a space shared between Ben, his peers and support staff and was the site at which they experienced and contributed to service culture. I was explored this through ethnographic 'hanging out' (Kusenbach, 2003, p. 463) which involved making fieldnotes relating to the practices, activity and customs of everyday life at the service, speaking to and asking informal questions to people and collecting archive material such as support plans.

Reflexivity
I undertook reflexive note-taking throughout fieldwork and analysis to consider what participants were communicating, whether methods could be adapted or introduced to enable stronger dialogue, and whether theories were relevant to their views and experiences. I considered my own position and perspective and how this could be impacting the research approach and process. I am a neurotypical person and former day-service support worker, meaning I had some assumptions and beliefs about day-service life and the experiences of autistic people with profound learning disabilities, such as ideas about the meaning of their relationships with support workers. At times I found myself drawn to areas of life seemingly unimportant to Ben, such as the conversation of support workers, and I explored this in relation to my previous role through reflexive note-taking. I kept in mind the question of how the research topic should be explored in a way that was relevant to Ben and his peers' experiences, perspectives and methods of communication, and attempted to re-focus my perspective accordingly.

These methods led to the development of two data sets, one concerning Ben's experiences and another concerning everyday life for members at the service, the site of service culture. The purpose of the analysis was to explore significant aspects of Ben's experiences and to see how they may have related to the broader experiences and practices of his peers.

Analysis
The first stage of analysis was done through the scope of lifeworld fractions. I did this by coding words, phrases or sections to the relevant fraction or fractions, though this could also be done with visual data, such as photos or video recordings, by coding sections or whole images. Fractions are 'mutually entailed, with overlapping… meanings…' (Ashworth, 2016, p. 23) and so data was often relevant to more than one and require coding as such. An example of fieldnotes coding:

He watches over her shoulder, directing his gaze at her hands and the man behind the counter (sociality and project). The tickets are printed and she shows Ben, clearly stating that we are in screen 4. He nods (sociality, embodiment and discourse).

Each fraction subsequently had a collection of data assigned to it (see Figs. 15.1, 15.2, 15.3 and 15.4):

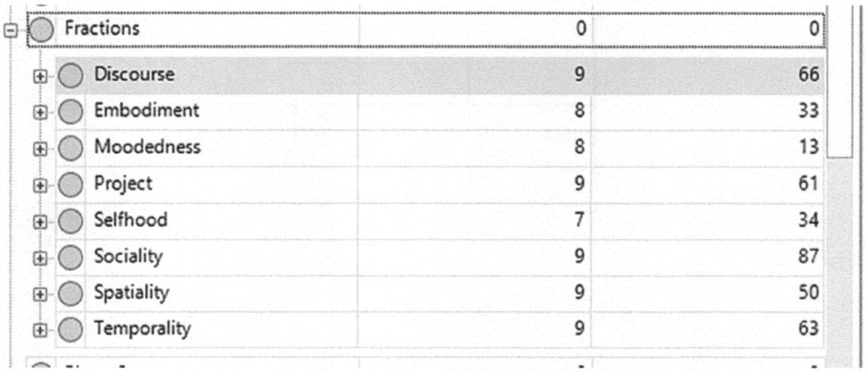

Fig. 15.1 An example of the fractions at the end of the initial stage. The numbers on the left indicate the amount of sources (e.g., a day of fieldnotes or a single interview), while the numbers on the right indicate the amount of individual parts of data (e.g., words, phrases or sections)

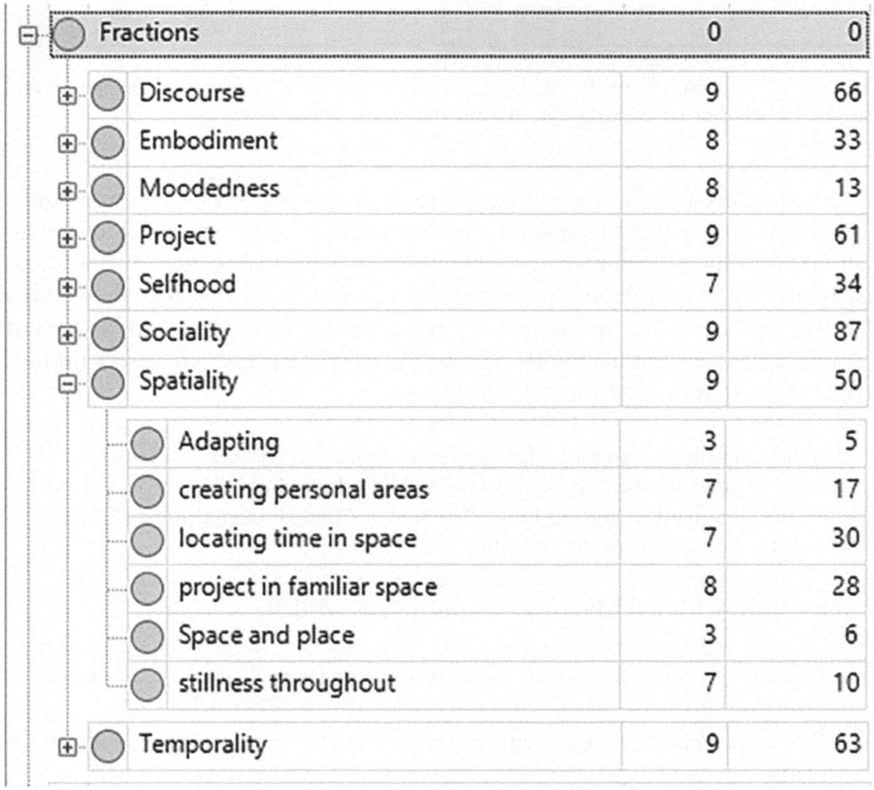

Fig. 15.2 A screenshot of an Nvivo analysis that illustrates six units of meaning within the spatiality fraction

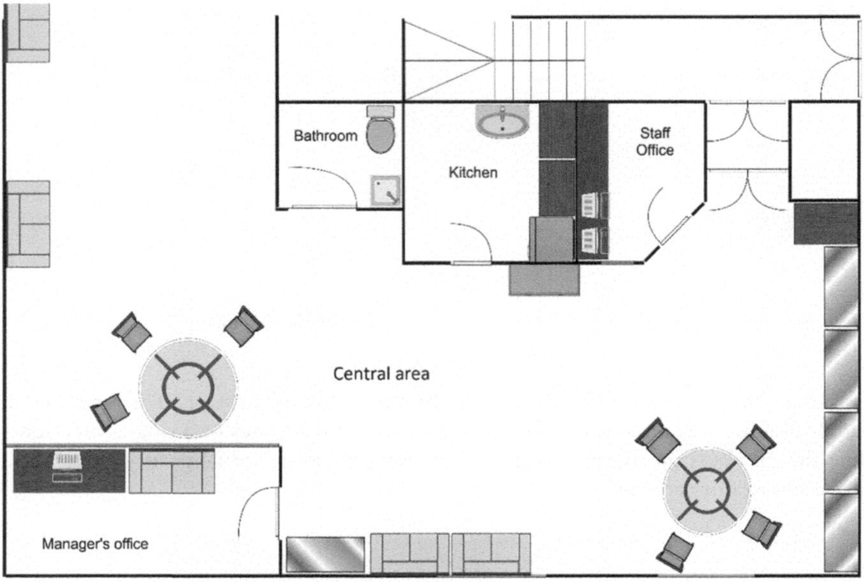

Fig. 15.3 An example of the fieldmap template developed to explore the spatial and temporal character of everyday life within the day-service

I then examined the data assigned to each fraction and suggested 'units of meaning', referring to my interpretations of what the experience meant to Ben. To do this, I attempted to orientate my perspective to that of Ben's and ask of the data: in relation to this fraction, what does this tell me of how Ben was experiencing the situation? I considered the data at hand, as well as insights gained during fieldwork. The below example details an observation of Ben after just arriving at the day-service:

> ...(Ben) continues watching the person complete the jigsaw. This person is vocalising quite loudly, 'eeees', 'ahhhhhs' and 'chup chup'. Emma brings out 2 bottles of squash saying nothing, Ben touches the summer fruits and Emma goes into the kitchen to get him this.

In relation to selfhood, the observation suggested that:

- Ben experienced himself as provided for in that drinks were brought to him.
- Ben experienced a sense of autonomy in the opportunity to choose a drink.

In relation to sociality:

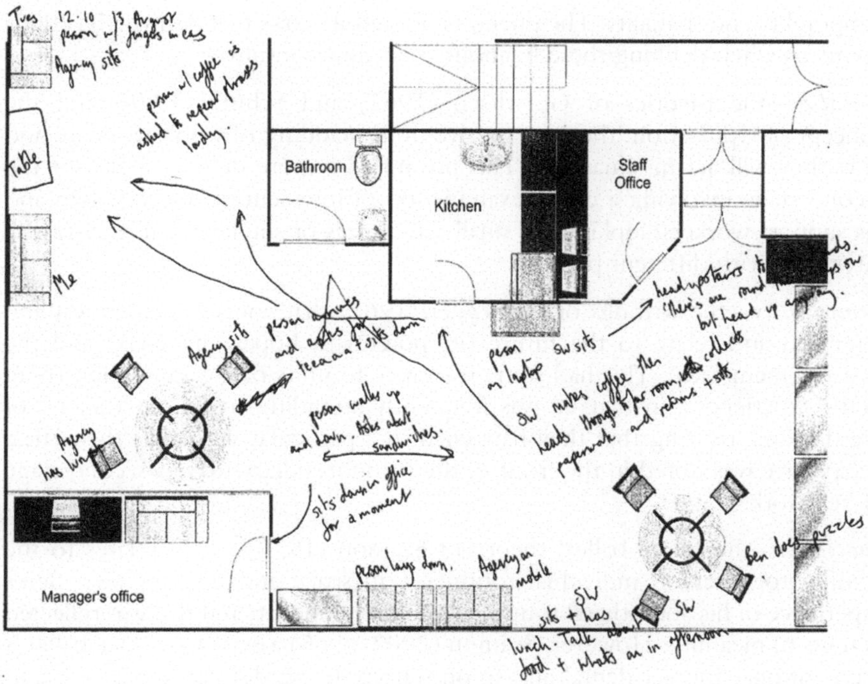

Fig. 15.4 An example of a completed fieldmap depicting ten minutes of activity in the day-service

- Ben was drawn to the activity of others as he observed them.
- Ben viewed support workers as providers (of choice or refreshments).

In relation to discourse:

- Ben expressed his choices through physical direction. He embodied his communication.

As the analysis progressed, units of meaning became more or less decided though never certain, particularly when analysis brought about new or alternative interpretations.

Connecting Analysis to Theory

The analysis will likely have highlighted one or more fractions as significant, which can guide a researcher to explore its foundational theories (see Ashworth, 2006, 2016) and to explore participants' experiences in new, complex and relevant ways. Researching with Ben, three fractions were suggested to be potentially significant after the initial analytic stage: sociality, temporality and spatiality. The phenomenological works that Ashworth (2006) points to when defining these fractions were thus consulted:

Sociality—the theories of Gurwitsch (1962) and Schutz (1976) and the concept of typification: the prereflective understanding of who others are and how they will act in situations. This prompted a more defined focus for the second stage involving a closer examination of interactions between Ben and his support workers, looking for small differences or similarities in the way he interacted with different people.

Temporality—the writings of Heidegger (1962) that concern protention and retention and relate to the future (its potentials, hopes and fears) and the past (its memories). This had some relevance to units of meaning concerning Ben's experiences, in that he was frequently attending to the next stages of the day and assuring that they happened in a particular way. This theoretical perspective was noted in the front of the fieldnotes diary for the second stage of fieldwork.

Spatiality—the 'place ballet' theory of Seamon (1979) which relates to the coming together of 'individual routines… in space and time' (p.58). Ben's experience of his and others' routines seemed significant, and this was reflected in units of meaning. However, Seamon (2002, p. 45) asserts that place ballet is unintentional and accidental and so on reflection this did not appear relevant to a day-service with planned activities and schedules. Further investigation was required to consider the concept in context.

I 'carried' these theories into the second stage of fieldwork by noting key ideas at the front of my fieldnotes diary and by focusing more closely on relevant aspects of Ben's day-service life, such as how Ben and his support workers engaged with each other in relation to sociality.

Stage Two

Fieldwork

I entered this second stage with ideas of what Ben was expressing about his experiences and the surrounding context, methods that were appropriate for his and his peers' communicative preferences, and particular fractions appearing most relevant to their experiences. I remained open to their views and experiences and whether the theoretical insights gained during analysis were still relevant. This was a case of further developing dialogue to explore previous and new possibilities.

The second stage involved six days (8.50am–3 pm) of fieldwork across three weeks. First stage methods continued though with a more defined focus that concerned:

- The ways in which Ben and his support workers interacted, including differences or similarities between support workers and how they sought to anticipate one another (sociality).
- Routines within the day-service and how these related to movement, environment and the activity (spatiality.
- The ways in which Ben looked to the future through planning, direction or action (temporality).

Analysis

In the second stage, spatiality and temporality were again suggested to be significant to Ben and his peers' experiences and the day-service context. Ben viewed schedules and routines as extremely important and prioritised related customs, which were also highlighted in units of meaning concerning the temporal and spatial order of the day-service and Ben's peers. However, contrary to the speculations I made in the first stage regarding the organisation of the day-service, second stage analysis suggested that people's movement and activity happened according to unwritten schedules and showed day-to-day life to be far less scripted than previously thought.

Consideration of the fractions led back to place ballet (Seamon, 1979), which refers to the ways individuals' routines can become experientially interrelated with others when connected to time and space. This creates shared customs and ways of experiencing a place, so if other people's routines change, a person may experience a disturbance of their own. This theory appeared highly relevant to the units of meaning relating to Ben's experiences, the customs of his peers and day-to-day life at the service. Exploring place ballet further identified a study involving a maps method (van Eck & Pijpers, 2017), which involves documenting qualitative, observational data on a bird's-eye view of an area or space over set periods of time. This method appeared as though it could capture aspects of day-service culture in a way relevant to Ben's experiences, the activity and customs of his peers, as well as the broader context of the day-service. Thus, moving to a third stage of fieldwork, I adopted a maps method with a guiding theoretical perspective relating to space, time and place ballet.

Stage Three and Onwards

The first two stages at Ben's service had highlighted areas of experiences important to him and his peers and suggested that a maps method, led by the theoretical perspective of place ballet, would be an appropriate way to explore this further. Thus, a maps method was used in ninety-minute periods across sixteen days of fieldwork, with a map completed every ten minutes.

This final stage produced a data set with a focus on the activity of everyday life at the day-service and its related spatial and temporal customs. I analysed this through theory-driven thematic analysis (Braun & Clarke, 2006) that used the concept of place ballet (Seamon, 1979, pp. 143–152) as this had been established as an appropriate frame of reference.

Discussion on the Approach

This discussion reflects on the ways this chapter's methodology enabled me to meaningfully involve Ben in research about a key part of his life, day-service culture. I do this by drawing from the phenomenological concept of the *while* (Heidegger, 1996) to consider the ways Ben participated, the ways his perspectives were responded to and how his input contributed to the project's development.

Heidegger (1996) used the term the *while* to refer to the in-the-moment nature of experience. Rix et al. (2020) drew from the *while* as a way to consider the inclusive nature of research, suggesting that, like experience, participation is in-the-moment and requires reflection in order to understand its meaning. They argued that inclusive researchers often position disabled people outside the *while* of research participation, during analysis for example. Participation thus requires reflection and consideration after the fact, including at what stages people could participate, how they could participate, and how this impacted knowledge construction. Below, Ben's participation in the *while* is considered through three key areas: dialogue, analysis and design.

Dialogue

Inclusive disability researchers have highlighted the importance of having adequate time to create meaningful dialogue (Brownlee-Chapman et al., 2018; Liddiard et al., 2019; Olsen & Carter, 2016). This is particularly apparent when researching with autistic people with profound learning disabilities (Bradley et al., 2015; Goode, 1994; Simmons & Watson, 2014, pp. 151–153), where long periods of engagement enable researchers to attend to 'actions, sounds and body language, but also to tempo, mood and potential significance of meaning' (Nind, 2018, p. 116).

Notable in researching with Ben was that it was not just the time spent with him in which dialogue emerged, but the *act* of repeatedly recording the minutiae of events and circumstances shared with him and subsequently reflecting on them through analysis. For example, Ben adjusted his socks at the same time, in the same place each week, and this gained significance because it had been noted down several times previously. Although appearing relatively trivial, Ben committed to this act and its importance was highlighted in each stage's analysis through relevance to spatial, temporal and embodied fractions. Within the approach's circular process, I suggest that meaningful dialogue was promoted, where, in the *while*, Ben could draw attention to aspects of day-service life and experience that were important to him, which could subsequently guide the research focus.

Analysis

Inclusive researchers have often struggled to include people with learning disabilities in analytic processes (Rix et al., 2020; Seale et al., 2015; Stack & McDonald, 2014), and this has been entirely unexplored in the context of autistic people with profound learning disabilities. There are significant hurdles in this regard, primarily relating to the knowledge capacities and communicative repertoires of people from this group, but also in relation to ethical questions it can raise: for example, Ben would experience upset if direct questions were asked of him (e.g., through creative methods) or if his usual, day-to-day activities were disrupted.

I undertook analysis away from Ben and the day-service he attended, creating 'a new source of participation' (Rix et al., 2020, p. 7). Here, Ben's fieldwork contributions mediated the *while*, though his participation and input were marginalised. Such marginalisation amplified my perspective, a dynamic that continues in this chapter. The circular multi-stage approach did, however, help to blur the boundaries of analysis and fieldwork, allowing Ben's perspective to shape analysis and its theoretical perspective, which subsequently shaped fieldwork and its methods.

The place ballet (Seamon, 1979) scope was a direct result of how space and time appeared significant to Ben in the *while* of participation; it had not been considered prior to researching with Ben. Importantly, the process was not tokenistic, it followed the principles of phenomenological research detailed by Berndtsson et al. (2007), which is to be guided by participants' experiences and circumstances and a relevant theoretical perspective.

Design

Involving community members in the design of research is important if it is to effectively explore salient aspects of their lives and circumstances (Cascio et al., 2020). Using lifeworld fractions provided the space and time in which to adopt and adapt methods in the *while*, in the ways Ben wanted to engage with me. This approach is not novel, merely following appropriate ways of engaging with autistic people with profound learning disabilities in research (Gjermestad et al., 2022). Where the *while* was extended, and where the approach showed innovation, was in how Ben's perspective led to the introduction of the maps method to explore day-service culture. It demonstrated that to greater or lesser extents, autistic people with profound learning disabilities can contribute to study design and that this can affect how their lives, circumstances and fellow community members are researched.

Limitations

The methodology detailed in this chapter requires an extended period of time, something that may not be readily afforded within the constraints of funders' expectations. The study's initial stages involved working with a single person and further research is required to understand how involving more would impact a study's timing and outcomes. Finally, ethical questions are inevitably raised around the extent to which researchers can interpret the views and experiences of autistic people with profound learning disabilities and this requires appropriate reflection and scrutiny. Presented in this chapter was an attempt to provide such scrutiny through its discussion on the *while* of participation, though this may be strengthened through formalised reflective processes or frameworks.

Conclusion: Reflecting on the Approach's Inclusive Merit

This chapter presented a methodology for doing research inclusively with autistic people with profound learning disabilities. This methodology incorporated the framework lifeworld fractions (Ashworth, 2016) and the approach thinking phenomenologically while doing phenomenology (Berndtsson et al., 2007). It was suggested that this methodology promotes five elements of doing research inclusively, which was evidenced through research with Ben in which he communicated his perspectives and experiences on important aspects of day-service life and how they should be studied. I suggest this phenomenological approach is useful in researching with autistic people with profound learning disabilities as:

- It helped me to identify perspectives on day-service culture significant to Ben and his peers, which to begin with I was not focusing on, and which are not represented in broader service culture literature.
- It introduced me to relevant theories and ways to research which I otherwise would not have come across.
- The fractions are, to greater or lesser extents, an accessible phenomenological concept, which I used as a PhD student with no previous phenomenological knowledge or experience.
- The extended periods of fieldwork and analysis gave me confidence and understanding in doing research using several methods.

References

Andrews, T., Hodge, N., & Redmore, N. (2019). The potential of the fractions of lifeworld for inclusive qualitative inquiry in the third space. *International Journal of Inclusive Education, 26*(2), 143–159.

Ashworth, P. (2006). Seeing oneself as a carer in the activity of caring: Attending to the lifeworld of a person with Alzheimer's disease. *International Journal of Qualitative Studies on Health and Well-Being, 1*(4), 212–225.

Ashworth, P. (2016). The lifeworld–enriching qualitative evidence. *Qualitative Research in Psychology, 13*(1), 20–32.

Berndtsson, I., Claesson, S., Friberg, F., & Öhlén, J. (2007). Issues about thinking phenomenologically while doing phenomenology. *Journal of Phenomenological Psychology, 38*(2), 256–277.

Bottema-Beutel, K., Kapp, S. K., Lester, J. N., Sasson, N. J., & Hand, B. N. (2020). Avoiding ableist language: Suggestions for autism researchers. *Autism in Adulthood, 3*(1), 18–29.

Bradley, E., Caldwell, P., & Korossy, M. (2015). Nothing about Us without Us: Understanding mental health and mental distress in individuals with intellectual and developmental disabilities and autism through their inclusion, Participation, and Unique Ways of Communicating.

Braun, V., & Clarke, V. (2006). Using thematic analysis in psychology. *Qualitative Research in Psychology, 3*(2), 77–101.

Brownlee-Chapman, C., Chapman, R., Eardley, C., Forster, S., Green, V., Graham, H., Harkness, E., Headon, K., Humphreys, P., & Ingham, N. (2018). Between speaking out in public and being person-centred: Collaboratively designing an inclusive archive of learning disability history. *International Journal of Heritage Studies, 24*(8), 889–903.

Cascio, M. A., Weiss, J. A., & Racine, E. (2020). Empowerment in decision-making for autistic people in research. *Disability and Society, 36*(1), 100–144.

Charlton, J. I. (2000). *Nothing about Us Without Us: Disability oppression and empowerment*. University of California Press. http://ebookcentral.proquest.com/lib/open/detail.action?docID=224299

Chown, N., Robinson, J., Beardon, L., Downing, J., Hughes, L., Leatherland, J., Fox, K., Hickman, L., & MacGregor, D. (2017). Improving research about us, with us: A draft framework for inclusive autism research. *Disability and Society, 32*(5), 720–734. https://doi.org/10.1080/09687599.2017.1320273

Cluley, V. (2017). Using photovoice to include people with profound and multiple learning disabilities in inclusive research. *British Journal of Learning Disabilities, 45*(1), 39–46.

Dahlberg, H., & Dahlberg, K. (2019). The question of meaning—A momentous issue for qualitative research. *International Journal of Qualitative Studies on Health and Well-Being, 14*(1), 1598723.

Doak, L. (2018). But I'd rather have raisins! Exploring a hybridized approach to multimodal interaction in the case of a minimally verbal child with autism. *Qualitative Research.* 1468794117752115. https://doi.org/10.1177/1468794117752115

Finlay, L. (2011). Phenomenology for therapists: Researching the lived world. Wiley.

Finlay, L., & Molano-Fisher, P. (2008). 'Transforming'self and world: A phenomenological study of a changing lifeworld following a cochlear implant. *Medicine, Health Care and Philosophy, 11*(3), 255–267.

Fletcher-Watson, S., Adams, J., Brook, K., Charman, T., Crane, L., Cusack, J., Leekam, S., Milton, D., Parr, J. R., & Pellicano, E. (2019). Making the future together: Shaping autism research through meaningful participation. *Autism, 23*(4), 943–953.

Gaudion, K. (2015). A designer's approach: Exploring how autistic adults with additional learning disabilites experience their home environment Ph. D. thesis, Royal College of Art, London].

Gillman, M., Swain, J., & Heyman, B. (1997). Life history or 'case' history: The objectification of people with learning difficulties through the tyranny of professional discourses. *Disability and Society, 12*(5), 675–694.

Gjermestad, A., Skarsaune, S. N., & Bartlett, R. L. (2022). Advancing inclusive research with people with profound and multiple learning disabilities through a sensory-dialogical approach. *Journal of Intellectual Disabilities*, 17446295211062390.

Goode, D. (1992). Who is Bobby? Ideology and method in the discovery of a Down syndrome person's competence. *Interpreting disability: A qualitative reader*, 197–212.

Goode, D. (1994). A World without Words.

Gurwitsch, A. (1962). The common-sense world as social reality: A discourse on Alfred Schutz. *Social Research*, 50–72.

de Haas, C., Grace, J., Hope, J., & Nind, M. (2022). Doing research inclusively: Understanding what it means to do research with and alongside people with profound intellectual disabilities. *Social Sciences*, *11*(4), 159. https://www.mdpi.com/2076-0760/11/4/159

Harré, R., & Gillett, G. (1994). *The discursive mind*. Sage Publications.

Heidegger, M. (1962). Being and time. 1927. Trans. John Macquarrie and Edward Robinson. Harper.

Husserl, E. (1970). The crisis of European sciences and transcendental phenomenology: An introduction to phenomenological philosophy. Northwestern University Press.

Koegel, L. K., Bryan, K. M., Su, P. L., Vaidya, M., & Camarata, S. (2020). Definitions of nonverbal and minimally verbal in research for autism: A systematic review of the literature. *Journal of Autism and Developmental Disorders, 50*(8), 2957–2972. https://doi.org/10.1007/s10803-020-04402-w

Kraus, B. (2015). The life we live and the life we experience: Introducing the epistemological difference between "Lifeworld"(Lebenswelt) and "Life Conditions"(Lebenslage). *Social Work and Society, 13*(2), 1–9.

Kusenbach, M. (2003). Street phenomenology: The go-along as ethnographic research tool. *Ethnography, 4*(3), 455–485.

Liddiard, K., Runswick-Cole, K., Goodley, D., Whitney, S., Vogelmann, E., & Watts MBE, L. (2019). I was excited by the idea of a project that focuses on those unasked questions. Co-producing disability research with disabled young people. *Children and Society*, *33*(2), 154–167. https://doi.org/10.1111/chso.12308

McCormack, N. (2017). Making memory sites: extending opportunities for people with profound and multiple learning disabilities to participate in life story work. *Ph. D. Thesis University of East London Art and Digital Industries*. https://doi.org/10.15123/PUB.6363

Merleau-Ponty, M. (1962). *Phenomenology of perception [Phénoménologie de la Perception]*. Routledge andKegan Paul.

Mietola, R., Miettinen, S., & Vehmas, S. (2017). Voiceless subjects? Research ethics and persons with profound intellectual disabilities. *International Journal of Social Research Methodology, 20*(3), 263–274.

Morris, J. (2003). Including all children: Finding out about the experiences of children with communication and/or cognitive impairments. *Children and Society, 17*(5), 337–348.

Nind, M. (2013). Inclusive research: Where does it leave people with PMLD? *PMLD Link.*

Nind, M. (2018). Multimodal listening and attending to children with complex disabilities in educational settings. In M. C. Twomey, Clare (Ed.), *Seen and heard: Exploring participation, engagement and voice for children with disabilities* (pp. 105 - 123). Peter Lang.

Nind, M. (2019). Thinking inclusively, acting inclusively, researching inclusively. *Journal of Special Needs Education in Ireland.*

Nind, M., & Vinha, H. (2014). Doing research inclusively: Bridges to multiple possibilities in inclusive research. *British Journal of Learning Disabilities, 42*(2), 102–109. https://doi.org/10.1111/bld.12013

Olsen, A., & Carter, C. (2016). Responding to the needs of people with learning disabilities who have been raped: Co-production in action. *Tizard Learning Disability Review, 21*(1), 30–38. https://doi.org/10.1108/TLDR-04-2015-0017

Palmer, C., & Walmsley, J. (2020). Are people with profound and multiple learning disabilities and their families welcome in the wider learning disability community? In M. Nind & I. Strnadova (Eds.), *Belonging for people with profound intellectual and multiple disabilities.* Routledge.

Redmore, N. (2023a). Day service cultures from the perspectives of autistic people with profound learning disabilities. *British Journal of Learning Disabilities*, *1*(12). https://doi.org/10.1111/bld.12545

Redmore, N. (2023b). Day service cultures from the perspectives of autistic people with profound learning disabilities. *British Journal of Learning Disabilities*, https://doi.org/10.1111/bld.12545 *early online access.* https://doi.org/10.1111/bld.12545

Redmore, N. (2023c). Proposing an academic space of 'autism and profound learning disability'. *The International Journal of Disability and Social Justice*, *3*(2), 91–106. https://doi.org/10.13169/intljofdissocjus.3.2.0091

Ridout. (2016). Well-being and creative methodologies as a tool for communicating with health and social care practitioners. In D. Milton & N. Martin (Eds.), *Autism and intellectual disability in adults* (Vol. 1). Pavilion Publishing and Media Ltd.

Rix, J., Garcia-Carrizosa, H., Hayhoe, S., Seale, J., & Sheehy, K. (2020). Emergent analysis and dissemination within participatory research. *International Journal of Research and Method in Education*, 1–16. https://doi.org/10.1080/1743727X.2020.1763945

Schutz, A. (1976). *Collected papers II: Studies in social theory* (Vol. 15). Springer Science & Business Media.

Seale, J., Nind, M., Tilley, L., & Chapman, R. (2015). Negotiating a third space for participatory research with people with learning disabilities: An examination of boundaries and spatial practices. *Innovation: The European Journal of Social Science Research*, *28*(4), 483–497.

Seamon, D. (1979). *A geography of the lifeworld: Movement, rest and encounter.* Routledge. https://doi.org/10.4324/9781315715698

Seamon, D. (2002). Physical comminglings: Body, habit, and space transformed into place. *OTJR: Occupation, Participation and Health*, *22*(1_suppl), 42S-51S.

Simmons, B., & Watson, D. (2014). The PMLD ambiguity: Articulating the life-worlds of children with profound and multiple learning disabilities. Karnac Books Ltd.

Skarsaune, S. N., Hanisch, H., & Gjermestad, A. (2021). Self-Determination: What can we Learn from Persons with Profound Intellectual and Multiple Disabilities? *Scandinavian Journal of Disability Research*. https://doi.org/10.16993/sjdr.830

Stack, E., & McDonald, K. E. (2014). Nothing about us without us: Does action research in developmental disabilities research measure up? *Journal of Policy and Practice in Intellectual Disabilities, 11*(2), 83–91.

Stewart, K., Bradshaw, J., & Beadle-Brown, J. (2018). Evaluating service users' experiences using talking Mats®. *Tizard Learning Disability Review, 23*(2), 78–86. https://doi.org/10.1108/TLDR-05-2017-0023

Teachman, G., McDonough, P., Macarthur, C., & Gibson, B. E. (2018). A critical dialogical methodology for conducting research with disabled youth who use augmentative and alternative communication. *Qualitative Inquiry, 24*(1), 35–44. https://doi.org/10.1177/1077800417727763

van Goidsenhoven, L., & De Schauwer, E. (2020). Listening beyond words: Swinging together. *Scandinavian Journal of Disability Research*. https://doi.org/10.16993/sjdr.756

van Eck, D., & Pijpers, R. (2017). Encounters in place ballet: A phenomenological perspective on older people's walking routines in an urban park. *Area, 49*(2), 166–173.

Vehmas, S., & Curtis, B. (2017). Profound intellectual disability and the bestowment view of moral status. *Cambridge Quarterly of Healthcare Ethics, 26*(3), 505–516.

Vehmas, S., & Mietola, R. (2021). *Narrowed lives: Meaning*. Stockholm University Press.

Walmsley, J., & Johnson, K. (2003). *Inclusive research with people with learning disabilities*. Jessica Kingsley Publishers Ltd.

CHAPTER 16

Narrative Research, Reflexivity, and AuDHD Lived Experiences

Lydia Stetson

Based on a True Story: Narrative Research and Reflexivity on AuDHD Experiences

At the core, narrative inquiry holds to the idea that individuals construct and communicate their life experiences through storytelling (Chase, 1995; Josselson & Hammack, 2021). For example, an AuDHD[1] doctoral student might endeavor to make sense of and speak about her experiences with narrative inquiry by sharing the story of how she found her way to qualitative research. In her story, the doctoral student will likely refer to circumstances that led her in one direction instead of another, the emotional impact of her decisions, how other characters influence her trajectory, and so forth. In this example, the author provides a version of events based on her positionality or frame of reference. The student's purpose in telling this story is not to claim that this is the only version of events in the time frame she described. Nor is it that others must view the same events similarly. Instead, the purpose of the narrative is to provide a glance into the student's contextualized self as an AuDHD doctoral student passionate about narrative inquiry.

[1] AuDHD refers to the combination of Autistic and Attention-Deficit/Hyperactivity Disorder (ADHD) experiences.

L. Stetson (✉)
Fielding Graduate University, Santa Barbara, California, USA
e-mail: Lydia.m.stetson@gmail.com

H. Bertilsdotter Rosqvist and D. Jackson-Perry (eds.), *The Palgrave Handbook of Research Methods and Ethics in Neurodiversity Studies*,
https://doi.org/10.1007/978-3-031-66127-3_16

As is possibly apparent, the AuDHD doctoral student I have briefly described is myself. Using myself as a subject falls squarely into the territory of autoethnography and strays from the chapter topic of narrative analysis or even reflexivity (Marshall et al., 2022). Nonetheless, using pieces of my narrative provides a convenient way to provide clarifying examples of my conceptualization and process of narrative inquiry. In this chapter, I do not seek to provide a definitive or exhaustive text on how to complete narrative interviewing and analysis. Rather, I offer my perspective on using, navigating, and embracing this approach to research as an AuDHD researcher with a passion for highlighting and affirming neurodivergent perspectives in research. More specifically, my aim with this chapter is to intersperse illustrative examples with a brief overview of narrative inquiry, discuss positionality and generalizability in research, explore ethical considerations, and provide a look at researcher reflexivity as of notable importance for AuDHD researchers and individuals doing research into AuDHD experiences.

A Brief Overview of Narrative Inquiry

Narrative inquiry is a research method particularly concerned with meaning-making and often involves questions of how people make sense of their experiences and histories (Josselson & Hammack, 2021). My experience with narrative inquiry is situated within the realm of clinical psychology. Nonetheless, narrative inquiry has a rich history and is applicable in many fields, including sociology, linguistics, cultural ecology, comparative history, education, and many more (Bruner, 2004; Clandinin et al., 2007; Johnstone, 2016). Narrative inquiry has broad application in social sciences as human beings have used storytelling to make sense of and share their lives with others throughout history (Bruner, 1990; Clandinin, 2006). The term narrative is difficult to define but may be conceptualized as a placeholder for life. That is, narratives are a way of observing, making sense of, and portraying life (Schiff, 2012). Therefore, researchers from many different disciplines have used narrative inquiry to deepen their understanding of how people experience their lives in various contexts, such as how women and girls develop a voice in male-dominated cultures (c.f. Gilligan, 1982) or how minority groups experience life within systems of oppression (c.f. Caine et al., 2017; O'Grady et al., 2018).

Life Stories

A fundamental aim of narrative inquiry is to elicit individuals' life stories to examine how each person draws meaning from their experiences and orients themselves contextually within the world (Josselson & Hammack, 2021; Murray, 2021). However, eliciting life stories does not necessitate prompting a person to recount their complete life history from birth until the present. Instead, life stories refer to subjectively profound or interesting periods or experiences in a person's life (Chase, 1995, c.f. Chapter 20, this volume). For example, consider the following excerpt of my life story regarding my foray into qualitative research.

> As I entered my clinical psychology doctoral program, my research experience was limited and entirely focused on quantitative inquiry. My few previous research activities included coding driving behaviors by looking at recorded videos and pushing buttons to facilitate participants' experiences in an audiology study. What little I knew of research was that I was respectful of but not particularly enthusiastic about quantitative research methods. So, when a faculty member encouraged me to consider that my clinical orientation (Humanistic-Existential) is well-suited to qualitative research due to comparable focus on context and meaning-making, I jumped at the suggestion in the hope that qualitative research would spark enough passion for research to allow me to get through my dissertation. While I am still making my way through my doctoral program and dissertation, my passion for qualitative research and narrative inquiry, more specifically, has only grown.

In this example, I situate my life story within the specific bounds of my experiences with research and, more specifically, my introduction to qualitative methods.

The Process of Narrative Inquiry

There are many ways to collect and analyze narrative data. Narrative data collection methods include interviews, focus groups, diaries, archives, letters, other documentation, or videos (Murray, 2021). My interest lies in collecting narrative data through interviews due to my values in sharing physical space and time with the individuals whose narratives I collect as a first step in honoring their stories. More specifically, a relational and empathic approach to narrative interviews is well suited to my interpersonal style. A relational empathic process to narrative interviewing involves careful listening and attunement to the emotions and subtext of the storyteller by using cognitive empathy (taking on another's perspective) and affective empathy (feeling another person's feelings) to deepen the researcher's insight and understanding of the narrator as a contextual and cultural being (Josselson, 2013). Additionally, relational empathic interviews allow the researcher(s) to note and follow up on emotional themes in real time. For example, if I told my brief example narrative in an interview, a relational and empathic interviewer might key into my word choices of passion and excitement and use these words to probe for deeper and richer stories and clarify why qualitative research interests me.

Furthermore, relational interviewing embraces the relationship between the researcher and the narrator, which influences how the narrator constructs and presents their account (Sefora & Ngubane, 2023). For instance, I shared my example narrative with myself when writing, and it is intended to serve as an illustrative example. Thus, I constructed this narrative in a manner that makes sense to me and in a way that is digestible to others who have little familiarity with me or with narrative inquiry. My account may have been similar but likely quite different in tone had I shared it in an interview with a narrative researcher. Finally, narrative interviews are starkly different from gathering demographic or biographical reports. Instead, narrative interviews should aim to elicit stories and not merely responses to basic questions (Chase, 1995). In contrast to reports, stories provide rich data that includes emotions, context, positionality, and identity. Had I provided a report rather than a story for my example, it would have focused on relatively objective information, such as that I am a student and am interested in a qualitative dissertation. While this information is accurate and congruent to my story, it lacks depth into my felt experiences and how I moved within the research context.

Just as there are varied approaches to narrative data collection, there is a range of narrative data analyses (Murray, 2021). For instance, an analysis may consist of work by the researcher or research team alone or involve the researchers in conjunction with participants, such as in participatory action research (Murray, 2021). My approach to narrative analysis is holistic, person-centered, and contextual. While the specific strategies for narrative analysis differ based on researcher experience and preference, a narrative analysis should be iterative and interpretive (Head, 2022). In other words, researchers should read narrative texts (i.e., interview transcriptions or archival narratives) multiple times, each time with a different lens or focus, to aid in developing a complete picture of the participant's gestalt, which is the structure or configuration of the parts that comprise a whole being (Sabar, 2013). Each time the researcher reads the narrative, they are working to integrate different pieces of the life story to create a cohesive understanding of the narrative. Foci of the recursive readings may include looking at the overall themes, voices or conflicts within the narrative, repeated themes and patterns in the data, and connections between the individual narrative and other narratives and with the broader literature base (Josselson & Hammack, 2021).

Throughout the analysis, the researcher(s) should note and reflect on how their positionality and context contribute to their interpretations of the participants' narratives (Head, 2022; Josselson & Hammack, 2021). That is, the researcher(s) should attend to how their understanding and meaning-making impact how they interpret meaning and sense-making within the narratives. For example, a clinical psychologist focusing on psychotherapy might emphasize my interest in interpersonal relationships as a critical factor in my path toward qualitative research. In contrast, a research psychologist might highlight my interest in depth of experiences. While qualitative research embraces the researchers' subjectivity, researchers must take care that their meaning-making does not overshadow or obfuscate the meaning-making of the participants through the practice of reflectivity, which is described in more detail in a later section (Josselson, 2007). In sum, there are diverse ways to

1. Identify the broad research question (the Big Q).
    a. The Big Q is a conceptual research question focusing on experiences that may be evident in life stories or narratives. Example: How do AuDHD experiences impact doctoral students' research methods selections for their dissertations?
    b. Note reflexivity.
2. Identify an experience-near or experiential question (the little q).
    a. The little q orients the participants to the research question and anchors the conceptual question in real-world narratives. Example: Tell me about your experience and process for choosing your dissertation topic and methods.
    b. Note reflexivity
3. Identify the recruitment population. Example: AuDHD doctoral students completing a dissertation.
    a. Note reflexivity.
4. Conduct and transcribe the interviews.
    a. Note reflexivity.
5. Complete iterative (repeated) readings of the interview transcripts focusing on:
    a. The gestalt, or holistic narrative theme.
    b. The voices or dialogues apparent in the interviews.
    c. Patterns in the interview/data.
    d. How the interview connects with theory.
    e. How separate interviews overlap or connect with each other.
    f. Note reflexivity throughout the readings.
6. Begin writing the manuscript
    a. Note reflexivity and include a brief section on researcher reflexivity and reflexive processes in the manuscript.

For more extensive examples and information on conducting and writing narrative studies, please refer to comprehensive texts on narrative inquiry (E.g., Clandinin, 2007; Josselson, 1996; Josselson & Hammack, 2021; Josselson & Lieblich, 1995).

Fig. 16.1 A brief outline for a narrative research project (outline based on Josselson & Hammack, 2021)

conduct a narrative study. However, a skeleton outline for a narrative study is shown in Fig. 16.1.

Positionality, Truth, and Ethical Considerations

While life stories are inherently distinctive to each person, their sociocultural context shapes their narrative, as humans are social and cultural beings (Craig & Huber, 2007; Johnson & Thacker Darrow, 2023). Culturally ascribed meanings inform the social categories that people inhabit and embody, and these categories influence the narratives people tell about their lives (Josselson & Hammack, 2021). For example, as an AuDHD person, it has been my experience that understanding and ascribing meaning to stimuli or events, such as a social cue or time management, is often different from how my neurotypical peers understand and respond to the same stimuli or events. In narrative inquiry, it is vital to understand and honor the narrators' lived experiences and sociocultural contexts or positionality as crucial facets of meaning-making and truth. Just as the participants' contexts and positionality influence their truths and meaning-making, so do the researchers' positionality and contexts (Johnson & Thacker Darrow, 2022). Therefore, ethical narrative researchers should engage in ongoing reflexivity practices, discussed further in a later section of this chapter, so that their stories do not end up overshadowing the participants (Josselson, 2007). The main takeaway from this section is that narrative analysis and interpretations should consider the subjective truths of the participants and the researchers (Chase, 1996; Josselson, 2007).

Importantly, truth in a person's narrative does not refer to a strictly factual or accurate telling of their life experiences. Instead, narrative truth refers to what makes sense given the individual's position in time and social and cultural domains (Josselson & Hammack, 2021). In other words, narrative inquiry, and qualitative methods more broadly, eschew the empiricist-positivist idea that a singular objective truth is nestled within a person's experiences (Gergen, 2014). This is not to say that truth is inconsequential to narrative inquiry. On the contrary, listening and looking for contextual truths in a narrative is essential to exploring and interpreting narrated life experiences (Head, 2022). For example, it is certainly not an objective fact that quantitative methods are uninteresting or lack excitement. However, my position as an individual who craves novelty and deep exploration of highly specific topics lends itself to my subjective truth that quantitative research methods are unengaging (to me) because they focus more heavily on broader, more generalized experiences and repetition or replicability. Thus, my truth that qualitative research provides the glimmer and spark I desire highlights the meaning and understanding of my narrative.

Nonetheless, it is vital to understand that researchers have an element of interpretive authority, meaning their subjective knowledge can and does shape their analysis of the narratives (Chase, 1996). Interpretative authority should add to, rather than detract from, the research. Therefore, it is crucial that ethical researchers acknowledge that their interpretations are a truth and refrain from asserting that their interpretation is the singular truth or only interpretation of their participants' life stories (Chase, 1996). Researchers can practice ethical interpretive authority by identifying what meanings they draw from the narratives and then checking their meanings against alternative interpretations found in the themes and meanings in the narratives, relevant theory and literature, and research collaborators (Olmos-Vega et al., 2022). For example, an AuDHD researcher might note that an AuDHD participant's narrative regarding their experience as a graduate student is superficially similar to the researcher's experiences. The researcher might reflect on similarities in a reflexivity journal and then invite alternative interpretations from other researchers with a range of neurodivergent identities. Finally, the researcher might explore how interpretations from the researcher and their collaborators are aligned or misaligned and use areas of alignment to inform ethical interpretive authority. Thus, it is imperative that researchers engage in reflexivity to address their subjective position within the study and that their interpretations stem from their place and context within the world (Bar-On, 1996; Josselson, 2007).

Researcher Reflexivity from an AuDHD Perspective

Reflexivity refers to understanding how the researcher's subjectivity impacts the collection, construction, interpretation, and dissemination of knowledge and meaning (Gemignani, 2017; Rankl et al., 2021). In contrast to quantitative methods, which aim to reduce bias by shunning subjectivity, researcher subjectivity is vital in qualitative research in which researchers are part of the meaning-making processes alongside the participants (Olmos-Vega et al., 2022). The contexts and experiences of both the researcher and participant shape how the participant creates, shares, and understands their narrative and how the researcher hears and interprets it. Additionally, in narrative interviews, the participants share their stories and construct meaning within the relationship between the researcher and the participant (Sefora & Ngubane, 2023). Thus, the interpretation of data involves ideas and meaning-making that the researchers have constructed are not purely objective facts that can be precisely replicated by others with different life experiences and histories (Chase, 1996; Murray, 2021). In addition to acknowledging how the researchers' positionality and subjectivity affect the study, reflexivity also involves using this awareness to honor and respect diversity and vulnerability in individuals' perspectives and experiences (Chase, 1996; Josselson, 2007). In other words, reflexivity is purposeful and aids in fostering an empathic research relationship with the participants and other researchers (Gemignani, 2016).

Reflexivity is interesting to consider from my AuDHD perspective. Namely, reflexivity requires robust introspective abilities to examine how personal context shapes meaning-making and interpretation of others' experiences (Rankl et al., 2021). Introspection and subjectivity are fascinating to me as an AuDHD individual as embracing the similarities and differences of my and others' experience comes naturally to me due to many experiences in which my interpretation did not align with others' perceptions of the same event. For example, my perception of "a few minutes" when referring to how long I will need to complete a task is often vastly different to how others interpret the idea of a few minutes. Subsequently, I often find myself preemptively stating that when I tell others I will need a few minutes to do something, I may need several or even hours to complete the task, but state that I will need a few minutes to convey my intention to start on the project very shortly and not later. Therefore, reflexively considering how and why my interpretation and meaning-making differ from participants or other researchers seems like a natural extension of my awareness of differences related to being AuDHD in a neurotypical environment (see Chapters 17, 18, this volume).

Additionally, from a practical standpoint, AuDHD and other researchers can use reflexivity practices to exercise accountability in tracking their responses to the data and ensuring their attention remains on the data related to the research question and not on themselves. Two common functional reflexive practices include collaboration groups or teams and journaling or memoing (Olmos-Vega et al., 2022; Rankl et al., 2021). Collaboration groups can benefit reflexivity by providing a space where each researcher shares their reactions to the data or other aspects of the research process to check in with the group. While collaborating with others, researchers explore whether their responses are idiosyncratic or align with co-constructed meaning and shared understanding of the research (Rankl et al., 2021). Journaling or memoing involves writing down, or notating in some fashion, the researcher's reactions to any aspect of a study, such as thoughts or emotions that arose for the researcher during an interview. Like collaboration groups, reflexive journaling allows researchers to practice attending to knowledge gaps, biases, and blind spots (Olmos-Vega et al., 2022).

To illustrate reflexivity in practice, consider the following excerpt from my reflexivity journal for a project as part of a research methods course. Specifically, I wrote this excerpt following a collaborative team meeting focused on the research course and a discussion for another class with a clinical focus:

> I was frustrated and disheartened by another case presenter's use of the Medical Model to conceptualize their case presentation on an autistic adult. The language felt diminutive and infantilizing, and I noted myself physically pulling back and looking away as the presenter described teaching their client about the client's experience as an autistic person. My heart sank as another audience member noted that autistic people are not usually as insightful or empathetic as the client presented with the case presenter. I felt frozen, wanting desperately to share my reactions but not having the words and feeling too paralyzed by the prospect of outing myself as autistic to these others I did not know well. … Maybe they wouldn't understand. Or worse, maybe their perception of me would morph into a diminutive and childlike caricature of an autistic person with communication deficits. If I felt this conflicted and frozen about addressing my experiences as an educated person in a supportive environment amongst peers and faculty, how do my interviewees feel speaking about their experiences when their experiences have involved actual (rather than imagined or predicted) misunderstanding and invalidation from others? How can [I] highlight my interviewees' voices in a way that is both true to them and understandable to a scholarly audience? I am in a privileged position of being heard, and I am obliged to use my voice, written or otherwise, to lift up the voices of others.

This excerpt was crucial for my ongoing work on my project for the research methods course as it spurred me to consider how my reactions might mirror or diverge from my practice participants. Furthermore, this excerpt and others allude to a few of the costs and benefits of being an AuDHD student and researcher in the field of clinical psychology. One notable cost to me is my perceived need to alter or modify how I speak depending on my context. For example, in the excerpt above, I stayed silent in a group of what I perceived to be neurotypical academics due to concerns about how they might perceive me. In contrast, I might have spoken out to a group of neurodivergent academics with whom I felt safe sharing my similar identity or experience. While this situation occurred in a conversation, it is conceivable that similar situations may arise in other areas, such as if and where I choose to publish my future work. This cost is also linked with the emotional cost of staying silent on my values. If I do not assert my values to others, am I passively agreeing with their values that do not align with mine? If so, what is my purpose or goal with research? However, this question segues to a notable benefit: as a researcher, I am well-equipped and positioned to amplify the voices of others.

Using Narrative Inquiry to Lift the Voices of AuDHD Individuals

Notably, neurodivergent populations have a history of being denied agency, being treated as less-than neurotypical others, and being funneled through harmful so-called best practices at the behest of the proponents of the medical model and not from the individual's own needs and desires (Rosenblatt, 2018). Because of this history, AuDHD individuals may be wary of participating in research. Nonetheless, research must attend to the voices of neurodivergent populations to promote change and accurate representation of neurodivergencies in society, clinical settings, and empirical literature (Bennett & Goodall, 2016; Rosenblatt, 2018). Narrative inquiry may offer a means of amplifying the voices of AuDHD individuals in research and exploring the social contexts that contribute to AuDHD meaning-making. Narrative inquiry is well suited for use with people or groups that the dominant society marginalizes or others (Creswell & Baez, 2021; Sefora & Ngubane, 2023). In particular, narrative inquiry embraces individuals' lived experiences and subjective truths within diverse cultural and social contexts (Josselson & Hammack, 2021). For instance, narrative inquiry could help illuminate the societal and individual experiences of neurodivergent individuals, such as accessibility in healthcare or education or the presence or lack of neurodiversity-affirming practices in a given environment. Furthermore, when enacted earnestly, reflexivity in narrative inquiry can assuage concerns that the researchers' experiences or agendas will overshadow or obscure the participants' life stories and meaning-making (Chase, 1996; Josselson, 2007). For example, a narrative researcher might use reflexivity to highlight their beliefs and biases regarding neurodivergent individuals and groups and subsequently bracket or set these aside to promote a continued focus on neurodivergent voices.

Concluding Remarks

Using narratives is one way that research can highlight and raise the voices of individuals, particularly individuals in groups or populations that society has othered or marginalized (Sefora & Ngubane, 2023). Narrative inquiry involves asking for life stories to deepen insight into people's lived experiences and idiosyncratic methods of meaning-making, and it differs from obtaining reports or highly generalized data (Chase, 1995). Additionally, reflexivity, a crucial component of narrative research, can assist in illuminating the researcher's positionality, relationship to the research question and participants, biases, stereotypes, beliefs, and assumptions (Gemignani, 2016; Olmos-Vega et al., 2022). Furthermore, reflexivity can be of particular importance for studies involving voices that have been underserved or overshadowed in research, as reflexivity helps to delineate between the voices of the participants and those of general society and previous literature or research. Thus, narrative inquiry offers a valuable way to lift the voices of AuDHD and other neurodivergent perspectives in the body of research that has historically overlooked them (Rosenblatt, 2018). This chapter is not a definitive or comprehensive text on the process and application of narrative inquiry. However, I leave you with a few tips worth considering when using narrative inquiry to explore neurodivergent experiences.

Researchers should:

1. Consider their subjectivity and how this relates to the research aims;
2. Engage in ongoing reflexivity practices through journaling and collaboration to ensure that their understanding aligns with and does not overshadow participants';
3. Seek and elicit the rich storied experiences of participants and avoid gathering only basic demographic information;
4. Critically engage with literature and their own research to evaluate whose voices are being heard and whose are kept silent.

References

Bar-On, D. (1996). Ethical issues in biographical interview and analysis. In *Ethics and process in the narrative study of lives*, (pp.45–59). Sage.

Bennett, M., & Goodall, E. (2016). Towards an agenda for research for lesbian, gay, bisexual, transgendered and/or intersexed people with autism spectrum disorder. *Journal of Autism and Developmental Disorders, 46*, 3190–3192. https://doi.org/10.1007/s10803-016-2844-z

Bruner, J. (1990). *Acts of meaning*. Harvard University Press.

Bruner, J. (2004). Life as narrative. *Social Research, 71*(3), 691–710.

Caine, V., Steeves, P., Clandinin, D. J., Estefan, A., Huber, J., & Murphy, M. S. (2018). Social justice practice: A narrative inquiry perspective. *Education, Citizenship and Social Justice, 13*(2), 133–143. https://doi.org/10.1177/1746197917710235a

Chase, S. (1995). Taking narrative seriously. In *The Narrative Study of Lives*, (pp.1–25). Sage.

Chase, S. (1996). Personal vulnerability and interpretive authority in narrative research. In *Ethics and process in the narrative study of lives*, (pp.45–59). Sage.

Clandinin, D. J. (2006). Narrative inquiry: A methodology for studying lived experience. *Research Studies in Music Education, 27*(1), 2–81. https://doi.org/10.1177/1321103X060270010301

Clandinin, D. J., Pushor, D., & Orr, A. M. (2007). Navigating sites for narrative inquiry. *Journal of Teacher Education, 58*(1), 21–35. https://doi.org/10.1177/0022487106296218

Clandinin, D. J. (2007). *Handbook of narrative inquiry: Mapping a methodology.* SAGE Publications, Inc., https://doi.org/10.4135/9781452226552

Craig, C., & Huber, J. (2007). Relational reverberations: Shaping and reshaping narrative inquiries in the midst of storied lives and contexts. SAGE Publications, Inc., https://doi.org/10.4135/9781452226552

Creswell, J. W., & Baez, J. C. (2021). Collecting data with marginalized populations. In *30 Essential skills for the qualitative researchers* (2nd ed., pp.142–148). Sage.

Gemignani, M. (2016). Toward a critical reflexivity in qualitative inquiry: Relational and posthumanist reflections on realism, researcher's centrality, and representationalism in reflexivity. *Qualitative Psychology., 4*(2), 185–198.

Gergen, K. J. (2014). Pursuing excellence in qualitative inquiry. *Qualitative Psychology, 1*(1), 49–60. https://doi.org/10.1037/qup0000002

Head, J. C. (2022). Waves of information, suffocated understanding, and the place for narrative inquiry in the "post-truth" era. *Journal of Constructivist Psychology, 35*(2), 491–510. https://doi.org/10.1080/10720537.2020.1727388

Johnson, M. C., & Thacker Darrow, N. E. (2023). Using intersectionality to explore narrative researcher identity development in narrative inquiry research. *Journal of Constructivist Psychology*. https://doi.org/10.1080/10720537.2023.2170503

Johnstone, B. (2016). Oral versions of personal experience: Labovian narrative analysis and its uptake. *Journal of Sociolinguistics, 20*(4), 542–560.

Josselson, R., & Lieblich, A. (1995). Interpreting experience: The narrative study of lives. SAGE Publications, Inc.

Josselson, R. (1996). *The narrative study of lives.* SAGE Publications, Inc.

Josselson, R. (2007). The ethical attitude in narrative research: Principles and practicalities. In D. Clandinin (Ed.), *Handbook of Narrative Inquiry: Mapping a Methodology* (pp. 537–566). SAGE Publications, Inc. https://doi.org/10.4135/9781452226552.n21

Josselson, R. (2013). Interviewing for qualitative inquiry: A relational approach. The Guilford Press.

Josselson, R., & Hammack, P. L. (2021). *Essentials of narrative analysis.* American Psychological Association.

Marshall, C., Rossman, G. B., & Blanco, G. L. (2022). Qualitative Research Genres. In *Designing qualitative research* (7th ed., pp. 1–17). Sage

Murray, M. (2021). Some ways of doing narrative research. In P. M. Camic (Ed.), *Qualitative research in psychology: Expanding perspectives in methodology and design* (2nd ed.) (pp. 101–121). American Psychological Association. https://doi.org/10.1037/0000252-006

O'Grady, G., Clandinin, D. J., & O'Toole, J. (2018). Engaging in educational narrative inquiry: Making visible alternative knowledge. *Irish Educational Studies, 37*(2), 153–157. https://doi.org/10.1080/03323315.2018.1475149

Olmos-Vega, F. M., Stalmeijer, R. E., Varpio, L., & Kahlke, R. (2022). A practical guide to reflexivity in qualitative research: AMEE Guide No. 149. *Medical Teacher,* https://doi.org/10.1080/0142159X.2022.2057287

Rankl, F., Johnson, G. A., & Vindrola-Padros, C. (2021). Examining what we know in relation to how we know it: A team-based reflexivity model for rapid qualitative health research. *Qualitative Health Research, 31*(7), 1358–1370. https://doi.org/10.1177/1049732321998062

Rosenblatt, A. (2018). Autism, advocacy organizations, and past injustice. *Disability Studies Quarterly, 38*(4). https://doi.org/10.18061/dsq.v38i4.6222Sabar; S. (2013). What's a gestalt? *Gestalt Review, 17*(1), 6–34.

Schiff, B. (2012). The function of narrative: Toward a narrative psychology of meaning. *Narrative Works: Issues, Investigations, and Interventions, 2*(1), 33–47.

Sefora, S., & Ngubane, S. N. (2023). Career development for students with disabilities in an open distance learning institution: A narrative inquiry. *Disability and Society, 38*(3), 445–459. https://doi.org/10.1080/09687599.2021.1946676

CHAPTER 17

On the Simple Joys of Bouncing: Duoethnography of Neurodivergent Researchers in Academia

Teddy G. Goetz and Noah Adams

Introduction

We are openly neurodivergent researchers with experiences of autism and attention deficit hyperactivity disorder (ADHD). In this chapter, we use duoethnography—"a collaborative research methodology in which two or more researchers engage in a dialogue on their disparate histories in a given phenomenon" (Sawyer & Norris, 2015, p. 1)—to make meaning from our experiences as individuals at the intersection of being neurodivergent and researching neurodiversity. We are particularly interested in the ways that academia structurally interfaces with neurodiversity and neurodivergent individuals' experiences and expectations of such environments (see also Chapters 16, 18, this volume).

In undertaking this exploration, we represent the experience of one individual who is autistic (Noah) and one who is ADHD (Teddy). In fact, these experiences are not mutually exclusive, with symptomatic and neurobiological overlap between the two (Antshel et al., 2013), and a significant population estimated to be both (Antshel et al., 2013; Rau et al., 2020), to the extent that it is sometimes abbreviated as "AuDHD." Executive function, for instance,

T. G. Goetz (✉)
Department of Psychiatry, University of Pennsylvania, Philadelphia, Pennsylvania, USA
e-mail: Teddy.Goetz@Pennmedicine.upenn.edu

N. Adams
Ontario Institute for Studies in Education, University of Toronto, Ontario, Canada

H. Bertilsdotter Rosqvist and D. Jackson-Perry (eds.), *The Palgrave Handbook of Research Methods and Ethics in Neurodiversity Studies*,
https://doi.org/10.1007/978-3-031-66127-3_17

covers a cluster of cognitive domains, such as activation, focus, effort, emotion, memory, and action (Brown, 2009), that are implicated in both ADHD and autism (Brown, 2009; Demetriou et al., 2018; Pellicano, 2012).

In this chapter, we will briefly start with some definitions, then alternate autoethnographic writing, interacting with each other's perspectives by noting shared and distinct experiences. We will conclude by discussing our experiences engaging in this generative project and what we hope this inspires in future investigative endeavors.

Definition of Neurodiversity

There are currently varied and sometimes conflicting accounts of neurodiversity, perhaps even within this book. We feel it necessary, and important, to define our own and thus the basis on which our discussion rests. Accordingly, we use the following definition; "Within the neurodiversity movement, all embodied diversity (including neurological differences) is accepted as a facet of human nature, whilst the concept of 'impairment' and related purist medical models are commonly denounced as normative" (McWade et al., 2015, p. 306). That is to say, *neurodiversity* expresses the idea that there are many different ways of perceiving the world, none of which are inherently superior or disordered. The term *neurodivergence* then refers to the ways that specific individuals "perceive, experience, and interact with the world" in a way that is distinct from societal norms (Hamilton & Petty, 2023, pp. 1–2; for more discussion of terminology see Adams et al., 2023). In this framework, *neurotypical* is used to describe non-neurodivergent experiences.[1]

Introducing Duoethnography as a Research Method in Autism and ADHD Research

There is little in the way of duoethnographic writing on the experience of autism or ADHD in academia. However, importantly, there is some precedent to the use of this methodology in duo (or multi)-ethnography in autistic autobiographical writings, though possibly not with this specific intention. Miller's *Women From Another Planet*, for instance, includes two chapters (*Beginnings*, and *Differences*) that put the texts' authors in conversation about being and communicating as an autistic person (2003).

Autoethnographies that explore the intersection of being both autistic and a researcher are somewhat more common. Autoethnographies that explore the academic discourse on this intersection are also increasing. Literature

[1] In this chapter, we will use "neurotypical normativity" to refer to societal assumptions of neurotypicality which accordingly render such experiences neutral, the invisible default. "Neurotypical privilege" will refer to the experience of not needing to consider one's position in the spectrum of neurodiversity due to having a neurotypical experience for which society has been structured (Bouckley, 2022); others have named it "the educational tyranny of the neurotypicals (Ito, 2018)".

on autistic phenomenologies—i.e., lived experiences, as opposed to external manifestations—is also growing. Raymaker (2017), for example, used evocative autoethnography to articulate and describe her experiences as an autistic, disabled "insider researcher," and so to push back in a way that pushed back against sociocultural narratives that expect autistic individuals to struggle to articulate themselves in speaking for themselves. Williams (2020) also published an amalgam of non-fiction and fictionalized autistic autoethnographies that drew from autistic participants' descriptions of their experiences with communication.

Collective autoethnography about autistic experiences to date includes Jackson-Perry et al. and and's (2020a, 2020b) text on sensory experiences and Bertilsdotter Rosqvist et al.'s (2020) on social spaces and community-building. Murray et al. (2023) engaged in an autistic/allistic discourse about individual life course experiences, as synthesized for common themes (e.g., the need for trust and reliability, the impact of context on regulation of emotion, sociability, and empathy) and divergent themes (e.g., primary sensory experience and social joining). Autoethnography of ADHD experiences, however, remains more limited (e.g., Ing & Mills, 2019; Salty et al., 2022, see also Chapter 12, this volume).

Why Duoethnography

We use duoethnography to articulate the experiences of individuals who are at the intersection of researching and being neurodivergent and to explore the ways in which these positionalities interface with academia. It is a very new methodology (Norris & Sawyer, 2012) which, alone amongst other qualitative methods, focuses on researchers, and the dialogue between them, as sites of information-gathering (Breault, 2016). It is thus an ideal method to use in exploring the divergent experiences of neurodivergent researchers within academia, where little such research exists.

Duoethnography is an extension and evolution of autoethnography, which "is a [qualitative] research method that uses personal experience… to describe and interpret… cultural texts, experiences, beliefs, and practices… [from the belief] that personal experience is infused with political/cultural norms and expectations (Adams et al., 2017, p. 1)." Both duoethnography and autoethnography serve to intimately illustrate concepts by grounding individual experiences within theoretical frameworks and, in doing so, respond to community stakeholders' calls to increase research that pertains to their lived experiences (Bertilsdotter Rosqvist et al., 2019; Fletcher-Watson et al., 2019). Duoethnography, however, takes autoethnography further and multiplies it by providing for a collaboration between two (or more) researchers to "juxtapose… and work in tandem to untangle and disrupt meanings about a particular social phenomenon (Burleigh & Burm, 2022)." While there is a young literature documenting autistic academics' experiences, ADHD experiences in academia remain relatively untouched, as does juxtaposition of neurodivergent experiences. Duoethnography as a methodology allows us to articulate and triangulate a common experience.

Data Collection

As a duoethnography, our primary data is composed of each researcher's individual experiences of being a neurodivergent individual who also conducts research on neurodiversity (Chang, 2013). We took turns writing two self-contained narratives on this experience, the first on our research process and the second on navigating our identities and roles as neurodivergent researchers. We purposely did not discuss or show each other what we would write before doing so. The narratives themselves were written in one-sitting and in a "free-flowing" manner, in order to best articulate our own internal methods of thinking. After the initial "burst" of writing, we went back and edited our own excerpts, while refraining largely from editing those of each other. Those edits we did make were in the form of suggestions and largely restricted to suggesting where something might be elaborated on. A Google Doc (Mountain View, CA) was used for this, which made co-working on the chapter much easier.

Analysis

As an emerging qualitative method of inquiry, the means of analysis of the data gleaned through duoethnography are not yet standardized (Breault, 2016). We opt here to simply compare and contrast our differing experiences of being neurodivergent individuals who also research neurodiversity; one from the perspective of an ADHD individual and one from an autistic. To do so, we each read and re-read the others' narratives and took notes on the aspects of neurodivergency that seemed shared—or different—between us. We then collaborated to provide a shared analysis of the experience of individuals at the intersection of being and researching neurodiversity. In doing so, we draw out common experiences between our two perspectives and provide the groundwork for more in-depth examinations.

Findings

Here we will present duoethnographic data, beginning with autoethnography about how ADHD impacts TGG's research process, followed by NA discussing how autism influences his research process, then TGG commenting on the experience of reading NA's and vice versa.

Neurodivergent Research Processes

ADHD. As I start to write this, I can't focus: I'm in a rolling office chair at work. It's objectively nice, offers adequate lumbar support. Yet, my legs are straightened in front of me under the desk, bouncing up and down, bumping into the wall periodically. For the past decade, my productivity has relied upon using a yoga ball as a chair. I forget about it until colleagues ask why I'm bouncing on video calls. My resting state is bouncing.

The graveyard of half-finished projects on my computer hard drive rolled over when I agreed to write this chapter. *Your day will come*, I assure them. I'm just waiting for the perfect weekend in which I have no obligations and wake up inspired to write one of those data sets into a paper! When that happens, the draft pours forth in a satisfying frenzy of emphatically loud typing. Unstoppable. Is there any difference between a paper written in eight straight hours based on data that sat untouched for two years and steady and deliberate progress on a project over the same period of time? To the journal reader, no. To the spouse, friend, or coworker witnessing the hyperfocus wormhole, absolutely. What magic! Not a researcher, but a machine! Perhaps coveting dazzling efficiency. To me, guilt covers neglected ideas as green biofilm, thicker and slimier each time I remember having forgotten and still lack the motivation needed to open the file and start. No matter what gets finished, I do not feel prolific or even celebratory—I have poor object permanence. *In press* may as well be in a closed drawer or cabinet: as good as forgotten.

I can't imagine spending a career studying one highly specific topic. I get bored. Or rather, I get distracted—I have variable attention (see Bertilsdotter Rosqvist et al., 2023a, 2023b; Hallowell & Ratey, 2021). What is as tantalizing as a new idea? It can make planning difficult. The proposal flows almost effortlessly. Data gathering is intoxicating—the only true novel information that exists in the world. As I said in middle school: science class was a misnomer—it, too, taught history. But where does one find the motivation to transcribe and code after conducting an interview? To revamp a rejected paper and reformat it for submitting to a new journal? There are so many steps between revelation and dissemination, each one treading through quicksand. My initial interest in research, as in medicine, was wanting to improve lives, but I fell in love with research as an adrenaline junky.

My friends in college and medical school could not understand what got me into the lab at 4:00am. My honest response of "data" was unable to convey my dopaminergic glee at the tantalizing possibility of discovering new information. Translating a few hours of work into bursts of numbers that promised a new understanding of the world was intoxicating. It felt like a slot machine that won more times than not; I learned over the course of five years in a wet lab that I didn't have the patience for techniques that didn't, that required more finesse. The microscopic began to threaten meaninglessness.

I switched to qualitative and mixed-methods research exploring human experiences and developing interventions to meet those expressed needs. Such work feels less dissociated and does not require a patience for repetition without payoff that I inherently lack. The stable nature of interview data is more robust to fits and starts—it is not subject to denaturing, swapped reagents, or mold contamination. It also runs on a different funding scale: there is less for me to manage. The intellectual stimulation to executive functioning self-discipline ratio is infinitely higher, allowing me to find a semblance of home, to imagine a career of success rather than self-sabotage—or, at least, to hope.

Autistic. Teddy has sent me an early draft of this article to look at. It's one long stream of text roughly broken up according to headlines that seem haphazard and temporary. I start by editing the abstract, before I even realize that there is more text. I am fixating and it occurs to me, after creating a wall of red track-changes, that Teddy might find this insulting. I realize that I'm not really getting anywhere, so I zoom out and start to work on the paper outline. Making sure that everything is in its right place, with page breaks, proper APA headings, and the margins just so. This is always a good place for me to start. It makes things feel orderly.

I set up my working environment in the same way. Sitting at a coffee shop on a comfortable couch and with my legs crossed or tucked under me. It always has to be the same, but not always in the same place. Right now, for instance, I'm in Glasgow, but usually I'm at a co-working space near my apartment in Berlin. I usually wear big headphones and recently switched to my first pair of active noise canceling headphones[2] (where have you been all my life!). When walking around I'm almost always listening to a podcast, but right now the words will distract me and only synthwave will do. In fact, at the moment I'm listening to Fleetmac Wood.[3]

The headphones are important because sounds and movements will distract me until they become unignorable and start to make me angry. Sudden high-pitched sounds are the worst. This way I can work without distractions, at least for a bit. Of course there are the distractions in my head. Sometimes it's still like a clear pond up there. I can get lost and hyperfocus on the work I'm doing. But sometimes I'm distracted with one of my special interests or, less charitably, obsessions. I listened to a podcast about UAF Flight 571 (Marshall, 2022) the other day. Now I can't stop my mind from wandering back and worrying at it. Like if I could figure out the logic of selflessness and heroism I could provide the formula to the world.

Concentrate Noah.

[2] Sony WH-1000XM4 for those who are interested.

[3] Check them out @ http://www.fleetmacwood.com/!

It bothers me that Teddy has filled in some areas and not others, created some headings, but left out expected ones. A section for our autoethnographies, but not data?[4] I attempt to impose some more order on it by placing the autoethnography titles under one subheading (Data) and starting a short section explaining our intent. That's a bit better. But still, where to start? In retrospect it makes sense that they would fill out the pieces that they know and leave the rest to me. This is collaboration or, apparently, duoethnography. And I recognize that I make the same assumptions with regards to my own writing, as we pass this back and forth. Nevertheless, I'm not good with a blank page and even worse with one that seems random. I need a model that I can extrapolate from. I often wonder whether it is possible for me to be creative or original, or if I am just very good, *very practiced*, at building seemingly new things from the bricks that others have left behind. I suppose it doesn't matter. My autoethnography is a good enough place to start. A corner that I can wedge the rest of the paper out of.

I keep bouncing back and forth between Teddy's section and mine. I can't say I speak for all autistics, but much of what others' write seems hauntingly familiar. The hyperfocus that results in writing screeds of information over a relatively short amount of time, followed by months of avoidance. The preference for late night work. I have never been what you might call a morning person. But, unlike Teddy, I love reformatting and editing papers. It's like pulling teeth to get it out, but going over and making it better is easy and what's more, fun. I feel like I'm achieving something and working towards a goal. That it's measurable in some definable way. I have always loved paperwork. Probably for the same reason I love data entry. I'm an excel-spreadsheet junkie. I find the idea of working steadily on a small but incremental contribution to a distant goal that I'm passionate about deeply satisfying. You see that tiny dot there? That's me!

That's an autistic thing too. That hyperfocus and fixation on 'special interests.' I have many special interests of course, but my interests in research methodology, as well as transgender and autistic subjects, probably serve me the best in an academic and research environment. I suppose in many ways this makes me the platonic ideal of a researcher. Diligent, hardworking, and without a good work/life balance. That's fine. But academia is more than that. There are hidden power dynamics and politics and the day-to-day realities of interpersonal communication in research groups and with colleagues. It would be an understatement to say that I am terrible at both. I lumber around like a bull in an emotional china shop. Stumbling into shelves and knocking over interpersonal teacups. People read intention into my misunderstanding or failure to pick up unspoken hints, but more often than not I just haven't noticed. On the occasion where I do, I have learned not to react for fear of assuming the wrong meaning and thus saying or doing the wrong thing. In

[4] Later when editing I will forget that I wrote this and haphazardly delete a heading that no longer seems adequate to me. I will wonder who put it there.

my calculation it seems better to gamble at people being confused than being sure that we will misunderstand each other. Most often though I don't even notice and so it will get back to me months, or even years later than people have assumed my terseness or lack of emotional response was a judgment of them. All in all, it seems better to just isolate myself from others and limit my interactions to email as much as possible.

Don't get me wrong. I have attempted to bridge this communication gap and explain when I notice that I am miscommunicating. It has been my experience however that neurotypical people tend to view this as 'making excuses' and to overestimate their ability to empathize with what it means for me to be autistic. Milton (2012) identify this as the double empathy problem; where communication breakdown occurs between autistics and allistic (non-autistic) people, but the autistic person is primarily or solely seen as the problem. So, all In all, it seems safer just to agree with them and stop speaking. Unfortunately, this isn't ideal for climbing the academic hierarchy which, admittedly, is something I have little to no interest in anyway. I do worry about my ability to navigate academic politics in hiring and staying employed though.

Navigating Identities and Roles as a Neurodivergent Researcher

ADHD. The main ghost that haunts my racing thoughts in the middle of the night is named "professionalism." An acceptably veiled term used to block access to expert status within medical and research from those of us for whom these fields were not made (e.g., Frye et al., 2020; McCluney et al., 2021). I am more familiar with this explicit phenomenon with regards to my transness, as my ADHD is generally obvious on observation to a trained eye, but is not written on ID badges and email signatures with my they/them pronouns–or stereotypically clocked in my short stature, youthful appearance, and patchy beard. I have a decade of research experience on transgender, non-binary, and/or gender expansive health, including developing the first animal model of gender-affirming hormone therapy as my master's thesis, and yet… I have been invited to speak at conferences with cis colleagues, only to later learn that while they will be reviewing the literature, I (the only speaker who conducted primary research on the topic) am to provide the "color commentary" discussing my personal experiences as a non-binary/trans person. I have swallowed my rage and pride, played the academia game, only to have my personal and professional expertise directly contradicted and undermined by my co-presenters.

With ADHD (as with my chronic illness), I am not tokenized the same way. Though I am open with colleagues about my lived experiences on principle–I refuse to contribute to stigma, to give myself permission to feel shame–those labels do not seem to color their every interaction with me. This is exceedingly ironic as my neurodivergence courses through my every thought and action in a way that my gender does not. Perhaps neurodivergence feels more nebulous to neurotypical brains—or the façade of neurotypical normativity remains particularly impermeable—and accordingly it's easier to consider manifestations as individual traits, or more often eccentricities, even failures.

My core ADHD fear as a doctor and researcher boils down to: ***am I interpreted as unprofessional?*** This blares in my brain as follows:

Sitting on a yoga ball chair improves my focus, but accentuates my baseline bouncing–in video meetings, should I forgo this aid to mask, try to shake and squirm entirely off-camera?

> Was I seen as too loud, too fast, too much when enthusiastically talking about my research with a potential collaborator or employer?
>
> Was I labeled rude when I was overstimulated by the noise and lights at a networking event and stopped engaging appropriately? Or, as aloof or arrogant when I then tried to hide in a corner alone to re-regulate, unable to suppress the wince on my face?

I have developed so many work-arounds for my profound difficulty holding multiple pieces of information in my awareness simultaneously–like responding to emails immediately, before they disappear from my awareness for good, and setting repetitive phone alarms to remind me of meetings–but sometimes I am interrupted or just too burnt out. How much erratic (dis)engagement can a work relationship tolerate before I am labeled unreliable or flaky? I fear that the times I'm able to meet expectations will be held against me when I fail to do so–evidence that I have chosen to fall short. Meanwhile, neurotypical privilege renders my efforts invisible.

How many typos, half-finished sentences, and formatting errors are acceptable in a paper draft? When writing (and editing), I jump around between sentences and sections, following the dopamine, allowing my thoughts to charge forward and flow out onto the page. Prolific! Yet, I leave fragments in my wake. Forcing myself to go sequentially leads nowhere: I get bored and bail. I feel self-conscious when the time-stamps on tracked changes and comments in word processing documents expose my process. There are currently three finished primary research papers that sit unpublished on my computer hard drive because of my physical inability to reformat them for a new journal's stipulations after our prior rejections. My shame prickles whenever I remember that manuscript graveyard; I have no idea what my co-authors–mentors and collaborators–think.

Navigating academic cultural expectations and values is not a neutral experience. How many times will I sit in a lecture by someone who does not share my lived experience, about my lived experience (and professional expertise), and hear incorrect (even offensive) information spouted as fact? How many times will my heart race, my face flush, my eyes burn, as I decide whether or not I may speak up *again* in that room? How many times does that trauma wreck my productivity for the remainder of the day? I am a scholar who studies my community's experiences, needs, and preferences and I work tirelessly to manifest that better world. Yet, trying to live up to the pressure of being the voice in those rooms, advocating for those who are not, takes a toll.

Autistic. "They say they want you successful, but then they make it stressful/ You start keepin 'up the pace, they start changing up the tempo" (Mos Def in Jackson, 2021).

Much of what Teddy has written above seems similar yet ever so slightly different from my own experiences. I don't, for instance, worry so much about the number of typos in a submission (especially since the almost universal availability of spell check). But I certainly self-police for whether my emotional and reciprocal expression appears correct to others. Was I too loud? Did I speak too much or too little? The rules seem capricious and unreliable.

It's a cliché, but I'm reminded of a scene from Star Trek Generations (Carson, 1994) where Data attempts to emulate a practical joke played on one crewmate by repeating it on another.[5] Logically, this makes sense. Data is simply recreating the joke played on Worf in an attempt to share the experience with his comrades. But it lacks context and, more importantly, the seemingly psychically communicated emotional subtleties that tell his colleagues where 'the line' is. As Teddy states, the person communicating in this way is perhaps seen as rude and unpredictable. The alternative is not to participate at all and thus be viewed as aloof, arrogant, and to be intentionally nonverbally communicating disdain and a lack of interest. It doesn't matter that I haven't said this or even, when asked, specifically refuted this intention. It seems that, no matter what I do, my social behavior is interpreted as unprofessional.

In fact, the very concept of professionalism represents a collection of unspoken and rarely articulated ideas about the 'proper' way to behave at work that are an outgrowth "of the values, actions, and emotions of straight white men (Mann, 2018)." Tami Jackson beautifully articulates how this especially affects black women who are "mentored to mask [their] personhood and humanity into a small, shallow, tight box… designed to suffocate [them] with white norms (2021)." These aren't, per se, challenges of the core values of professionalism, such as respect for patient welfare and a commitment to being on time (Lucey & Souba, 2010). Rather the challenge is to the 'neutral' baseline on which professional appearance and communication is determined.

I have, nevertheless, developed a series of adaptations to manage my own workflow and communication with others. First and foremost, I strongly prefer online communication if possible. This allows me time to consider and frame my responses in a way that neurotypicals can understand. I use an email client to collect and access all my email addresses into a central location and I treat the inbox as a list of things that need to be acted on. Everything else is immediately moved to 'archives.' I am always in search of 'inbox zero,' and achieve it 1–2 times a month. Like Teddy, I have a mountain of unfinished projects. Though, over time, I have become better at picking out and pursuing projects

[5] I highly recommend Gil Goletski's Bodies with purpose: An exploration of the intersection of autistic and transgender coding in Star Trek (2019).

to completion. Part of this is learning to say no more often and part is knowing that the time for old projects often comes back around again. I've developed a digital filing system for those that get shelved to allow me to easily access them again in the future. I also often offer unfinished projects to others if I see that they are doing something similar. Better that they do something with it than it just sits on my hard drive forever.

Analysis

Teddy's Reflection on Noah

Reading Noah's reaction to getting my chaotic seedling of a document made me laugh out loud. "Haphazard and temporary" is how all of my work starts—I chafe against strict formats, I struggle with being told what to do. I cannot relate to his reflection, "I need a model that I can extrapolate from." Rather than hurt, I was profoundly relieved when Noah reorganized everything. It did not change my writing process at all (I continued to bounce between sections) and saved time and frustration on the back end. Trying to tidy up in that way is exhausting and difficult for me—I couldn't care less and don't even really see it.

Noah's description of needing specific sounds or quiet for different tasks resonates with mine. For my brain, the most taxing cognitive functions require silence, but I have specific playlists that I play while writing if I need a stimulus to prevent me from just zoning out thinking for the next hour. This week my hyperfixation has been Rachel Carson after learning more about her work and life in an audiobook that I was reading. I wonder about the structure of Noah's headphones. I can't handle over-the-ear ones. I bop my head too much and they wiggle and drive me nuts. Also they squish my ears into my glasses. I need soft-tipped earbuds (never the hard plastic!).

Noah's invocation of the bull in the china shop metaphor interests me, because I've often been described as moving that way (impulse to muscle faster than it's possible to notice the person or end table that is already occupying that space. I don't resonate as much with Noah's interpersonal iteration. I sometimes bowl people over with overly direct responses, but navigating social (and really political) dynamics doesn't mystify me the same way that it seems to Noah. I don't like feeling like a cartoon spy playing laser limbo, but with intention I'm usually able to avoid getting burned.

I share Noah's quest for 'inbox zero,' but am eternally out of reach of it. Part of it is that certain messages are not truly crossed off my to-do list. The larger factor is the digital manifestation of object impermanence. The same reason I forget to close kitchen cabinets and leave items lined up on the top of my desk instead of in a drawer—for me, out of sight is out of mind and I don't want to lose things that I think might be important in the future. I have a comprehensive email inbox folder filing system, but it's an archive: I only move messages into storage when I don't think I will actually need them, but keep them for posterity just in case. I wonder if Noah has the same problem or if his filing system feels dynamic, accessible, always within reach. I wonder if he trusts himself to know where he leaves things.

Noah's Reflection on Teddy

As I read through Teddy's reflections on their processes as a neurodivergent researcher I am struck by the similarities to my own experiences. We share a non-linear thinking style, but it is articulated in different ways and with unique dimensions. We are alike in that we have both consciously—and probably unconsciously—adapted to working in and with the neurotypical world. But the ways in which we adapt sometimes differ. Teddy bounces. I burrow into a topic until I can't hear anything around me, preferably while sitting on a weatherbeaten sofa in the corner of a coffee shop. We both worry about the acceptability of our adaptations and the degree to which they are tolerated by our colleagues. I deal with this by avoiding them, but I imagine Teddy doesn't have that luxury.

I am trying hard here, as we agreed, to not read Teddy's response before finishing my own, but I did spy the last line, in which they muse on my own relationship to object permanence. That is, do I so meticulously construct my 'intellectual filing system' because I don't trust myself to know where I put things. The short answer is that I don't. Out of sight, out of mind. I structure things in a consistent and patterned manner so that, even when I forget about the whole, I can find the individual components and trace things from there.

Teddy shares their need to work on new projects in order to maintain interest and enthusiasm. I have certainly had these moments but, overall, I prefer to slowly and methodically work through the intricacies of data, entering things in spreadsheets, and organizing them. And while Teddy mentions difficulty in returning to and revising articles, I thoroughly enjoy it… as long as I can understand and accept the revisions suggested, which is far from guaranteed. I suspect my problem here is more one of stubbornness than attention though.

Both Teddy and I appear to have learned to direct our academic work towards our strengths and away from the areas that we find intolerable or impossible. Though our directions might be different, the goal is the same and often the 'work-arounds' and adaptations we develop are as well. Likewise, we share concern that 'professionalism' is used to disparage the work of researchers who diverge from the mean. Researchers who bounce. It never seems to occur to those defining normality that they are, perhaps, the abnormal ones. Teddy observes that academic cultural expectations and values are not neutral, and I add that they are also invisible to the very people who are centered by them. I am fascinated by their observation that their trans identity is more often the source of academic marginalization and tokenization than their neurodivergence though.

Then there is the shame. The fear of being first noticed as different and then excluded for it, no matter the quality of your work or time spent on it. And the fear that your non-linear style of thinking is bad, or somehow wrong, simply because it is different. And the burn out when we are forced to perform in a personally abnormal manner for too long.

Discussion

The process of creating this duoethnography, our first, was illuminating. The focus of this chapter is on juxtaposing our experiences of neurodivergence *vis a vis* being both neurodivergent individuals and researchers of this subject. This exploration led us, unexpectedly, to find common experience between our respective methods of navigating neurotypical normative academic environments. We both, for instance, have non-linear thinking styles, issues with object permanence, and deploy multiple adaptations—particularly organizational—to integrate into the neurotypical world. Our differing neurodivergencies (ADHD and autism) go some way to explaining the distinctions between how we experience and do these things, but the similarities are, at least to us, surprising.

The key similarity between our experiences appears to be the overall sense of hypervigilance we have over our actions and behaviors, and the ways in which they are perceived by the neurotypical world. Teddy with their bouncing, and Noah with his uncertainty regarding others' expectations. While we focus on moving toward our neurodivergent strengths, this vigilance has a continuous and corroding impact on our sense of self and certainty as individuals and academics. It is a situation aptly described by the double empathy problem, whereby neurodivergent individuals bear the higher burden of rectifying communication breakdowns with neurotypicals (Milton, 2012). But this is also a problem of masking, that is the multiple strategies used by neurodivergent people to hide their diversity, which has been found to have a profoundly negative self-impact (Miller et al., 2021).

Masking, however, is often perceived as necessary to compete in the highly structurally neurotypical world of academia, where neurodivergent behaviors (e.g., rocking, stimming, needing adaptations) may receive harsh criticism and even censure from neurotypical peers. In this context, neurotypical behavior is seen as normative and natural, whereby many neurodivergent ones are perceived, without evidence, as intentional, problematic, and even harmful. Indeed, research suggests that neurotypicals may be much more vulnerable to making snap negative decisions about neurodivergent peers—specifically autistics—based solely on social and non-verbal differences and despite 'functioning' status (Sasson et al., 2017).

In fact, this is an expression of ableism in academia more broadly, where disclosure by and representation of disabled academics continues to lag behind the student population (Brown et al., 2018). Nevertheless "'coming out' and the complex issue of representation for disabled academics is a double-edged sword involving an intricate web of negotiations, rarely with the promise of a positive career outcome (Gillberg, p. 13)."

A recent German study found that the unemployment rate amongst autistics is roughly 5 times the general population (25.2%), that these periods average 23 months, and that "interpersonal problems [are]... the main reason for contract termination (Espelöer et al., 2023, p. 731)." Higher levels of education did not protect against this. This study is typical, with others finding even higher rates of unemployment (Ohl et al., 2017). Whilst information on unemployment of ADHD individuals appears to be scarcer, one study did note that these individuals had a 17% lower income and more days of unemployment per year than neurotypicals (Jangmo et al., 2021). Accordingly, neurotypical bias against neurodivergent individuals' ways of thinking and interacting with the world appears to carry a real potential to negatively impact our career possibilities and outcomes.

One potential answer to these dilemmas may be to support neurodivergent individuals in conducting and leading research on neurodiversity. Doing so would create an opening for neurodivergent researchers in the academy, increase knowledge on our experiences, and provide better access to what can be a very hidden population. At its most basic, doing so would entail acknowledging that neurodivergent individuals can act as competent and objective researchers on subjects that center our own experiences. From a more structural perspective, adaptations to Research Ethics Boards and a greater appreciation for neurodivergent communication styles in the academy, on project teams, and in meetings would be necessary.

Conclusions

Engaging in ADHD/autistic duoethnography about our experiences as neurodivergent scholars was both revelatory and healing. We were able to glean a better understanding of how our own minds worked and plumb the richness of divergent and overlapping experiences that openly naming—and celebrating!—neurodiversity offers. Explicitly articulating our chafing against neurotypical normative social contracts and sharing that with a neurodivergent colleague combatted the shame that we feel for such, usually internalized, struggles; this Google Doc was a space in which we found community. We hope this project underscores the value of collaboration and knowledge exchange between individuals with distinct neurodivergent experiences and inspires future projects utilizing similar methodologies.

References

Adams, T. E., Ellis, C., & Holman Jones, S. (2017). Autoethnography. In J. Matthes, C. S. Davis, & R.F Potter (Eds.), *The international encyclopedia of communication research methods* (pp. 1–11). Wiley. https://doi.org/10.1002/9781118901731.iecrm0011

Adams, N., & Goetz, T.G., Walsh, R., & Derrick, Z. (2023). DoubleQueer: Being neurodiverse and gender diverse neurodiversity and gender diversity. In T. Goetz, & A. S. Keuroghlian (Eds.), *Gender-affirming psychiatric* care. American Psychiatric Association Publishing.

Antshel, K. M., Zhang-James, Y., & Faraone, S. V. (2013). The comorbidity of ADHD and autism spectrum disorder. *Expert Review of Neurotherapeutics, 13*(10), 1117–1128. https://doi.org/10.1586/14737175.2013.840417

Bertilsdotter Rosqvist, H., Hultman, L., Österborg Wiklund, S., Nygren, A., Storm, P., & Sandberg, G. (2023a). Intensity and variable attention: Counter narrating ADHD, from ADHD deficits to ADHD difference. *British Journal of Social Work, 53*(8), 3647–3664. https://doi.org/10.1093/bjsw/bcad138

Bertilsdotter Rosqvist, H., Hultman, L., Österborg Wiklund, S., Nygren, A., Storm, P., & Sandberg, G. (2023). Naming ourselves, becoming neurodivergent scholars. *Disability and Society*. https://doi.org/10.1080/09687599.2023.2271155

Bertilsdotter Rosqvist, H., Kourti, M., Jackson-Perry, D., Brownlow, C., Fletcher, K., Bendelman, D., & O'Dell, L. (2019). Doing it differently: Emancipatory autism studies within a neurodiverse academic space. *Disability and Society, 34*(7–8), 1082–1101. https://doi.org/10.1080/09687599.2019.1603102

Bertilsdotter Rosqvist, H., Örulv, L., Hasselblad, S., Hansson, D., Nilsson, K., & Seng, H. (2020). Designing an autistic space for research: Exploring the impact of context, space, and sociality in autistic writing processes. In H. Bertilsdotter Rosqvist, N. Chown, & A. Stenning (Eds.), *Neurodiversity studies: A new critical paradigm* (pp. 156–171). Routledge.

Bouckley, C. (2022, February 24). Neurotypical privilege in the labour market. *LSE Business Review*. https://blogs.lse.ac.uk/businessreview/2022/02/24/neurotypical-privilege-in-the-labour-market/

Breault, R. A. (2016). Emerging issues in duoethnography. *International Journal of Qualitative Studies in Education, 29*(6), 777–794. https://doi.org/10.1080/09518398.2016.1162866

Brown, N., & Leigh, J. (2018). Ableism in academia: Where are the disabled and ill academics? *Disability and Society, 33*(6), 985–989. https://doi.org/10.1080/09687599.2018.1455627

Brown, T. E. (2009). ADD/ADHD and impaired executive function in clinical practice. *Current Attention Disorders Reports, 1*(1), 37–41. https://doi.org/10.1007/s12618-009-0006-3

Burleigh, D., & Burm, S. (2022). Doing duoethnography: Addressing essential methodological questions. *International Journal of Qualitative Methods, 21*. https://doi.org/10.1177/16094069221140876

Carson, D. (Director). (1994, November 18). *Star trek: Generations* [Film]. Paramount Pictures.

Chang, H. (2013). Individual and collaborative autoethnography as method. In S. H. Jones, T. E. Adams, & C. Ellis (Eds.), *Handbook of autoethnography* (pp. 107–122). Left Coast Press.

Jackson-Perry, D., Bertilsdotter Rosqvist, H., Annable, J. L., & Kourti, M. (2020). Sensory strangers: Travels in normate sensory worlds. In H. Bertilsdotter Rosqvist, N. Chown, & A. Stenning (Eds.), *Neurodiversity studies: A new critical paradigm* (pp. 125–140). Routledge. https://doi.org/10.4324/9780429322297-12

Jangmo, A., Kuja-Halkola, R., Pérez-Vigil, A., Almqvist, C., Bulik, C. M., D'Onofrio, B., Lichtenstein, P., Ahnemark, E., Werner-Kiechle, T., & Larsson, H.

(2021). Attention-deficit/hyperactivity disorder and occupational outcomes: The role of educational attainment, comorbid developmental disorders, and intellectual disability. *PLoS ONE, 16*(3), 1–16. https://doi.org/10.1371/journal.pone.0247724

Demetriou, E. A., Lampit, A., Quintana, D. S., Naismith, S. L., Song, Y. J., Pye, J. E., Hickie, I., & Guastella, A. J. (2018). Autism spectrum disorders: A meta-analysis of executive function. *Molecular Psychiatry, 23*(5), 1198–1204. https://doi.org/10.1038/mp.2017.75

Espelöer, J., Proft, J., Falter-Wagner, C. M., & Vogeley, K. (2023). Alarmingly large unemployment gap despite of above-average education in adults with ASDH without intellectual disability in Germany: A cross-sectional study. *European Archives of Psychiatry and Clinical Neuroscience, 273*(3), 731–738. https://doi.org/10.1007/s00406-022-01424-6

Fletcher-Watson, S., Adams, J., Brook, K., Charman, T., Crane, L., Cusack, J., Leekam, S., Milton, D., Parr, J. R., & Pellicano, E. (2019). Making the future together: Shaping autism research through meaningful participation. *Autism, 23*(4), 943–953. https://doi.org/10.1177/1362361318786721

Frye, V., Camacho-Rivera, M., Salas-Ramirez, K., Albritton, T., Deen, D., Sohler, N., Barrick, S., & Nunes, J. (2020). Professionalism: The wrong tool to solve the right problem? *Academic Medicine, 95*(6), 860–863. https://doi.org/10.1097/ACM.0000000000003266

Gillberg, C. (2020). The significance of crashing past gatekeepers of knowledge: Towards full participation of disabled scholars in ableism academic structures. In N. Brown, & J. Leigh (Eds.), *Ableism in academia: Theorising experiences of disabilities and chronic illnesses in higher education* (pp. 11–30). UCL Press. https://doi.org/10.2307/j.ctv13xprjr.7

Goletski, G. (2019, November 3). Bodies with purpose: An exploration of the intersection of autistic and transgender coding in Star Trek. *Medium*. https://medium.com/@ggoletski/bodies-with-purpose-an-exploration-of-the-intersection-of-autistic-and-transgender-coding-in-star-49070f5ed391

Hamilton, L. G., & Petty, S. (2023). Compassionate pedagogy for neurodiversity in higher education: A conceptual analysis. Frontiers in psychology, 14, 1093290. https://doi.org/10.3389/fpsyg.2023.1093290

Hallowell, E.M., Ratey, J.J. (2021). ADHD 2.0: New science and essential strategies for thriving with distraction-from childhood through adulthood. Ballantine Books.

Ing, C., & Mills, J. P. (2019). 'Hey, look at me': An auto ethnographic account of experiencing ADHD symptoms within sport. *Qualitative Research in Sport, Exercise and Health, 11*(2), 274–283. https://doi.org/10.1080/2159676X.2017.1405362

Ito, J. (2018, September 6). The educational tyranny of the neurotypicals. *Wired*. https://www.wired.com/story/tyranny-neurotypicals-unschooling-education/

Jackson, T. (2021, March 18). Operationalizing justice: The problem with 'professionalism.' *The Inclusion Solution*. http://www.theinclusionsolution.me/operationalizing-justice-the-problem-with-professionalism/

Jackson-Perry, D., Bertilsdotter Rosqvist, H., Annable, J. L., & Kourti, M. (2020). Sensory strangers: Travels in normate sensory worlds. In H. Bertilsdotter Rosqvist, N. Chown, & A. Stenning (Eds.), *Neurodiversity studies: A new critical paradigm* (pp. 125–140). Routledge.

Lucey, C., & Souba, W. (2010). Perspective: The problem with the problem of professionalism. *Academic Medicine, 85*(6), 1018–1024. https://doi.org/10.1097/ACM.0b013e3181dbe51f

Marshall, S. (Executive Producer). (2022, October 31). Flight 571: Survival in the Andes with Blair Braverman. [Audio podcast episode]. In *You're Wrong About*. https://www.youtube.com/watch?v=kTmcNAjn9U8

Mann, S. (2018, August 2). The problem with "professionalism." *Bust*. https://bust.com/feminism/194931-problem-with-professionalism.html

McCluney, C. L., Durkee, M. I., Smith, R. E., II., Robotham, K. J., & Lee, S. S. L. (2021). To be, or not to be... Black: The effects of racial codeswitching on perceived professionalism in the workplace. *Journal of Experimental Social Psychology, 97*(104199), 1–12. https://doi.org/10.1016/j.jesp.2021.104199

McWade, B., Milton, D., & Beresford, P. (2015). Mad studies and neurodiversity: A dialogue. *Disability and Society, 30*(2), 305–309. https://doi.org/10.1080/09687599.2014.1000512

Miller, D., Rees, J., & Pearson, A. (2021). "Masking is life": Experiences of masking in autistic and nonautistic adults. *Autism in Adulthood, 3*(4), 330–338. https://doi.org/10.1089/aut.2020.0083

Miller, J. K. (2003). *Women from another planet*. AuthorHouse.

Milton, D. (2012). On the ontological status of autism: The Double empathy problem'. *Disability & Society, 27*(6), 883–887. https://doi.org/10.1080/09687599.2012.710008

Murray, D., Milton, D., Green, J., & Bervoets, J. (2023). The human spectrum: A phenomenological enquiry within neurodiversity. *Psychopathology, 56*(3), 220–230. https://doi.org/10.1159/000526213

Norris, J., & Sawyer, R. (2012). Toward a dialogic method. In J. Norris, R. Sawyer, & D. Lund (Eds.), *Duoethnography: Dialogic methods for social, health, and educational research* (pp. 9–40). Left Coast Press.

Ohl, A., Sheff, M. G., Small, S., Nguyen, J., Paskor, K., & Zanjirian, A. (2017). Predictors of employment status among adults with autism spectrum disorder. *Work, 56*(2), 345–355. https://doi.org/10.3233/WOR-172492

Pellicano, E. (2012). The development of executive function in autism. *Autism Research and Treatment, 2012*(146132), 1–8. https://doi.org/10.1155/2012/146132

Rau, S., Skapek, M. F., Tiplady, K., Seese, S., Burns, A., Armour, A. C., & Kenworthy, L. (2020). Identifying comorbid ADHD in autism: Attending to the inattentive presentation. *Research in Autism Spectrum Disorders, 69*(101468), 1–12. https://doi.org/10.1016/j.rasd.2019.101468

Raymaker, D. M. (2017). Reflections of a community-based participatory researcher from the intersection of disability advocacy, engineering, and the academy. *Action Research, 15*(3), 258–275.

Salty, K. I., Gobernatz, A. E., & Close, E. W. (2022, September 13–14). *ADH... disorder? Discoveries on ADHD and physics learning from collaborative autoethnography* [Conference session]. Proceedings of the Physics Education Research Conference, Grand Rapids, Michigan.

Sasson, N. J., Faso, D. J., Nugent, J., Lovell, S., Kennedy, D. P., & Grossman, R. B. (2017). Neurotypical peers are less willing to interact with those with autism based on thin slice judgements. *Scientific Reports, 7*(40700). https://doi.org/10.1038/srep40700

Sawyer, R.D., & Norris J. (2015). *Duoethnography: Understanding statistics*. Oxford Academic. https://doi.org/10.1093/acprof:osobl/9780199757404.001.0001
Williams, G. L. (2020). We're all strangers here. *Anthropology and Humanism*, *45*(1), 123–129. https://doi.org/10.1111/anhu.12270

CHAPTER 18

The (In)Accessibility of UK Universities' Ethics Application in Humanities and Social Sciences

Katie Munday

Introduction

Research ethics frameworks aim to ensure that researchers maintain high ethical standards, integrity, and confidentiality, protecting the researcher and their participants, as well as the reputation of their affiliated university (Bryman, 2012). Ethics policies within UK universities outline that all research and innovation activities undertaken by staff and students require ethical review (this includes undergraduate students who speak to human participants as part of their dissertation). Ethics applications within Humanities and Social Sciences (HSS) in UK universities should be flexible enough to work across different disciplines but precise enough to keep individuals safe and ensure good quality outputs (Federman et al., 2013). Despite the importance of ethics applications and the practices they uphold, the process of obtaining a favourable opinion may contain barriers which disproportionately impact neurodivergent students. In this chapter, I explore the barriers in formal research ethics assessments for neurodivergent individuals in UK universities, as well as the informal researcher ethics around ableism and inaccessibility. I reflect on my own experiences and those shared with me online. From this, I suggest recommendations for UK universities to make the process more accessible and give advice for other Autistic ADHD researchers. The aim of this work is to begin to deconstruct the UK ethical application process to make it

K. Munday (✉)
University of Portsmouth, Portsmouth, England
e-mail: Katie.Munday@myport.ac.uk

H. Bertilsdotter Rosqvist and D. Jackson-Perry (eds.), *The Palgrave Handbook of Research Methods and Ethics in Neurodiversity Studies*,
https://doi.org/10.1007/978-3-031-66127-3_18

more accessible to *all* students and researchers. The examples and experiences discussed in this chapter are related to autism and/or ADHD, with a focus on my Autistic and ADHD experiences. It is worth noting that 'neurodivergent' covers a much broader range of neurologies and experiences, with overlapping but non-identical access issues in the ethics application process.

UK University Ethical Standards in the Humanities and Social Sciences

Humanities and social sciences (HSS) research students and faculty in UK universities must abide by a strict ethical code of practice whilst conducting research involving human participants. These ethical standards are based on the Belmont Report (National Commission for the Protection of Human Subjects of Biomedical and Behavioural Research, 1979), which outlines boundaries between practice and research as well as how to gain informed consent, create risk assessments, and recruit participants. Over 30% of UK universities also follow the ethical guidelines set out by a dedicated research integrity organisation, the UK Research Integrity Office (UKRIO). UKRIO aims to enhance the international reputation of UK research, as well as protect the public by encouraging a strong professional ethos based on integrity, accountability, and continued professional development. Under UKRIO guidance, universities and their researchers (students *and* staff) can reflectively strengthen best practice. UKRIO suggests that universities provide researchers with varied development opportunities to enable them to conduct research to the required ethical standards.

The ethical guidelines used in higher education in the UK are also beholden to national and international law. University-based researchers must show a knowledge of these laws and how they apply to their specific area of study and methodology. A researcher who works with children and young people, for example, would need to recognise the United Nations Conventions of the Rights of the Child (UNICEF, 1989) to protect the interests of young participants. Researchers who collect sensitive data must follow the European guidelines of General Data Protection Regulation ([GDPR], EU, 2016/679) which outline the safe retention, reuse, and destruction of personal information. GDPR, with regard to research, protects raw data, however, when storing personal data (whether anonymised or not) participants can request that their data be destroyed (GDPR, EU, 2016/679). Since the UK left the European Union in 2020, researchers must now follow the guidelines of UK GDPR set out under the Data Protection Act (Department for Digital, Culture, Media & Sport, 2018).

Where relevant, researchers in UK universities must also follow the requirements of professional bodies in their field of research and the professional regulating bodies of which they are members (Universities UK, 2018). All research involving National Health Service (NHS) patients and/or their relatives requires ethical submission to the appropriate NHS Research Ethics Committee (REC). Studies outside of the NHS may need to be reviewed by other external bodies including the National Social Care Ethics Committee; Her Majesty's Prison and Probation Service (for research involving prisoners) and the Ministry of Defence REC (for research relating to military personnel). In these cases, researchers may need to submit an ethics form to their university (as is standard in most UK university-based research) as well as gaining external ethical opinion, which must be shared with their faculty ethics committee. All faculty and students involved in research are accountable to their research volunteers and funders, to their profession, the university, and to society (City University of London, 2021).

Methods

This chapter is a reflection on (in)accessibility, using my own lived experience of applying for ethical approval as a postgraduate researcher in Humanities and Social Sciences in the UK. I also share the views and opinions of other Autistic ADHD individuals within this chapter to better understand the current state of ethics applications in UK universities. Information was gathered from an online Autistic researchers' group as well as Twitter and Facebook; answers to which have been pseudonymised and they/them pronouns have been applied. My Facebook and Twitter posts were posted on 17th April 2023 and read: "Do you have any recommendations for UK universities to make the ethics application process easier for Autistic ADHD individuals? I'm writing a book chapter on this." These reflections are on formal and informal ethical issues which make university-based research more difficult for neurodivergent people.

Autistic ADHD Experiences of UK University Ethics Applications

My undergraduate dissertation project, in the School of Education and Sociology, was literature-based and therefore did not need ethical approval. There was an opportunity to engage in research with human participants, although students were discouraged to do this due to the complexity of researching with young people. In my dissertation, I explored how to support lesbian, gay, bisexual, and transgender Autistic school students. My Master of Research (MRes) dissertation collected the narratives of gender diverse Autistic people and their thoughts on research on our lived experiences (Munday, 2022). All MRes research students needed ethical approval which covered anything from archival and literature-based work to research with children and young people. These applications were complexified by inconsistent COVID-19 restrictions in the UK,[1] which changed the parameters of health and safety within

[1] COVID-19 restrictions started in the UK in March 2020 and stopped Oct 2022. Restrictions included mandated lockdowns, restriction of movement, mandatory face coverings, and personal space.

participant-based research. Many of my cohort started as full-time campus-based students but, as it became more obvious that COVID-19 would be a long-term issue, many of us changed to part-time and studied from home.

Thankfully, I started my MRes with some knowledge of formal research ethics gleaned from an undergraduate psychology module, in which I collected and analysed data for a short research assignment on verbal fluency. Whilst this gave me some experience of empirical research, I needed a much deeper understanding of formal ethical standards for my MRes. I was not prepared for the formal barriers within ethics assessments, including a plethora of paperwork, unclear expectations, and miscommunication from faculty. Nor did I expect the informal issues which underpin many of these processes including perceived neurodivergent 'bias' and positionality, and the perception of vulnerability and naivety of Autistic participants.

Paperwork, Paperwork, and More Paperwork

My MRes project consisted of interviewing 13 participants who were Autistic, either formally or self-diagnosed, and transgender and/or non-binary (Munday, 2022). I recruited 8 participants for video-conferencing interviews and a further 5 for written interviews. Due to the nature of my study, I needed to create many different forms including consent forms, debrief sheets, and participant information sheets. The creation of these forms ran alongside the formation of interview questions, risk assessments, scripts for oral consent, and a recruitment advert (with plans of where and when to recruit). Many ADHD people (including me) cannot efficiently move between different tasks due to differences in working memory, response, and attention (Bueno et al., 2017). This issue with task-switching means that an ethics application can quickly become overwhelming, especially when being undertaken for the first time. As Sharon, a neurodivergent postgraduate, suggested: "I was expected to magically know the conventions" having never engaged in an ethical application before. Thankfully my university had lots of explanative notes in their one-size-fits-all paperwork, but I remained confused around some of the nomenclature and unexplained acronyms.

When I asked for clarity, I was often told by faculty that all the information I needed was on Moodle[2] and that I simply needed to look for it. However, navigating online systems is difficult for me especially as their design elements make them inaccessible to most students and faculty (Martin, 2020), let alone those of us who communicate differently. When I started my postgraduate in 2020, some accessibility features were added to the university's Moodle site,

[2] Moodle is an online university portal used throughout the UK—it is an online space where students can check their schedule and information around exams and course work. Usually, lecturers share notes, videos, and other resources there too.

including dark mode,[3] and the ability for screen readers to make sense of the text layout. Around a third of Autistic people have co-occurring disabilities (CDC, 2022), which mean many of us have issues with accessing, processing, and recalling written information. Having an online portal where information was stored haphazardly with little or no thought for accessibility made 'looking' for information incredibly difficult. Even when resources were clearly signposted, I was sent on a magical mystery tour for resources which did not exist.

When I started my course, before COVID-19 lockdowns, I wanted to offer participants face-to-face interviews, for which I needed to write a risk assessment. The links to find the university's risk assessment documents led nowhere and the person in charge of creating the documents could not be contacted. This was exceedingly stressful for me and maybe for other neurodivergent researchers due to our differences in processing speed and time management, and challenges with asking for clarification (Van Hees et al., 2015; White et al., 2016). When I did eventually speak to my supervisor and other faculty members (I emailed a lot of people for clarity), it was assumed that I was 'okay' to carry on face-to-face work without risk assessments.

The ethical responsibility for my work, as well as any risk assessments which I may or may not have needed, was unclear throughout my course. The university's guidelines state: "The supervisors of undergraduate and taught postgraduate students take the responsibility of principal investigator" (University of Portsmouth, 2020) however, my course was not a taught course, so it was unclear who the principal investigator was. I could not ask for clarity from my course liaison tutor, as they left in the first year and were replaced by someone who never got in contact with me. There were just too many occasions on which I needed to ask for help but wasn't sure who to ask or how to ask for it. Consistently having to translate expectations leaves many neurodivergent students to expend significant amounts of energy to organise admin, keep to timelines, and understand who we need to communicate with (Pellicano, 2012). Miscommunication can be an issue for any student/researcher: however, it is more difficult for neurodivergent people due to issues with working memory, response speed, and other areas of executive functioning (Gobbo & Shmulsky, 2014).

Whilst the skillset required for ethics applications is challenging for many in academia, it can be particularly difficult for neurodivergent people as we often have an uneven cognitive profile, in which we have great strengths in certain areas and great challenges in others (Jones et al., 2009). When asking for support with administration, we can be met with incredulity, especially if we are considered to have academic rigour, as it is considered implausible that we can write and present information well but need help with 'simple' paperwork.

[3] Dark mode is a display setting for user interfaces including smartphones or laptops. The display background is black, with white text. This mode can be helpful for people who are dyslexic, colourblind, or have processing differences.

Not only is this a false representation of Autistic ADHD embodiment,[4] it also undermines the significant amounts of effort which neurodivergent students and researchers put into academic work. As Bella, someone who was unable to access university, shared: "One of the main reasons I never went to uni: I can't do paperwork without support and support I am sadly without."

The Luck of the Draw

Students in the UK can apply for support through the Disabled Students Allowance (DSA), if they have a mental health problem, long-term illness, or any other disability (Department for Education and Skills [DfES], 2009). DSA can support Disabled students with specialist equipment, non-medical support (such as a mentor or specialist note taker), travel expenses to university and placements, and other disability-related study support (DfES, 2009). DSA support can only be granted if the applicant has a formal identification for their need, meaning fewer than a third of Disabled students access the funding (Grant et al., 2023; Holmes, 2022). I applied for DSA in the final year of my undergraduate once I had my formal autism identification, up to that point I was unable to access support.

Having access to a mentor for the final part of my undergraduate degree, and throughout my MRes, meant that I could access specific help with regard to ethical application. Thankfully, my mentor had just finished her sociology PhD, so I was able to ask her questions and speak at length about the application process throughout our sessions together. Sam, a participant in Munday (2022), also shared their 'luck' in university:

> We [me and the other neurodivergent students] all had a lot of extensions. We all struggled a lot with actually getting the work done on time, and to understand what's being asked of us. I don't know whether that's because of what kind of course we were doing, or it was just really complete luck of the pot of who ended up teaching us and was in the well-being service at the time. Yeah, they did a much better job than I've ever experienced in any prior education I've ever been in and yeah, I just feel sad at how that wasn't normal for most people and I'm just like 'how did I just suddenly get this random jackpot here?'

Sam shared that they felt they had hit the 'jackpot' with their faculty and that they felt supported. They reflected that they felt sad that this was not usual for other research students on their course. Many neurodivergent students and faculty researchers cannot access tailored support for their ethics application. It is understandable that some students may feel like Sheila, a neurodivergent postgraduate, who shared: "My ethics application in the UK nearly broke me, it is such an unfriendly process."

[4] Autistic ADHD embodiment means how we understand, respond to, and move around in the world as Autistic ADHD people.

Miscommunication

The turnaround for feedback for my ethics application was unclear, which left me in limbo—I could not progress with my time-constrained project due to administrative backlog and faculty miscommunication. These miscommunications can be viewed as an example of the Double Empathy Problem (Milton, 2012) in which the breakdown in reciprocity and mutual understanding between Autistic and non-Autistic people is situated within *both* individuals. Unfortunately, the blame for communication breakdown follows many neurodivergent people across our lives, including in academia. Many academics have limited understanding of Autistic and ADHD embodiments, meaning neurodivergent students and researchers must constantly self-advocate to ensure that our supervisor or colleagues are aware of—and responsive to—our needs (Elias & White, 2018; Jones, 2023). Self-advocacy becomes more complex with regard to disclosure; if you do not disclose you are less likely to get support, if you do disclose you risk ignorance and stigma following you throughout your academic studies and/or career (Jones, 2023).

Even if we do disclose and self-advocate, faculty may remain reluctant to provide explicit instructions for ethics applications—we are expected to find our own way of doing things whilst miraculously following a framework and hierarchy which is never communicated to us (Martin, 2020). The *correct* person and advice to follow is not always made clear, as Samantha, a neurodivergent Graduate, shared:

> The process, when I applied, was very long with lots of different people needing to get back to you without a timeframe. So, the entire process is really overwhelming [and] could prevent people for applying for university in the first place. The process at the time required individuals to keep logging in and checking for updates on their application. This can be really anxiety-provoking.

Samantha explained that a lack of clear communication and expectations made them feel overwhelmed and confused. Having to correspond with different faculty meant that they had the same conversations several times, and lost track of who they needed to speak to about specific issues, and within what timeframe. Similar issues were found in Martin (2020) with one neurodivergent participant sharing: "Administrative barriers are the most immediately stressful. Complex and convoluted online systems and programmes, labyrinthine networks of people that you need to call or email to get things done" (p. 297). The lack of timeframes around feedback on their university's online portal meant Samantha felt the constant need to check their ethics status. Interestingly, Samantha also considered that the application process itself may put off potential students from applying to university altogether.

Neurodivergent 'Bias' and Positionality

Institutional ableism is often exemplified by a culture in which normativity dictates processes, and divergence is not accepted, let alone celebrated (Kattari et al., 2018, see Chapters 7, 8, 9, 13, 17, 25, this volume). Unfortunately, institutional ableism is deeply ingrained in academia, often meaning that neurodivergent students and faculty must do *more* work for the same outcomes as our non-Disabled peers (Botha, 2021; Brown & Leigh, 2018). Neurodivergent researchers are often assumed to be biased when working with members of our own community, we are considered to lack epistemic authority over our own culture, history, and knowledge (Fisher, 2023; Hacking, 2009; Hens et al., 2019). These ideas affected the way I framed my ethics applications as I was made aware that first-hand accounts of Autistic (and otherwise neurodivergent) people are often considered biased and inaccurate simply due to our neurology. Positionality and reflexivity is an important part of *all* research and should not be dependent on the identity of individual researchers (Braun & Clarke, 2019).

Throughout my MRes, I reflected on my own experiences, both professional and personal, as I "did not feign the distanced impartiality of an outside observer: [as] it would erase my own queer Autistic voice to do so!" (Sparrow, 2020, p.164). My lived experience informed *all* aspects of my study including ethical considerations, as I wanted to ensure that participants felt listened to, as well as keeping them safe. From my own experience, I knew that certain subjects could be triggering, so I chose a narrative methodology which allowed participants to share what they wanted to. I also created a robust ethics entry around accessibility of interviews to reduce the stress of participants. Participants were told of my identity upon recruitment, and some commented that they felt more comfortable talking to a researcher who shared parts of their identities (Munday, 2022). My position also influenced how I recruited participants, as I recruited through online gender diverse Autistic groups which I have been a member of for several years. I may have unintentionally sought out participants who reflected my own experiences and views, however, it is essential to remember that no two Autistic experiences are the same (Mendes & Maroney, 2019). I felt there was a greater need for me to reassure the ethics committee that I was underpinning my work with research created by non-Autistic researchers. However, this may reflect my anxiety around being 'taken seriously,' as opposed to explicit instruction from the ethics committee.

Feelings of anxiety and inadequacy can be felt elsewhere in ethical applications: as we must illustrate how we ensure the safety and wellbeing of participants but not whether we are protected from reading deficit-based work (Botha, 2021). When researching within disability studies and autism research, we must often work from diagnostic manuals and theoretical underpinnings which can be traumatising to read. Consistently reading that you are considered to have *disorders, deficits, inflexibility, failures,* and *difficulties* is distressing and yet does not appear to be a consideration for ethics committees in HSS in the UK. Ethics applications are created and upheld by neuronormative systems of academia which do not consider neurodivergent researchers or those we research with.

Perceptions of Autistic Participants

During my ethics application, there were several barriers to working with Autistic adults, due to the assumptions that Autistic people are either incompetent or too vulnerable (Hacking, 2009; Hens et al., 2019). It is important that *all* participants in research are protected. However, harmful stereotypes still appear to affect the ability of researchers to work with Autistic participants. Unfortunately, this perpetuates autism research, which focuses largely on non-Autistic people's experiences of Autistic people. Our teachers, parents, carers, family, and therapists are spoken to more than we are, often leaving our voices unheard (Fisher, 2023; Milton & Bracher, 2013; Roche et al., 2020). Focusing exclusively on the experiences of others ignores the most important part of the autism community: Autistic people. In trying to protect us, ethics processes can unintentionally silence us, especially those of us who are non-speaking, learning disabled, or come from marginalised backgrounds. The stigma which sees all Autistic people as 'too vulnerable' or lacking mental capacity is something we must fight against within and without academia (see Chapter 10, this volume).

Recommendations for UK Universities to Make the Ethics Application More Accessible for Autistic ADHD Researchers

Many of the issues I faced in applying for ethical approval were reflected in the experiences and recommendations of those who shared their experiences online. Recommendations to improve ethical assessment processes specifically include creating comprehensible guidelines, offering individualised support, and allowing researchers access to examples of past ethics applications and forms.

Comprehensible Guidelines

The literature around ethical applications can be unclear, and it can be difficult to understand the expectations of paperwork and practice. Neurodivergent students have reported that they benefit greatly from structure and concrete instructions (Cai & Richdale, 2016; Knott & Taylor, 2014). Universities, and all those they support, would benefit from:

- Virtual meetings in which students (neurodivergent and otherwise) can ask specific questions about their specific application.

- Clearer guidelines on how to tailor the ethics forms to different types of study, e.g. qualitative or quantitative, literature-based or participatory.
- University websites which act as a 'one-stop shop' which would inform students on what they'll need to gain approval, what clearances are needed, a timeframe and contactable person(s) to go to for further clarification. These steps would be based on the information the student/researcher inputs.

Offering Individualised Support for Those Applying for Ethical Approval

Autistic ADHD individuals often benefit from clear expectations and guidelines, and individualised support. A mentor, specifically someone who is neurodivergent or has experience working alongside neurodivergent people, could support an Autistic ADHD student with all aspects of the application. Support could be in the form of:

- Interpreting feedback.
- Completing revisions.
- Signposting to useful faculty members and resources.
- Keeping the student/researcher to time with gentle reminders.
- Creating a longer timeline together which gives structure but allows for flexibility.

The final point not only supports the executive function of neurodivergent individuals, but it also appreciates that many of us have co-occurring disabilities and chronic illnesses which affect our energy levels and work output. Creating plans for dips in energy and function is an important part of supporting neurodivergent students and researchers.

Allowing Researchers Access to Examples of Past Ethics Applications and Forms

Examples of recent previous successful ethics applications and connected forms (such as participant information sheets and debrief forms) would allow Autistic ADHD individuals to understand the expectations of ethical applications in their university. During my ethical application process, I was given unclear guidelines on word counts or referencing expectations. My ethics proposal was given a favourable opinion with minor revisions, however, there were comments that what I had done was 'too much.' If I had been given more information upfront, including examples, then I could have used this energy elsewhere. Examples and templates would support Autistic ADHD individuals understanding of:

- What information to put under what heading.

- How much information is 'just right' to be successful.
- Understanding phrasing and language expectations.
- Word counts—even if these exist as a general range to aim for.
- How to answer open-ended questions.
- Jargon and acronyms.
- Resources to draw from.

It is important to note that changes which would make HSS ethics applications in the UK more accessible for neurodivergent individuals would make things easier for *all* students and researchers.

Advice for Autistic ADHD Individuals Undergoing Ethics Application in UK Universities

Undertaking Higher Education can be difficult for those of us who are under-represented in the formulation of curriculum, including those of us who are multiply neurodivergent. I was formally diagnosed as Autistic in the final year of my undergraduate degree which meant I was unsupported during most of my undergraduate degree. Even with support, there is an often-chaotic relationship between my Autistic and ADHD embodiment. Ideas which would support my Autistic sensibilities—lists, organisation, clear communication, and asking lots of questions—would not necessarily suit my more impulsive eleventh-hour ADHD tendencies! There was a constant need for me to re-evaluate how I was working, as it would have been easy to become overwhelmed. Things which may help other Autistic ADHD students/researchers include looking to others for support, giving yourself plenty of time to apply, and asking for examples of previous applications.

Ask for Support

It sounds like obvious advice, however, experiencing consistent invalidation and ignorance can leave many of us doubting our skills of self-advocacy. This is your research so *take up space*, this is your time to ask questions, and look for support. Things which may help:

- Ask as many questions as you need until it becomes clear—do not be afraid to ask more than once or to ask several faculty members (they are contacted by many people, so almost certainly won't speak to each other about it).
- Make sure you have support around you within and outside of the university. Faculty can help with research specific issues, family and friends can help with emotional and physical support (support with other elements of your life can leave the executive function for the ethics application process).

- Your supervisors or research lead are there to support you—ask them questions and get their advice.
- If your supervisor is not responding to your attempts to contact them, contact the person in charge of dissertations in your department, or 'postgraduate research' if you are a research student.
- Create or join a neurodivergent research group, either face-to-face or online. Connecting with peers means that you can find answers together rather than struggle independently. These groups can also help with body-doubling, where you keep each other accountable whilst studying together or regularly checking in with one another.

The ethics process in HSS in the UK can be overwhelming, however, there are people who can help—disability and academic skills services, your tutor, supervisor or research lead, and your colleagues. Your ideas and your work are worth other people putting their time and effort into.

Give Yourself Plenty of Time

Again, this sounds obvious, but this is not something which comes naturally to many of us. The following ideas may help with time management:

- Look up timeframes in advance.
- Make a list of all the forms you need at the start, download them, and work through them with plenty of time (more than you think you will need). The application process can be easier to contend with when it is broken down into steps. These steps give the opportunity for many 'jumping in' points or the ability to do everything all at once, depending on your study style.
- Complete the ethics application as soon as possible—this allows your supervisor or research lead to support you if needed. Peer-review can also take time, so starting early helps.
- Make a note of when the ethics committee closes (some close over the summer in the UK). It can take up to three months to get feedback from ethics applications, so ensure to factor in closures and feedback time.
- Look at the steps at the beginning of a course as they can take time, for example, if you need a DBS (disclosure and barring service) check to work with children or vulnerable adults, this can take several months and is worth doing early.

Plans can be created alongside your supervisor or research lead, with whom you can have 'check in' sessions or emails to check on your progress. Having someone to keep you accountable can be helpful. It can feel overwhelming looking at everything which needs to be submitted, breaking this down into chunks makes it easier to process and ticking things off your list can be satisfying. Remind yourself as often as you need to about the timelines. For me, this looked like a wall planner with a sticker system, which clearly showed my progress (I went for cute glittery stickers, but plain ones work just as well!).

Ask for Examples

Examples can give a clear indication of expectations, including word counts, use of language, and referencing. Fortunately, the ethics application at my university gave maximum word counts and had clarifying points for each question. This allowed me to get on with most of the work by myself. However, this is not a practice used by all UK universities in HSS. I emailed several researchers from other universities about their ethical and methodological approaches and was pleasantly surprised by their response. There is nothing quite like connecting with people who share your academic interests. I was worried about reaching out but kept reminding myself that the worse thing people could say to me was 'no' or simply skip over my email.

There are many places where you can find examples to help you create a more robust application. Your supervisor or research lead should be able to help you find previous successful ethics applications and may even share those from their own work. You can reach out to researchers who have done similar work to yours, just be aware that some of the information they share may be country or institution specific. Combining approaches from examples and advice from researchers in similar areas, as well as your own contributions, will help create a strong ethics application. If your application goes through the first time, or comes back with a small amount of feedback, then you have more time to give to your research project. Also, many of the parts you use in your ethics application may be used in your final research thesis, so time spent on the application will benefit you when you come to write your final thesis.

Conclusion

Ethical frameworks aim to ensure that researchers maintain high ethical standards to protect the researcher, participants, and the affiliated university. To gain ethical approval, all researchers must create and organise multiple pieces of paperwork, usually in a short timeframe. For neurodivergent researchers, little individualised support may be available around translating nomenclature. Breakdowns in communication between neurodivergent and neuronormative individuals, mixed with the power dynamics of university hierarchies, mean that neurodivergent researchers are often blamed for misunderstandings (Milton, 2012). Self-advocacy becomes more difficult when our support needs are questioned, or universities simply do not believe we exist in their institutions (Martin, 2020). This stigma puts undue pressure on neurodivergent researchers to consistently adapt to neuronormativity. Even when support is in place, it is not always appropriate due to the misunderstanding of support needs in UK universities (Cage & Howes, 2020).

This chapter has identified some ideas which may help neurodivergent people when applying for ethics in HSS in UK universities including looking to others for support, giving yourself plenty of time to apply, and asking for examples of previous applications. Universities should also improve their accessibility within ethics by creating comprehensible guidelines, offering individualised support to those applying for ethical approval, and allowing researchers access to examples of past ethics applications and forms. Many of these recommendations for university ethics applications processes would improve outcomes for *all* students—a university degree should be evidence of knowledge, *not* a test of resilience.

References

Botha, M. (2021). Academic, activist, or advocate? Angry, entangled, and emerging: A critical reflection on autism knowledge production. *Frontiers in Psychology, 12*. https://doi.org/10.3389/fpsyg.2021.727542

Braun, V., & Clarke, V. (2019). Reflecting on reflexive thematic analysis. *Qualitative Research in Sport, Exercise and Health, 11*(4), 589–597.

Brown, N., & Leigh, J. (2018). Ableism in academia: Where are the disabled and ill academics? *Disability and Society, 33*(6), 985–989. https://doi.org/10.1080/09687599.2018.1455627

Bryman, A. (2012). *Social research methods* (5th Ed.). Oxford University Press.

Bueno, V. F., da Silva, M. A., Alves, T. M., Louzã, M. R., & Pompéia, S. (2017). Fractionating executive functions of adults with ADHD. *Journal of Attention Disorders, 21*(11), 944–955.

Cage, E., & Howes, J. (2020). Dropping out and moving on: A qualitative study of autistic people's experiences of university. *Autism, 24*(7), 1664–1675.

Cai, R. Y., & Richdale, A. L. (2016). Educational experiences and needs of higher education students with autism spectrum disorder. *Journal of Autism and Developmental Disorders, 46*(1), 31–41.

City University of London (2021). *Research ethics approval process*. https://www.city.ac.uk/research/support/integrity-and-ethics/ethics/approval

Department for Education and Skills. (2009). *Disabled students' allowance – DSA*. https://www.gov.uk/disabled-students-allowance-dsa

Elias, R., & White, S. W. (2018). Autism goes to college: Understanding the needs of a student population on the rise. *Journal of Autism and Developmental Disorders, 48*(3), 732–746.

Federman, D. D., Hanna, K. E., & Rodriguez, L. L. (2013). *Responsible research: A systems approach to protecting research participants*. Institute of Medicine: Committee on Assessing the System for Protecting Human Research Participants.

Fisher, J. (2023). Epistemic injustice: The silenced voices. *Mental Health Nursing*. https://doi.org/10.1111/inm.13163

Gobbo, K., & Shmulsky, S. (2014). Faculty experience with college students with autism spectrum disorders: A qualitative study of challenges and solutions. *Focus on Autism and Other Developmental Disabilities, 29*(1), 13–22.

Grant, A., Williams, G., Williams, K., & Woods, R. (2023). Unmet need, epistemic injustice and early death: How social policy for Autistic adults in England and Wales fails to slay Beveridge's Five Giants. *Social Policy Review, 35*. https://doi.org/10.51952/9781447369219.ch012

Hacking, I. (2009). Autistic autobiography. *Philosophical Transactions of the Royal Society Biological Sciences, 364*(1522), 1467–1473. https://doi.org/10.1098/rstb.2008.0329

Hens, K., Robeyns, I., & Schaubroeck, K. (2019). The ethics of autism. *Philosophy Compass, 14*(1). https://doi.org/10.1111/phc3.12559.

Holmes, C. (2022). Report into Disabled Students Allowance. https://lordchrisholmes.com/report-disabled-students-allowancedsa/

Jones, S. C. (2023). Advice for autistic people considering a career in academia. *Autism, 27*(7), 2187–2192. https://doi.org/10.1177/13623613231161882

Kattari, S. K., Olzman, M., & Hanna, M. D. (2018). "You look fine!" Ableist experiences by people with invisible disabilities. *Affilia, 33*(4), 477–492.

Knott, F., & Taylor, A. (2014). Life at university with Asperger syndrome: A comparison of student and staff perspectives. *International Journal of Inclusive Education, 18*(4), 411–426.

Martin, N. (2020). University through the eyes of autistic students and staff. In D. Milton (Ed.), *The neurodiversity reader* (pp. 289–310). Pavilion.

Mendes, E. A., & Maroney, M. R. (2019). *Gender identity, sexuality and autism: Voices from across the spectrum*. Jessica Kingsley Publishers.

Milton, D. E. M. (2012). On the ontological status of autism: The 'double empathy problem.' *Disability and Society, 27*(6), 883–887.

Milton, D. E. M., & Bracher, M. (2013). Autistics speak but are they heard? *Medical Sociology Online, 7*, 61–69.

Munday, K. (2022). Stories from under the "double rainbow": Trans and non-binary autistic narratives [unpublished postgraduate thesis]. https://doi.org/10.13140/RG.2.2.34886.75843

National Commission for the Protection of Human Subjects of Biomedical and Behavioral Research. (1979). The Belmont report: Ethical principles and guidelines for the protection of human subjects of research. *U.S. Department of Health and Human Services*. https://www.hhs.gov/ohrp/regulations-and-policy/belmont-report/read-the-belmont-report/index.html

Pellicano, E. (2012). The development of executive function in autism. *Autism Research and Treatment*. https://doi.org/10.1155/2012/146132

Regulation (EU) 2016/679 of the European Parliament and of the Council of 27 April 2016 on the protection of natural persons with regard to the processing of personal data and on the free movement of such data. https://www.legislation.gov.uk/eur/2016/679/contents

Roche, L., Adams, D., & Clark, M. (2020). Research priorities of the autism community: A systematic review of key stakeholder perspectives. *Autism, 25*, 336–348. https://doi.org/10.1177/1362361320967790

Sparrow, M. (2020). *Spectrums: Autistic transgender people in their own words*. Jessica Kingsley Publishers.

UNICEF UK. (1989). *The United Nations convention on the rights of the child*. https://downloads.unicef.org.uk/wp-content/uploads/2010/05/UNCRC_PRESS200910web.pdf?_ga=2.78590034.795419542.1582474737-1972578648.1582474737

University of Portsmouth (2020). *Ethics policy*. https://policies.docstore.port.ac.uk/policy-028.pdf?_ga=2.34442469.554403068.1579529397-2119224136.1579529397

Van Hees, V., Moyson, T., & Roeyers, H. (2015). Higher education experiences of students with autism spectrum disorder: Challenges, benefits and support needs. *Journal of Autism and Developmental Disorders, 45*(6), 1673–1688.

White, S. W., Elias, R., Salinas, C. E., Capriola, N., Conner, M., Asselin, S. B., et al. (2016). Research in developmental disabilities students with autism spectrum disorder in college: Results from a preliminary mixed methods needs analysis. *Research in Developmental Disabilities, 56*, 29–40. https://doi.org/10.1016/j.ridd.2016.05.010

CHAPTER 19

Meaningful Engagement with Neurodivergent Public Work

Alyssa Hillary Zisk

Introduction

Some researchers have connections within the communities we research. Others may not have these personal connections, but still want to use publicly available work (e.g., blogs, videos, or podcasts) in their research. This engagement can be an important way of including the experiences, theories, and work of community members who work outside of academia within academic conversations. It can also be extractive, taking community knowledge and independent scholars' work and erasing its source. This chapter will explore and illustrate ways of engaging with neurodivergent work beyond academia. I hope this chapter can serve as a precedent, a sort of citational seed (Liborion et al., 2022), to be an academically published reference that neurodiversity researchers who want to cite blogs can use to justify doing so.

The first thing to recognize is that attempting to meaningfully engage with neurodivergent work beyond academia is working against current norms. Doing so isn't necessarily against any *rules*—I have never been called on citing blogs, videos, or fanfiction—but it is against the grain of how academic research is typically done, and that means it takes *work* to meaningfully achieve (Hawkins et al., 2021).

It is, generally speaking, easier to follow an existing model (or to take pieces from a few different models) than to do something without a model. Even

A. H. Zisk (✉)
AssistiveWare, Amsterdam, The Netherlands
e-mail: alyssazisk@uri.edu

H. Bertilsdotter Rosqvist and D. Jackson-Perry (eds.), *The Palgrave Handbook of Research Methods and Ethics in Neurodiversity Studies*,
https://doi.org/10.1007/978-3-031-66127-3_19

when we, personally, do not know about any of the potential models, what we're trying to do probably isn't totally new. Claiming newness or firstness often requires stringent limitations on what we count as knowledge or as relevant, which perpetuates the kind of exclusion we want to avoid while making that exclusion harder to perceive (Liborion & Li, 2022).

Existing Forms of Engagement

Neurodivergent Researchers Doing Neurodiversity Research

There is considerable community favor for autistic autism researchers and neurodivergent neurodiversity researchers in general (Eb, 2015; theoriesofminds, 2015, 2017). There are now more examples of this than I can reasonably put in a chapter that is mostly Not About This, but Michelle Dawson's body of work (e.g., Dawson & Fletcher-Watson, 2022) is an historically important one.

Neurodivergent researchers doing neurodiversity research get neurodivergent ways of thinking about neurodivergence into the academic record. At best, rather than taking a seat at a table still run by neurotypical professionals and academics, this can lead to at least a small table of our own—if we get support for our research programs, or when we sit on the editorial boards of journals (Nicolaidis, n.d.; *Ought*, 2019–2023).

I am not aware of formal studies examining this, but in my experience, openly neurodivergent neurodiversity researchers seem to be at least somewhat more likely to engage in the other practices described below. When we do this, we need to make sure to maintain proper ethical and research standards–while academic establishments may view our work with greater suspicion due to our disclosed neurotypes, we are often operating from a position of more trust *within* our communities due to our identities.

Reprinting and Sometimes Expanding Work From Outside Academia in Academic Journals

Reprinting neurodivergent work from outside academia gets that work into the academic record. This is useful, but most relevant to people who, for whatever reason, are either unable or unwilling to cite work from beyond academia. Also, blog preservation is an Actual Issue (Peoples, 2010). Autistics.org, for example, is now only available through the internet archive. Reprinting in an academic journal with permission from the blog author is one way of preserving at least a few key blog posts that are considered foundational documents.

This is also a patchwork, band-aid style solution. It's typically applied to blog posts that have already attracted significant attention, and the academic journals that *do* this are fairly low in number, with limited capacity. *Autonomy*, which did this (Arnold, 2019), now seems to *also* only be available through the internet archive (Autonomy, n.d.).

The article in *Disability Studies Quarterly* (Yergeau, 2013) that I know comes significantly from a blog post (Yergeau, 2011) is important, one I regularly refer to: however, Yergeau is also an academic.

Bibliographies and Other Lists

One of the challenges of meaningfully including and engaging with neurodivergent work is in finding it. This can be true even when we have community connections. For example, at different times, I primarily engaged in blogging (Blogspot and Tumblr), on the site formerly known as Twitter, and in certain Facebook groups. Each of these communities had its own norms and its own overlapping but non-identical participants. I am, by and large, not aware of the people who are currently theorizing on TikTok or other audiovisual-heavy platforms. When I am aware of people who are primarily active in modalities other than text, it is usually because I have been directed to that work by other people who are active in both those communities and at least one of mine.

Direct person to person recommendations of work can be the most personalized. We can (and often do) share links with each other based on known interests. But there are also times when we organize lists which can help support engagement with particular kinds of knowledge.

This can include broad lists, aiming to help people find a wide variety of work from members of a particular group or addressing a particular topic. Broad lists of autistic work include Bagg's list of books by autistic authors (2006). There is also now an Autism Books by Autistic Authors Project (n.d.) containing over 1000 books ranging from new 2024 releases to out of print books referenced in Bagg's list. On a similar note, there are several listings of autistic bloggers, YouTubers, or pages (Bristol Autism Support, 2021; Judy, n.d.; Neuroclastic, n.d.), as well as a listing of autistic researchers (Thom-Jones, n.d.). These blog listings are particularly useful to people who either are not familiar with existing Autistic blogger communities or are aiming to look at a wider variety of blogs, as searching for autistic blogs can in many cases lead to *autism* blogs which are written by family members, researchers, or professionals who are not themselves autistic. Similar issues are apparent in searches for ADHD blogs and are likely present for a variety of neurodivergences.

These broad lists, however, are often very long, and shorter guided lists also have their place. A shorter guided list does not aim to cover a topic or category thoroughly. Instead, it is usually an introduction to one topic or a few related topics, with explanations for why each piece in it fits the list. The Framework for Open and Reproducible Research Training (FORRT) has a Team Neurodiversity (n.d.) with one such annotated reading list as a project in early stages. I also published an annotated list of sources from beyond academia which are relevant to critical autism or neurodiversity studies (Zisk, 2023). My list aimed to highlight key pieces from diverse creators whose other work may also be worth considering when seeking to engage with neurodivergent neurodiversity work, as well as to capture items which may not be as widely known (e.g., no 'Don't Mourn for Us', despite its definite importance; Sinclair, 1993) and some prior archival efforts.

Community-Based Participatory Research

Community-based participatory research (CBPR) is one way to ensure people can be involved in directing research whether or not going through all the formal education currently needed to lead it is a practical possibility. The Academic Autism Spectrum Partnership in Research and Education (AASPIRE) is a community-based participatory research partnership born of the need for autistic-led research (Raymaker, 2020, see also Chapters 4, 25, this volume). That is, AASPIRE comes from the same needs for autistic-led research community members described in #AutChat conversations (e star, 2022; Eb, 2015; theoriesofminds, 2015, 2017). AASPIRE is, to the best of their knowledge (Raymaker, 2020) and mine, the first use of CBPR with autistic people. More have followed, including additional partnerships involving members with multiple roles: AASPIRE is co-directed by an autistic academic and a parent-academic (Raymaker, 2020), and I serve as both an Additional Collaborating Academic and a community partner working with Dr. Amy Donaldson (Donaldson et al., 2023). Some autistic people have suggested using research to verify community knowledge (Eb, 2015; see also Chapter 24, this volume). This requires finding out what the community's knowledge is. Qualitative CBPR studies are an option here (e.g., Donaldson et al., 2021).

Neurodivergent Review

At varying stages of research projects, it is possible to ensure that members of relevant groups review the work (see also Chapters 4, 10, this volume). Sometimes, this is done on an individual level, analogous to authors of fiction seeking beta readers. This can overlap with CBPR, though any consultants who are checking over the work and making suggestions at set points rather than throughout the research may more accurately be considered neurodivergent reviewers than community members of a research team. The same project can have both community representatives on the team *and* review from external neurodivergent people. One reason a research team might choose to do this is if a particular person who wants to be a community representative doesn't have the time to dedicate to that role—someone else could fill that role, and the original person could be a reviewer or consultant with a smaller time commitment at certain set points.

Neurodivergent review is possible at any point where review occurs, including grants, requests for proposals, and journal articles. However, the examples of systematic neurodivergent review that I am aware of are both for autism journals. *Autism in Adulthood* and *Ought: The Journal of Autistic Culture* both systematically ensure autistic people review papers. AASPIRE is part of how and why *Autism in Adulthood* exists to have this policy (Raymaker, 2020). There may be other journals that do so. This is a good thing! I have served as a reviewer for both of these journals in the past, and if I am asked I will likely do so again. I have also served as an aphantasiac reviewer of a paper about aphantasia, though I (1) cannot tell you what paper this was, and (2) have no idea if this was intentional on the part of the editors.

There are multiple advantages of autistic peer reviewers for autism research, and similarly of other neurodivergent people for the research relevant to their neurotypes. One is that we can often catch the most obvious neurotypical (mis)interpretations that come from people interpreting behavior as if they were the ones who took the action, rather than in the reality that someone significantly different from them took the action (Hillary, 2020; Milton et al., 2022). Getting autistic reviewer perspectives while there's still time to edit or reject a paper helps reduce these misinterpretations before publication (Raymaker, 2020). One illustration of the need for autistic review comes via an observation of what happens without it. Some autistic people have noted that they prefer reading autistic perspectives and summaries of existing research to reading academic publications directly due to the pathologizing misinterpretations common in the academic literature (Eb, 2015).

Another advantage of autistic reviewers for autism research (and similar policies for other neurotypes) is that authors submitting to a journal known to have this sort of policy are at least intellectually aware that members of the group they're studying will read their work and comment on it. It's a reminder that we are, in fact, in the audience.

Blog-like Work as a Data Set

Sometimes, people use blogs, YouTube videos, Facebook pages, or similar types of neurodivergent work as data sets to address a variety of research questions. Examples include Angulo et al. (2019) study of autistic perspectives on neurotypicality, Coles-White, Jones, and Saka's (2023) study of perspectives on camouflage from Black Autistic women, Guberman's (2023) study of epistemic resistance on Twitter, Harmens, Sedgwick, and Hobson's (2022) work on women's experiences during identification and diagnosis, and Seidmann's (2021) on blogging, bloggers, and autistic space itself. Most of this research is similar to interview-based qualitative research, and researchers may in some cases (e.g., Seidmann, 2021) conduct qualitative analysis on both the blog data *and* on interviews with some or all creators behind the analyzed blogs. An advantage of analyzing blog data rather than (or in addition to) interview data is that it does not require bloggers to repeat what they have already written, reducing the work they have to do.

The Work of Ethical *Engagement*

Researchers in general are expected to conduct ethical research. For work with human participants, this generally means alignment with certain principles and may also mean following guidelines set by laws, supervisors, and/or review boards. For example, authors may be required to declare that their work followed the principles of the Declaration of Helsinki (World Medical Association [WMA], 2013), or to have a committee review the ethics of their proposed research before conducting it. Whether or not formal review is required, researchers have an obligation to consider the ethics of our work, both in how we do it and in the effects it may have. Community membership may engender more trust from fellow community members, but it does not automatically make our work ethical—we still need to think about ethics and make sure our work is ethical.

Ethics Review and Vulnerability

Some forms of engagement with neurodivergent people and neurodivergent work may require formal review. In principle, this review is a good thing. However, neurodivergent academics have encountered reviewer comments suggesting that all autistic people be categorically considered decisionally impaired or that 'there is not adequate evidence that the self-reports of individuals on the autism spectrum are valid or reliable' (Anonymous qtd. in Raymaker, 2020, see also Chapters 10, 18, this volume). In these cases, ethical review becomes a discriminatory and exclusionary barrier to meaningful engagement with neurodivergent people. Research ethics require additional protections for vulnerable participants, but it is important to consider situational vulnerability, rather than beginning and ending with the idea that certain categories of people are always vulnerable (Gordon, 2020). Disabled people are vulnerable to exploitation in a variety of ways, and our presence in research often triggers additional scrutiny in ethics assessments (Scully, 2014). However, a great many disability-related vulnerabilities are at least somewhat situational, rather than categorical, particularly under a social or relational model of disability. We therefore must be careful not to increase situational vulnerability to certain forms of harm or exploitation in the name of mitigating inherent categorical vulnerability. Otherwise, categorical vulnerability may be used to justify paternalistic research and interventions (Mackenzie, 2014). Instead, responses to and protections for vulnerability should ideally support autonomy (Mackenzie, 2014).

One way that protections for categorical vulnerability can aggravate situational vulnerability is when ethics boards treat community partners in research as identical to research subjects—while ethical engagement and compensation for community partners is important, these are not in fact identical situations requiring identical protections. Community partners are ideally in a position to determine research questions and procedures, meaning that they are *necessarily* involved before a particular research procedure is submitted for ethical approval. Requiring ethics approval for the involvement of community partners then leads to either the need for an additional submission about forming the community-based participatory research team (likely the best solution for a research team facing this challenge) or a situation in which research questions and procedures are significantly determined before partners can be involved (limiting the truthfulness and effectiveness of the claim to community-based research).

Another way that protections and autonomy can be in tension is when work with blog data is considered to be a form of human subjects research, thereby requiring confidentiality. The best ways to handle blog and social media work in research are situational and subject to significant discussion—there are times for anonymity and heavy disguise of work (Bruckman, 2002), but there are also times when anything except credit and citation would constitute theft and erasure of disabled people's work (e.g., Jackson et al., 2022).

Tensions of Credit, Confidentiality, and Erasure

When working with blog or social media data, community opinion seems to tend toward the idea that researchers should engage with autistic writing on social media and blogs—but also that they should ask permission, and that people should have choices about non-use, anonymity, pseudonymity, and citation (Solveig et al., 2022; Williams, 2019). That is, rather than a default of taking the names off public writing but leaving quotes intact, which leaves people identifiable but not credited—the worst of both worlds, but approved by some ethics boards (e.g., Harmens, Sedgwick, & Hobson, 2022). It is best to ask the people whose work you are considering using. A listing of public neurodivergent work may be a place to *find* the work, but researchers would in this case still be responsible for seeking consent and for respecting people's choices around credit versus confidentiality.

Tensions of credit, confidentiality, and erasure overlap with citational politics. Citational politics is the politics of how sources get cited in academic work (Hawkins et al., 2021). Citational politics looks at how citational practices make disciplines seem whiter, more male, more cisgender, more global north (Ahmed, 2013, 2020; Liborion & Li, 2022, see Chapter 22, this volume)—and more neurotypical. Different disciplines have different patterns of what gets counted as a source and about what should be cited, but there are some patterns that are common across many disciplines. Reading lists and reference sections are full of white men (Ahmed, 2013, 2017), for example.

These 'standard' citational tendencies are present—and a problem—in neurodiversity work[1]. Neurodiversity work is too large for any one discussion to cover it, so choosing a focus and then specific things to engage with within

[1] Neurodiversity work is distinct from neurodivergent work, in that neurodiversity work addresses the topic of the work, whereas neurodivergent work is about its authorship. The terms can and often do apply to the same work, but the emphasis differs and each term can apply even when the other does not.

(or related to) that focus is a practical requirement. It's also a political decision, and in practice, discussions of neurodiversity often leave people behind (e.g., described in Baggs, 2014). Academic discussions of neurodiversity are no more immune to this than neurodiversity work beyond academia. In addition, academic discussions tend to prioritize other 'traditionally academic' work as sources to cite, whether by not engaging with broader work or by engaging but *failing to cite it* (Jackson et al., 2022; Liborion & Li, 2022).

These tendencies lead to a wide variety of community knowledge either being ignored or being taken with incorrect or missing credit—so, this is not only a philosophical ethical issue, but a very practical one. They are compounded when multiply marginalized figures in neurodiversity movements are also not academics and write primarily in non-academic spaces, as in Kassiane Asasumasu's coining of 'neurodivergent' on About.com (Asasumasu, 2015a), not in an academic paper, and her explicit intentions in coining the term (Asasumasu, 2015b). Singer has been mistakenly attributed with coining 'neurodivergent' on the basis of her early use of 'neurodiversity' (Botha et al., 2024). Looking for academic sources, rather than references to the blogs, social media posts, and email lists where neurodiversity work has often actually happened makes it much harder to follow who actually said what, when, where, and why.

Citations of blogs may not be typical—but typical ways of doing things lead people to repeat those same ways, or one way of study being dominant can lead people to think it's the only way—the published literature is its own domain of im/perceptibility (Liborion et al., 2022). Besides, the common reference formats support blog citation, and I've yet to be challenged when I simply *do* it.

Patterns of exclusion in citation can be present even when researchers theoretically want to do otherwise, because doing otherwise takes Significant Effort. It also tends to require significantly atypical choices about citation—seeking out non-academic sources to cite is one example of an atypical choice that can lead to crediting a wider range of creators. However, with effort, it can be done. This chapter, for example, currently cites 52 sources—of which 25 are not from books, academic journals, or the World Medical Association. Similarly, while describing the authorship of a source such as an *entire journal* is complicated, about half the academic sources where it makes sense to describe the authorship include openly neurodivergent authors.

Additional Labor and Who Is Asked to Do It

People who blog and make public posts have already done the work of putting together whatever they have shared. They have already self-published, with both the speed and the control over their own content that provides (Nelson, 2022). Asking them to put it together again in a formal research study is, in fact, asking them to do more work even if they retain the same level of control over what they say, write, or share. Depending on the extent to which existing works directly or tangentially address a specific research question, asking for this labor may be warranted. However, we need to be clear that it is a request for additional labor.

A preference for anonymity may also be a reason to consider this labor. However, depending on how well known a person's blog and social media work may be, this rephrasing or recreation may not even prevent them from being identifiable to other community members (e.g., Donaldson et al., 2021). If confidentiality isn't going to be maintained effectively with any disguise strategy that leaves quotes intact (Bruckman, 2002), then we need to seriously consider whether or not there's any advantage to a new qualitative study worth requiring participant time and effort over citing existing public writing.

Seeking consent from creators ahead of using their work is labor for researchers. However, the extent to which it is truly ***additional*** labor is questionable. Similar alternatives include conducting an entire qualitative study, which also requires the work of seeking consent. Disguise methods also require labor to implement, and every creator who chooses citation and credit is a creator whose work the researchers do not need to disguise.

Compensation

Researchers typically get some sort of compensation for our work. The exact nature may still be exploitative, but if research is part of our job, then there is generally some incentive or compensation around the fact that we are Doing Research (c.f. Chapters 4, 5, 6 this volume). For research participants, compensation is an option but not generally a requirement, and incentives that would constitute full payment for their time may in some cases be criticized as undue influence (Largent & Lynch, 2017), particularly with groups who are considered categorically vulnerable (whether or not that is an accurate consideration).

For other activities associated with research, compensation may not typically be considered or available—but when including non-academic community members for whom these activities are ***not*** part of their regular jobs, it should be. For people with certain kinds of academic and academic-adjacent jobs, peer review is part of the job, and expectations about productivity may even recognize this reality. For everyone else, peer review is just extra unpaid labor, which it is still not fair or reasonable to expect of disabled people. Autism in Adulthood recognizes the problem—they're looking into ways to compensate this labor for autistic reviewers—but they do not currently do so (Nicolaidis, n.d.). The Community Against Prejudice Towards Autistic People (CAPTAP) also notes the importance of compensating autistic reviewers in their draft guidelines (Botha et al., n.d). I am not aware of other journals which are looking into this issue for reviewers.

When compensation is offered in consultation or advisory roles, underpayment is an issue (Wells, 2023), just as it is when considering compensation for research participants. Payment which is not on par with the pay for people with academic or clinical credentials is an equity issue for neurodivergent reviewers, consultants, and community partners.

Recommendations

There is value to engaging with neurodivergent work beyond the academy. There are existing frameworks for how this can be done, from which we can borrow the pieces that work well. There are also challenges in ensuring that the engagement is both meaningful and ethical. Overall, I suggest the following:

- Consider neurodivergent work from beyond academia (such as blogs, videos, or podcast) as sources of theory and in some cases as primary sources, not only as sources of data.
- Actively look for these sources; when relevant consider seeking them out *before* looking for (neurotypical-authored) academic sources that repeat community knowledge.
- Ask neurodivergent creators if they would like their work to be used in research, and if so whether they prefer credit and citation, anonymity, or pseudonymity[2]. Note that in many cases, blogs or other work may already be written under a pseudonym (e.g., Chavisory, n.d.), and that pseudonymity and citation may not be mutually exclusive.
  - Allow sufficient time for creators to think about this question and respond ahead of any deadlines.
  - You may need to come to an agreement with a supervisor or review board about the inclusion and/or level of disguise used for creators who do not or cannot respond.
  - Ideally, neurodivergent participants would not *need* to initiate the conversation about asking to be cited rather than anonymized, and the option would be given from the beginning. However, if the option is for whatever reason *not* initially available (e.g., concerns from an ethics committee, or joining a research project after its initial approval), then a request for credit from a neurodivergent creator should lead to returning to the ethics committee to support the request.
  - Look into funding mechanisms to appropriately and equitably compensate neurodivergent collaborators for whom any research work is not part of their jobs.

[2] For this chapter, I did ask living people whose non-academic blog posts or Twitter threads I cited if it was acceptable to do so. I also asked some but not all creators of academic or academia-adjacent blogs.

References

Ahmed, S. (2013, September 11). Making feminist points. feministkilljoys. https://feministkilljoys.com/2013/09/11/making-feminist-points/

Ahmed, S. (2017, December 19). Diversity work as complaint. feministkilljoys. https://feministkilljoys.com/2017/12/19/diversity-work-as-complaint/

Angulo, H., Chan, M., & DeThorne, L. (2019). Life is a stage: Autistic perspectives on neurotypicality. *Autism in Adulthood, 1*(4), 276–285. https://doi.org/10.1089/aut.2019.0024

Arnold, Larry. (2019). Autonomy, the critical journal of interdisciplinary autism studies. *Autistic Community and the Neurodiversity Movement: Stories from the Frontline*, 211-220. https://doi.org/10.1007/978-981-13-8437-0_15

Asasumasu, Kassiane [sherlocksflataffect]. (2015b, June 12). PSA from the actual coiner of "neurodivergent". Lost in my Mind TARDIS. https://sherlocksflataffect.tumblr.com/post/121295972384/psa-from-the-actual-coiner-of-neurodivergent

Asasumasu, Kassiane [sherlocksflataffect]. (2015a, June 12). How... how are we supposed to know... [Anonymous ask]. Lost in my Mind TARDIS. https://sherlocksflataffect.tumblr.com/post/121326525344/how-how-are-we-supposed-to-know-you-actually

Autism books by autistic authors project. (n.d.) https://autismbooksbyautisticauthors.com/

Autonomy. (n.d.) *Autonomy, the critical journal of interdisciplinary autism studies.* https://web.archive.org/web/20210729152553/http://www.larry-arnold.net/Autonomy/index.php/autonomy/index

Baggs, A. M. (2006). Autistic authors booklist and facts. *Autism Information Library*. [archive] https://web.archive.org/web/20110620082141/http://archive.autistics.org/library/booklist.html

Baggs, M. (2014, May 30). Your politics have a problem when they contradict the real-life experiences of the people they're supposed to be about. Ballastexistenz. https://ballastexistenz.wordpress.com/2014/05/30/your-politics-have-a-problem-when-they-contradict-the-real-life-experiences-of-the-people-theyre-supposed-to-be-about/

Botha, M., Chapman, R., Giwa Onaiwu, M., Kapp, S. K., Stannard Ashley, A., & Walker, N. (2024). The neurodiversity concept was developed collectively: An overdue correction on the origins of neurodiversity theory. *Autism*. https://doi.org/10.1177/13623613241237871

Botha, M. Friedman, S., McHaffie, S. & Waldock, K. E. (n.d.). Draft Journal Guidelines. https://livekentac-my.sharepoint.com/:w:/g/personal/kew48_kent_ac_uk/Eefd6awE8_ZAnZJDRAPURc4BQxXwb66nKrnQl1tYmykMZw?rtime=U8ioJEUh3Eg

Bristol Autism Support. (2021, 17 March). A list of autistic bloggers and YouTubers https://www.bristolautismsupport.org/actuallyautistic-blog-youtube-bloggers/

Bruckman, A. (2002). Studying the amateur artist: A perspective on disguising data collected in human subjects research on the Internet. *Ethics and Information Technology, 4*, 217–231. https://doi.org/10.1023/A:1021316409277

Chavisory. (n.d.). *Chavisory's Notebook*. https://chavisory.wordpress.com/

Coles-White, D., Jones, L., & Saka, R. (2023, November 17). Social media POV: Black autistic female perspectives on the use of camouflaging. American Speech Language Hearing Association Conference. Boston, MA,

USA. https://asha2023-asha.ipostersessions.com/?s=FE-B9-7B-8C-6E-B6-C2-98-53-89-46-C3-9C-AA-0D-0D

Dawson, M., & Fletcher-Watson, S. (2022). When autism researchers disregard harms: A commentary. *Autism, 26*(2), 564–566. https://doi.org/10.1177/13623613211031403

World Medical Association [WMA]. (2013). Declaration of Helsinki. https://www.wma.net/policies-post/wma-declaration-of-helsinki-ethical-principles-for-medical-research-involving-human-subjects/

Donaldson, A. L., Corbin, E., & McCoy, J. (2021). Everyone deserves AAC: Preliminary study of the experiences of speaking autistic adults who use augmentative and alternative communication. *Perspectives of the ASHA Special Interest Groups*, *6*(2), 315–326. https://doi.org/10.1044/2021_PERSP-20-00220

Donaldson, A. L., Corbin, E., Zisk, A. H., & Eddy, B. (2023). Promotion of communication access, choice, and agency for autistic students. *Language, Speech, and Hearing Services in Schools*, *54*(1), 140–155. https://doi.org/10.1044/2022_LSHSS-22-00031

e star. (2022, August 23). Interacting with research (August 14, 2022). #Autchat. https://autchat.com/interacting-with-research-august-14-2022/

Eb. (2015, December 6). September 29, 2015: Research. #AutChat. https://autchat.com/september-29-2015-research/

Framework for Open and Reproducible Research [FORRT]. (n.d.) Team Neurodiversity. https://forrt.org/neurodiversity/

Gordon, B. G. (2020). Vulnerability in research: Basic ethical concepts and general approach to review. *Ochsner Journal, 20*(1), 34–38.

Guberman, J. (2023). # ActuallyAutistic twitter as a site for epistemic resistance and Crip futurity. *ACM Transactions on Computer-Human Interaction, 30*(3), 1–34. https://doi.org/10.1145/3569891

Harmens, M., Sedgewick, F., & Hobson, H. (2022). The quest for acceptance: A blog-based study of autistic women's experiences and well-being during autism identification and diagnosis. *Autism in Adulthood, 4*(1), 42–51. https://doi.org/10.1089/aut.2021.0016

Hawkins, K., Liu, R., Liborion, M., & Rivers, M. (2021, June 10). The researchers that search engines make invisible. *Civic Laboratory for Environmental Action Research (CLEAR).* https://civiclaboratory.nl/2021/06/10/the-researchers-that-search-engines-make-invisible/

Hillary [Zisk], A. (2020). Neurodiversity and cross-cultural communication. In H. Bertilsdotter Rosqvist, N. Chown, & A. Stenning (Eds.), *Neurodiversity studies: A new critical paradigm* (pp. 91–107). Routledge. https://doi.org/10.4324/9780429322297-10

Jackson, L., Haagaard, A. & Williams, R. (2022, April 19) Disability dongle. Platypus: The castac blog. https://blog.castac.org/2022/04/disability-dongle/

Judy (n.d.) Actually autistic blogs list. *An Autism Observer*. https://anautismobserver.wordpress.com/

Largent, E. A., & Lynch, H. F. (2017). Paying research participants: The outsized influence of "undue influence". Irb, 39(4), 1. https://www.ncbi.nlm.nih.gov/pmc/articles/PMC5640154/

Liborion, M. & Li, R. (2022, March 2). Citational politics in tight places. *Civic Laboratory for Environmental Action Research (CLEAR).* https://civiclaboratory.nl/2022/03/02/citational-politics-in-tight-places/

Mackenzie, C. (2014). The importance of relational autonomy and capabilities for an ethics of vulnerability. Vulnerability: New essays in ethics and feminist philosophy, 33–59.

Milton, D., Gurbuz, E., & López, B. (2022). The 'double empathy problem': Ten years on. *Autism, 26*(8), 1901–1903. https://doi.org/10.1177/13623613221129123

Nelson, J. J. (2022). An analysis of self-published novels by autistic authors as a form of advocacy. *Ought: The Journal of Autistic Culture*, *3*(2), 5, p. 10–29. https://doi.org/10.9707/2833-1508.1090

Neuroclastic. (n.d.) Non-Speaker and Speller Links. https://neuroclastic.com/directory-of-nonspeaker-pages-blogs-media/

Nicolaidis, Christina. (n.d.) Autism in Adulthood: Information for Autistic Reviewers. https://s3-us-west-2.amazonaws.com/clarivate-scholarone-prod-us-west-2-s1m-public/wwwRoot/prod1/societyimages/aut/Reviewer_instructions_post_invitation_2019.pdf

Ought: The Journal of Autistic Culture. (2019–2023). https://scholarworks.gvsu.edu/ought/all_issues.html

Peoples, L. F. (2010). The citation of blogs in judicial opinions. *Tulane Journal of Technology and Intellectual Property*, *13*, 39–80. https://journals.tulane.edu/TIP/article/view/2590

Raymaker, D. M. (2020). Shifting the system: AASPIRE and the loom of science and activism. Autistic community and the neurodiversity movement: Stories from the frontline, 133–145. https://doi.org/10.1007/978-981-13-8437-0_10

Scully, J. L. (2014). Disability and vulnerability: On bodies, dependence, and power. Vulnerability: New essays in ethics and feminist philosophy, 204–21.

Seidmann, V. (2021). On blogs, autistic bloggers, and autistic space. *Information, Communication and Society, 24*(15), 2277–2292. https://doi.org/10.1080/1369118X.2020.1754878

Sinclair, J. (1993). Don't Mourn for Us. *Autism Network International*. https://www.autreat.com/dont_mourn.html

Solveig ⚧—Autistic Acceptance Worldwide [@autisticb4mmr], e* says hi (also at @endeverstar.bsky.social) [e star; @endeverstar], & Dr. Alyssa [Zisk, Alyssa H., @yes_thattoo]. (2022, August 14). In short, get permission first. [Twitter thread]. https://twitter.com/autisticb4mmr/status/1558911323399827456

theoriesofminds. (2015, December 13). Dec 13, 2015: Research. #AutChat. https://autchat.com/dec-13-2015-research/

theoriesofminds. (2017, June 2). May 28, 2017: Research (transcript). #AutChat. https://autchat.com/may-28-2017-research-transcript/

Thom-Jones, S. (n.d.). Autistic Researchers Researching Autism. *Autistic Professor*. https://www.autisticprofessor.com/autisticresearchers

Wells, A. (2023, August 7). I don't want a seat at your table: Co-production in mental health services. The National Survivor User Network (NSUN). https://www.nsun.org.uk/i-dont-want-a-seat-at-your-table-co-production-in-mental-health-services/

Williams, Rua M [@FractalEcho]. (2019, Nov 3). Q2: I find the emerging practice of "scraping" publicly posted data for research purposes troublesome. [Twitter thread.] https://twitter.com/FractalEcho/status/1191103123684806661

Yergeau, M. R. (2011, December 27). That's just your autism talking (and other phrases that shouldn't appear in an autism essay). autistext: on autism, rhetoric, & ELO. https://autistext.com/2011/12/27/thats-just-your-autism-talking-and-other-phrases-that-shouldnt-appear-in-an-autism-essay/

Yergeau, M. [Remi]. (2013). Clinically significant disturbance: On theorists who theorize theory of mind. *Disability Studies Quarterly*, *33*(4). https://doi.org/10.18061/dsq.v33i4.3876

Zisk, A. H. (2023). Critical autism studies beyond academia: An annotated list. *Ought: The Journal of Autistic Culture*, 5(1), 6. https://doi.org/10.9707/2833-1508.1144

CHAPTER 20

Participatory Methods: Researching Autistic Sexuality, Intimacy, and Authenticity

David Jackson-Perry

Introduction

The past decade has seen increasing interest in participatory frameworks within autism research, alongside a certain pressure to integrate participation and collaboration with autistic people into research processes (Bertilsdotter Rosqvist et al., 2019; Crane et al., 2018; Woods & Waltz, 2019; Hobson et al., 2023; for discussion, see also Chapter 22, this volume). However, while early-career researchers may be enthusiastic about using participatory approaches, few resources give detailed descriptions as to how participatory methods have been implemented in research (Fletcher-Watson et al., 2018; Pickard et al., 2021).

To address this gap, I draw extensively on my doctoral research for this chapter. (Jackson-Perry, 2023). However, at the time I identified (to myself and to participants) as a neurotypical researcher: I have since received a diagnosis of ADHD and reflect on this where it seems relevant for what follows. I start with a brief overview of the study and its aims before giving a detailed account of the participatory process employed during the doctoral study on which this chapter draws. I describe the five principal steps I undertook: formal and informal community engagement; the creation of a research website; the recruitment and involvement of the Research Advisory Group; an online survey, and finally written interviews. The emphasis here is on practical,

D. Jackson-Perry (✉)
Queen's University, Belfast, UK
e-mail: djacksonperry01@qub.ac.uk

H. Bertilsdotter Rosqvist and D. Jackson-Perry (eds.), *The Palgrave Handbook of Research Methods and Ethics in Neurodiversity Studies*,
https://doi.org/10.1007/978-3-031-66127-3_20

methodological questions of the early and middle parts of the research process, up to and including data collection: findings have been published elsewhere (Jackson-Perry, 2024; see also Chapter 22, this volume).

Study Overview

The doctoral research from which this chapter has been adapted explored the barriers and opportunities experienced by autistic people in their intimate and sexual lives. It used a critical autism studies (CAS) framework and drew on constructivist grounded theory for data analysis. The research questions were:

1. What are the challenges and opportunities faced by autistic people as they develop their sexual and intimate lives?
2. What are the processes by which they overcome those challenges and maximize those opportunities?

The 16 autistic participants, aged between 22 and 54 years old, and with a wide range of gender and sexual/affective orientations, took part in a total of 24 semi-directive, mostly written qualitative interviews.

I had retained four principles as being central to this research process, which I based on readings within critical autism studies and critical disability studies more broadly:

> Consultation and engagement: both formal and informal, with autistic individuals and communities throughout the research process.
> Relevance: questions investigated, and the results and reflections to which they lead hold the potential to contribute to positive impact for the people concerned, within academic autism discourse and/or in practice.
> Accessibility: both the research process and its results are to be as accessible as possible for participants and autistic individuals and communities more broadly.
> Empowerment: a move "to increase the influence of relatively powerless sectors of society" (Hahn, 2002, p. 181) is "not in the gift of the researcher" (Mercer, 2002, p. 237). However, by rejecting medical or deficit readings of participants' accounts, and respecting the points above, I sought to "facilitate this process" (Oliver, 1992, p. 111) as far as possible.

These principles were all the more important to me as I am not autistic. At the time I thought of myself as a neurotypical researcher. This may—or not—be pertinent as I now realize that I was groping, as far as was possible to me at the time, toward a form of what Raymaker (2020, p.138) calls "scientific activism," defined as work seeking:

- To encourage the inclusion of people on the autism spectrum in matters which directly affect them.
- To include people on the autism spectrum as equal partners in research about the autism spectrum.
- To answer research questions that are considered relevant by the autistic community.
- To use research findings to effect positive change for people on the spectrum.

Formal and Informal Community Engagement

Fletcher-Watson and colleagues (2018, p. 944) note that "engagement with the community in general (e.g., via social media) and consultation with relevant individuals or community organizations" is one of various "manifestations of participatory research." Indeed, consulting one or more autistic people does not, as Fetcher-Watson and colleagues (2018, p. 950) make clear, give "non-autistic researchers access to a consistent 'community view.'" I therefore found it important, as Milton and Green (2024, p.3) have since advised, to "(E)ngage with the autistic community at an early stage of the research process, as not doing so may lead to some costly regrets" (see also Chapters 4,10, 19, this volume). The need for building bridges between autistic communities more broadly and researchers is also noted in a review of researchers' views and experiences of participatory research processes (Pickard et al., 2021).

Previously, the only autistic people with whom I had had any sustained contact (to my knowledge) were the children and adolescents I supported—largely concerning sexuality—in their homes or schools. Of the 140 references in my 2016 master's dissertation, only three (apart from the posts from the website I was analyzing) came from sources I knew to represent autistic people or organizations. This reflected both the reading list for the master's program in 'autism spectrum disorders', which drew mainly on non-autistic third-party, professional, and caregiver accounts, and my ignorance at that point of other, more critical bodies of literature.

I then became aware of CAS, first through reading Worlds of Autism (Davidson & Orsini, 2013), and of autistic academics and activists who have been debating various aspects of autism, including knowledge creation (Arnold, 2012; Bertilsdotter Rosqvist, 2014; Chown et al., 2017; Milton, 2014; Milton & Bracher, 2013; Murray et al., 2005; Yergeau, 2013). I engaged with lay and academic autistic communities, online and through attending conferences. This positively impacted the research process in various ways. It led to possibilities for collaboration, facilitated calls for participation as I was familiar to a number of people who were interested, and brought a level of transparency that I hoped might mitigate some of the "distrust from autistic persons toward researchers" (Woods and Waltz, 2019, p. 2). Finally, observing and engaging with exchanges of ideas within autistic communities sensitized me to questions of importance to them, perhaps mitigating the risk of finding myself in an academic silo, only addressing questions that were considered to be of interest in research in my own discipline (Arnold, 2012).

I also joined the Critical Autism Network (CAN, 2020) and the Participatory Autism Research Collective (PARC, 2020, no page), attending—and presenting at one of—their conferences. PARC aims "to bring autistic people, including scholars and activists, together with early-career researchers and practitioners who work with autistic people." My engagement evolved and snowballed, deepening organically. With some of the people I met in those contexts, we organized *Intimate Lives*? the first conference in the UK specifically exploring the intersection of autism, sexuality, and gender. With a group of mostly autistic academics, activists, and lay-people, we initiated a loosely formed writing collective that went on to produce several publications (Bertilsdotter Rosqvist & Jackson-Perry, 2020; Bertilsdotter Rosqvist et al., 2019; Jackson-Perry et al., 2020). This form of 'reflexive cross-neurotype' collective writing is seeing an increasing interest within Neurodiversity Studies (see Chapters 7, 8, 9, 14 this volume).

While I have only briefly described my community engagement here, it was an intense part of my life for three years. While not a simple undertaking, from which I felt the need to take a distance as I wrote up my thesis, the stimulation, critical insight, support, and friendship that community engagement brought has been essential to my development in the field, as a researcher, and as a human being.

The Research Website

Another important early step was setting up a research website[1] in 2018. The rationale behind the website was as follows:

1. Transparency: the "history of distrusting researchers among 'minority communities" (Raymaker & Nicolaidis, 2013, p. 169) had been confirmed by many of the autistic people I met. It was important to me that anyone interested in being involved in this study, as advisors or as participants, could easily access information about me and the research. Through the site, potential participants could learn about the study-and the person and intentions behind it—without needing to contact me directly, deepening the notion of informed consent.
2. Accessibility: "what happens to the products of research" (Zarb, 1992, p. 128) is an important question. Too often, communities and individuals cannot access research results, either 'consumptively' (the physical ability to access research results that are often unavailable without academic affiliation) or 'transformatively' (that results, once accessed, are *useful*

[1] Autismsexualityresearch.com.

and *usable* to those who need them) (Brewer et al., 2014). With the website, I sought to make results accessible to whoever might be interested, whether or not they were part of the study.
3. Impact: extending the previous point, by impact here I mean encouraging other novice researchers to consider using participatory methods in their work. Doctoral students interested in participatory methods may wonder how to create "an active engagement with autistic communities" (Fletcher-Watson et al., 2018, p. 950), and those methods may be perceived as overly complex, risky, or too resource-costly to undertake (Pellicano et al., 2014). I hoped that having a window onto my process via the website might contribute to others considering participatory methods, whether to borrow from, build on, or reject the way I proceeded (Fig. 20.1).

Site content evolved as the research process unfolded. Initially, it contained some information about myself and the study, a statement clarifying some of the principles of my approach, and a link to my academic profile page, which also gives access to publications or conferences I have been involved in. As I looked for members for the Research Advisory Group (RAG—see below) I added the call for advisors and linked to the site from various online media. Once the RAG was in place, and in response to RAG members' suggestions that communicating synchronously across three time zones would be complicated, I added a private forum to the site to facilitate communication between members of the advisory group and myself: this provided a way for advisors to 'meet' that would have otherwise have been impossible in such a large group from different countries and regions. Calls for participants for the survey and in-depth interviews were also linked to the site, and sections provided information about them. These sections were later used to give provisional results, and I informed participants and advisors as they were updated.

Several early-career researchers also contacted me, with various questions about the site and the research process more generally. This suggests that the potential for impact within academia is indeed there, as I had hoped. It also perhaps confirms that despite the arguments made for participatory research

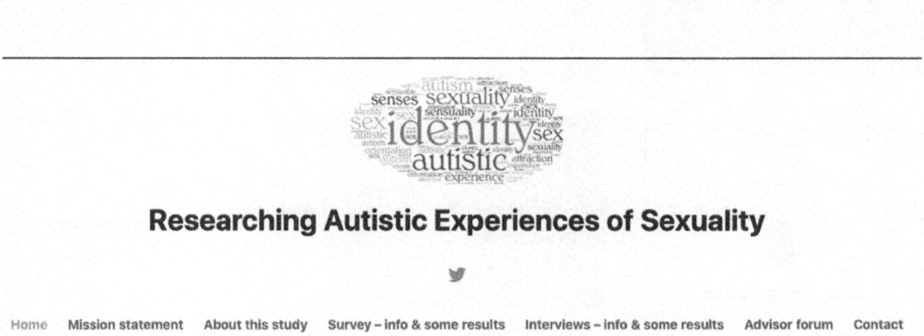

Fig. 20.1 The header for the research website

methods, there remains a need for more detailed guidance as to how exactly to go about them, beyond relatively vague statements of intent.

If I were to do it again, I would give the site a far more central role in the research process, even if it meant cutting out other research steps. To have been able to produce, for example, video presentations of results in lay language would have further increased the impact and accessibility of this study. The relatively high time investment could have been amortized by using the site for potential participants to access consent forms and information packs, reducing administrative burden elsewhere.

The Research Advisory Group

One of my earliest decisions was to seek input from a research advisory group (RAG) of autistic people, and I set out to recruit them with a call on Twitter (now known as X). The RAG's role was to be largely decided by the group itself, although their provisional remit was "to guide the research process, and to ensure that this study is respectful, accessible, and relevant to autistic individuals and communities." (see also Chapters 10, 19, this volume) (Fig. 20.2).

The inspiration for the RAG was Martin (2015), who worked with three "adults with Asperger's Syndrome" in research seeking to determine the support requirements of others. Martin initially contacted her members of her group through a regional Asperger's support group, and they met face-to-face at key points during her research project. Following Martin, my initial call sought advisors in the UK for four to five physical meetings during the year.

I very quickly received community feedback on this. Some people told me that the imposition of a face-to-face format was inaccessible either for reasons of mobility or the social challenge of meeting as a group. I also received

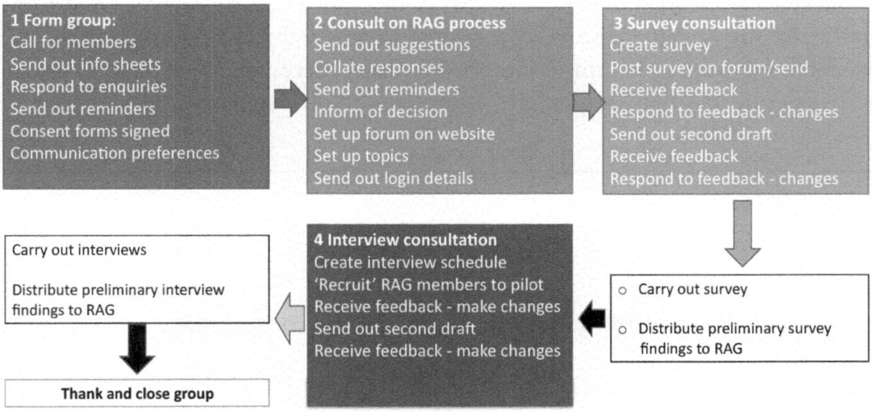

Fig. 20.2 RAG process

emails questioning the logic behind an exclusively UK-based group. Therefore—despite considerable logistical misgivings–I opened the possibility of participation to an international audience and indicated that the format would be decided by the group once it was formed. The resulting RAG was made up of nine autistic adults from the UK, Germany, and North America.

The early days of my Ph.D. were characterized by an often-paralyzing sense of illegitimacy as the neurotypical researcher that I still considered myself to be, and that I presented myself as being in all my interactions with participants and RAG members. Now that I have a formal diagnosis of ADHD, I wonder if having been able to present myself as 'neurodivergent but not autistic' might to some extent have mitigated this. As an ADHDer, I am of course no closer to having insight into autistic subjectivity (and indeed, I now realize that it conceivably worked against me regarding accessibility, as I describe below). However, from a neurodiversity perspective, being in some ways a "cousin" (Kathy Grant, in Sinclair, 2005, no page) may have mitigated the discomfort I often felt, although I cannot be sure. However, when I set up the RAG, I had neither received a diagnosis nor self-identified, and whether for this or any other reason I did not feel confident enough to ask potential members for information about themselves: even hitting 'send' on an email to RAG members sometimes came with much angst, and the concern that I was imposing on them. Therefore, the only information on individual members of the RAG is that which they spontaneously provided.

With a large, international group, face-to-face meetings were not possible. After discussion, some advisors decided to communicate individually with me, others through the advisors' forum on the website, and the rest with a combination of these options. Setting up a forum on the website again required a higher time investment on my part than I had expected: time management difficulties were a signature of this part of the research process. This was largely due to my own inexperience, to having to find solutions to problems that I had not expected, and that are well outside my comfort zone, such as managing a website.

Indeed, the creation and management of the RAG was complex, time-consuming, and often emotionally draining. Each individual item in Table 20.1 represented hours of administrative work, all in parallel to data collection and analysis: to avoid tokenism, an advisory group cannot be a discrete element, but runs parallel to, interacts with, and impacts other steps.

I had set no criteria as to how I would include RAG members: all those who completed the process of receiving information and giving consent were included. While my thesis supervisors had approved this, they also warned me that it would come with drawbacks, which was the case. Initially intending to recruit two to six advisors in the UK, I ended up with nine people in different time zones, making it necessary for me to think about ways to communicate that would take this into account. In hindsight, limiting membership to one geographical location and fewer members, like Martin (2015), would

Table 20.1 RAG members

	Location	*Age*	*Diagnosis*	*Education/ work*	*Preferred communication*	*Other information (as given by members*
Advisor 1	USA	47	As adult	Ph.D.	Email & forum	Male. Has a son with diagnosis of autism
Advisor 2	UK	26	Yes (age unknown)		Email & forum	Female. Identifies as queer/bisexual
Advisor 3	UK	25	As adult		Email	Female. Married to asexual-identified autistic man
Advisor 4	USA		As adult	Ph.D.	Email & forum	Female
Advisor 5	UK	29	Asperger's	Researcher Ph.D. candidate	Totally flexible	Female
Advisor 6	Germany		Self-identified	Medical statistician	Email & forum	Female
Advisor 7	UK	30's	Yes (age unknown)	Computer programmer	Forum	Bi/pan-sexual, non-binary
Advisor 8	UK	30's	Yes (age unknown)	Ph.D. candidate	Email & forum	Female
Advisor 9	UK		Yes (age unknown)	MSc	Email & forum	Male

have been more manageable within the constraints of a doctoral thesis, if less diverse.

Nicolaidis and colleagues (2019, p. 12) note that "community partners often feel that instruments developed for general populations are not accessible, usually due to confusing or imprecise language or ableist assumptions." This was indeed a major area of input from the RAG, both for the survey (see below) and the interview (adapted from McAdams, 2008; see below). I will give one particularly representative example of each here. My original question in the survey (to explore which areas related to.

sexuality respondents would like to see explored) was initially:

- **A** - Please outline one question/area you would like to see researched about sexuality **(Open answer)**
- **B** - Please outline a second question you would like to see researched about sexuality **(Open answer)**

I had wanted to avoid being directive but received several comments on this from advisors. One said:

> It's rather daunting to have a "go for it" section without an easy out. I'm guessing there are some established areas that might be worth putting down for those who could be overwhelmed by a blank canvas... I personally don't know what areas are being studied.

Another, more specifically, suggested:

> It might be worth re-formatting this question to include areas of research as examples, I have no clue what I would answer for this and there is too much choice. If I was writing this question myself, I would give 5-10 examples of research areas and ask people to number 1-3 their most important, with a comment box below if they feel you've missed anything.

The final wording provided six possible areas to be numbered from one to six in order of importance for respondents, with a free response question at the survey's end.

In the written interviews I had phrased a question on sensory experience as follows:

> A sensory experience. Please describe an experience that sticks out in your mind that was in some way connected to your sensory desires or needs. Was it pleasurable, painful, exciting, difficult, comforting? Were you alone or with someone else? How did you feel during and after this experience?

One advisor found the interview "very interesting to complete and quite therapeutic." noting that she could "see the aim is to encourage people to explore their experiences quite freely." However, she went on to say that "information overload," for her, made "questions with a long preamble or a series of questions within a question...disorienting." Her proposition was this:

A sensory experience: Please describe an experience that sticks out in your mind that was in some way connected to your sensory desires or needs. Think about:

- How it felt (for example: pleasurable, painful, exciting, difficult, comforting).
- Whether you were alone or with someone else.
- How you felt during this experience.
- How you felt after the experience.

The final wording of this question was very close to her suggestion. Both my original questions and advisors' comments highlight the tension between wanting to leave respondents as much room as possible to respond in their own way, and the inaccessibility that this might bring by being over-broad in my questions, confirming the importance of "providing enough specificity, even with open-ended questions; and ensuring questions are sufficiently concrete" (Nicolaidis et al., 2019, p. 11). This feedback, along with the relatively short responses given by both advisors who piloted the interview, also led me to add a 'warning' to the interview information sheet: "some of the questions are quite 'broad' and open-ended. This is so that you can talk about subjects that are important to **you**, and not to limit you to what **I** think might be important."

I continued to work in parallel with my Ph.D., and this, combined with my inexperience, meant that I found it difficult to predict and manage the duration of each step. Data collection and analysis—which I had expected to take nine months—lasted almost two years. This seemed too long a period to expect advisors to remain engaged. I was feeling increasingly uncomfortable with communicating with them and felt guilty that I was asking too much of them. I sent a final email to advisors thanking them, stating that I would no longer be contacting them but inviting them to contact me if they had any comments, and directing them to the research website for updates.

The question of time management was a prickly one throughout. From the word go, some members were highly responsive, contributing a great deal of input. However, others were not, and I found myself caught between the need to send reminders to people and the risk of overwhelming or harassing them. This concern was compounded by some advisors having physical and mental health difficulties at various points during the process. In retrospect I would make this part of initial discussions regarding the RAG functioning—how frequently would people like to be reminded, at what point do I stop sending reminders and take silence as a definitive though implicit 'no more please'. I was encouraged to continue contacting advisors with updates when they did not respond by several messages similar to the one below, when an advisor said that they were having health difficulties and were unsure about being able to participate further, but added:

> Thank you so much for checking back in, I was so caught up in my own world due to the recent (mental health) flare that I didn't even have it on my radar anymore. I hope that I can still help with the project in the near future, but right now there's a lot of personal stuff that will be taking priority. Please don't stop keeping me updated / checking in, though. I really appreciate it!

My concern about the level of participation I could reasonably expect from advisors was aggravated by several structural difficulties. One was that I was not able to pay RAG members for their time. While there is "an ethical issue in asking that participants give their time without compensation" (MacLeod et al., 2014, p. 415), I was also advised that an ethical issue of another kind—potentially inducing people to participate if they were paid—was problematic for my ethics committee (see Chapters 10 and 18, this volume for more about ethics applications). I contacted Autistica, an autism research charity in the

UK, which supported the formation of autistic research advisory groups. Far from presenting an ethical difficulty, for Autistica, payment of advisors was a *condition* of their support (which, incidentally, reflected my own view). However, no funding mechanism being in place at that time via Autistica for doctoral research, I was in a catch-22 in this regard.

I was also advised by my supervisors that *too much* input from advisors (particularly co-analysis of data and publications) would bring into question whether the thesis could be considered my own work, a condition for Ph.D. theses in the UK. At the time I was frustrated with this comment. However, with hindsight I am grateful: involving third parties in analysis and publication would have added yet another element to a process which was already, I now see, overly ambitious within the time, resource, and requirement constraints of a Ph.D.

The Survey

There was no information available at the time I was writing concerning which areas of sexuality constituted research priorities for autistic people. I therefore carried out a survey, in which I stated that I sought to understand:

> what aspects of sexuality people (responding to online research) would like to see explored, and the most accessible way to do this. The results of this survey will guide the next stage of my research, the questions I explore, and the way I explore them.

I recruited respondents through snowballing via research advisors and acquaintances; direct emailing to national, local, and student organizations active in autism advocacy and services; publication on Twitter; a link to the Facebook page of the Critical Autism Studies network and on a Facebook page created for the purpose of distributing the survey, and via link to the survey on the research study website. The call used four different color schemes, as suggested by a RAG member. The survey had 12 questions, and I had responses from 567 people from 27 countries in two weeks. I then published a last post, thanking those that had participated and directing anyone interested to the research website for preliminary results (Fig. 20.3).

The results relevant here are, firstly, the research areas of most interest to respondents (Table 20.2).

Based on responses to the survey, I had expected my interviews to be concerned largely with sensory experience. However, this was not the reality of interview participants' responses (see below for discussion).

The second result of interest here was that most respondents preferred a guided interview format, and to respond to questions in writing and in their own time. This is supported elsewhere, with Howard and Sedgewick (2021), finding that particularly when communicating with unfamiliar people, written communication was the preferred option for most of the autistic people

Fig. 20.3 The call for survey participants on Twitter and Facebook

Table 20.2 Research priorities (as a percentage of respondents selecting each option as either their 1st or 2nd choices)

Sexuality and sensory experience	*44.7%*
Intimate relationships'	39.4%
Social barriers to sexuality	35.5%
Sexual education	30.6%
Sexuality and gender roles	26.2%
Constructing sexual identity	24.9%

solicited. I therefore proposed this as one amongst other possible formats for the in-depth interviews, and written responses were indeed the most popular format for interview participants.

In retrospect, if I had to do it all again, I would not carry out a survey, for several reasons. Firstly, my aim with this thesis was to gather as much depth and complexity as possible in responding to my research questions: clearly, a survey is not an appropriate tool in this case. Secondly, I realize that carrying out a survey to understand areas of research that are important to people, and then interviewing a different set of people, is not a good fit with inductive analysis in general or grounded theory in particular, in which areas to be studied are led (here) by interview data. Finally, the information regarding preferred forms of participation was already available (see, for example, Howard & Sedgewick, 2021; Benford & Standen, 2011).

The Interviews

In-depth interviews produced the majority of data here. Fifty-one people responded to the call on Twitter, and 16 completed the pre-interview process (reading the detailed information sheet and signing the consent form) and went on to be interviewed (Fig. 20.4 and Table 20.3).

This resulted in 24 interviews: seven participants responded to a second follow-up interview, and one completed a third, allowing me to revisit certain themes and check my understanding of previous responses. For the first interviews, 13 participants chose an asynchronous written interview, two a Skype interview, and one instant messaging. People choosing Skype/instant messaging were sent questions in advance. For the second interviews, one person (who had previously chosen written questions) chose a Skype interview, and the others (including one who had previously chosen Skype) chose written questions. The only person to do a third interview did so in writing,

Why ask autistic people about sexuality, intimacy, and sensory experience?

- There is very little research about sexuality and intimacy from an autistic perspective.
- In a survey of 567 autistic adults, sensory experience was the most popular choice for future research about sexuality and intimacy.

Who can take part?

- Autistic adults interested in answering questions about sexuality, intimacy, and sensory experience.

What can you expect?

- To answer some questions about your sexual, sensory, and intimate experiences. This can be done via Skype or I can send you a list of questions to answer in writing, as you prefer. This first phase will probably take 1 – 2 hours.
- To take part in a second, follow-up interview probably lasting about half an hour.
- As a 'thank you' you will receive a £25 voucher of your choice for each interview.

More information?

Here: www.autismsexualityresearch.com
Or email me: djacksonperry01@qub.ac.uk
Or DM me on twitter @IntersectQUB

Fig. 20.4 The call for interview participants on Twitter

Table 20.3 Country of origin, age, diagnostic status, sexual and gender identity (as given by participants: pseudonyms were chosen by participants if they so wished)

Name	*Country*	*Age*	*Diagnosis/ self-ID*	*Diagnosis/ID age*	*Sexual identification*	*Gender identification*
Charlotte	UK	32	DX	32	Queer	Cis woman
John	UK	32	DX	4	Neuro-queer/ Grey	Cis male
Sara	UK	36	DX	33	Bisexual/ Polyamorous	Gender-queer
Charles	USA	42	DX	22	Hetero/ Demi-ace	Cis male
Finn	Australia	22	DX	19	Bisexual	Neutrois, AFAB
A	UK	23	DX	19	Asexual	Non-binary
Beatrice	UK	25	DX	23	Bisexual	Cis female
Julie	UK	30	DX	28	Lesbian	Woman
Mia	UK	40	Self-ID	40	Bisexual	Cis female
Hazel	UK	23	DX	18	Heterosexual	Cis female
Mike	Canada	54	DX	54	Hetero (?)	Male
James	UK	20	DX	14	Bisexual	Man
Gillian	UK	41	DX	38	Heterosexual	Female
Oliver	UK	27	DX	5	Pansexual	Cis male
Karen	UK	47	DX	44	Lesbian	Non-binary
Susan	UK	22	DX	14	Pansexual	Female

as they had for the two other interviews. All non-written interviews were transcribed prior to analysis. The average length of written interviews was 2602 words. Two Skype interviews ran two hours and ten minutes and the third just under 45 minutes; the instant messaging interview lasted approximately 30 minutes. In July 2019, I sent an email to participants with some preliminary results, and in April 2020, I sent a final email thanking participants and directing them to the research website for further information.

The interview schedule was adapted from a life story interview schedule, initially used in personality psychology, and shared with me by Marianthi Kourti, a Ph.D. colleague with whom I have worked on several writing projects. While I wasn't seeking to analyze my participants' personalities, McAdams' (2007, p.1) interview schedule aimed "to understand the different ways in which people in our society live their lives and the different ways in which they understand who they are." This ambition intersected with my research questions and with my understanding of the importance not just of events themselves, but of the 'meanings given to events and the actions/interactions/emotions expressed in response, along with the context in which those response and events occur' (Corbin, 2009, p. 38). After piloting the interview schedule with members of the RAG, I made further adaptations (some of which I described above) before sending interviews out.

I conclude this section on the interviews I carried out with a brief reflection regarding the balancing act between accessibility and the gathering of rich data. Firstly, most participants chose written interviews as being the most accessible for them (c.f. Chapter 21, this volume). I kept many of the questions as open as possible (to allow participants to explore avenues that *they* felt to be important, rather than that I had decided were important), while still giving enough guidance to make responding accessible. This meant that sometimes important elements 'emerged', such as when one participant said "*There were two times I can think of when I was raped*" without giving any further information or context. Had this been in a face-to-face format, I feel I would have been able to infer whether this could have been safely developed, as well as to whether support should be offered. However, to do this in a written follow-up seemed intrusive and perhaps risky. There was clearly a tension between gathering deep, rich data, and making it possible for people to safely participate at all.

Secondly, while I had thought a great deal about how I could be as accessible as possible for participants, it had not occurred to me to worry about my *own* potential difficulties: what is accessible for participants is not necessarily accessible for the researcher (c.f. Chapter 13, this volume). Susan chose to be interviewed using instant messaging, and so we exchanged in short sentences in real time, similar to using WhatsApp. For her, this was the only format that would have made the interview accessible. I found it frustrating, overwhelming, and exhausting. Susan was used to typing very rapidly and replied to my questions immediately and incisively. However, I couldn't process her responses rapidly enough to reply, either to prompt or to ask further questions, to the extent that at one point she wrote "*what is the problem??*" I ended up cutting the interview relatively short with a sense of defeat.

In our interaction, what was accessible for Susan was inaccessible for me. Since carrying out this research, I have received a diagnosis of ADHD, which perhaps goes some ways to explain the incompatibility that Susan and I experienced. Hume (Chapter 13, this volume) notes that it is important to proactively propose accommodations to all participants rather than wait for them to ask for them. In this case, I had done this. However, what I had not seen coming was my own incapacity to use a 'quick-fire', single-sentence format, either from a motricity point of view (of not being able to type fast enough), or from a processing perspective.

That access for one person might be 'access denied' for another was explored briefly in the authors' reflections on the process behind a prior publication (Jackson-Perry et al., 2020). While it is theoretically satisfying to find support for the importance of context when considering accessibility, as a research experience, it was extremely disagreeable. Obviously, the burden of finding an accessible mode of communication for participants is squarely on the researcher's shoulders, which in this case I was able to do only to a limited extent. However, this situation gave me the uncomfortable but useful opportunity to find myself in the shoes of those who are excluded from research participation for questions of access.

Limitations and Strengths

Principal limitations throughout are, firstly, that some participants were likely to be part of my social media network, and secondly that we cannot be sure that all respondents meet the diagnostic criteria for autism. My Twitter profile at the time read: *Queer-hood | Researching autistic accounts of sexuality—intimacy | Critical Autism Studies | Sexual health | HIV @CHUVLausanne | He/him* which is likely to bias the profiles of those who follow me. I have no way of knowing how people accessed the call for participants. However, it is possible that responses may have come from a broader section of the autistic population than if I had recruited via existing 'gatekeepers' such as clinics, a point also noted in the few studies that used a similar methodology (see, for example, Byers et al., 2013).

A further limitation throughout this study is a lack of inclusion of those who might be, for example, learning disabled: this limitation is often stated by researchers within autism studies, but is one in which little progress has been made (Fletcher-Watson et al., 2019; see Chapter 15, this volume, for an exception). Indeed, RAG members here had high education levels, which may be itself be a limitation. Further, as I described above, I did not feel legitimate at the time to ask many direct demographic questions of either RAG members or interview participants, and so have no data concerning the ethnic diversity—or otherwise—of these people.

Conclusion

The process I have described here was not the simplest or the quickest way of carrying out participatory research. I discussed above the 'casual engagement' with autistic community members that I experienced early in the research process: one of these was a meeting with Dr. Damian Milton in 2017 at the 3rd International Disability Conference in Amsterdam at which we were both giving presentations. His advice to me, given the constraints of a Ph.D., was something like this: "even if you only do one small thing that is participatory during your study, and you do it well, this is already more than most people are doing and is better than nothing." This was the most useful advice that I could have received at that time, and certainly one I would pass on to anyone else doing a Ph.D.. Unfortunately, it was advice that I did not have the confidence to follow at that time: I felt so unsure of myself that I was always looking at how to do *more*, rather than how to do less, but better. Despite having no regrets as to the route I took, as I read this back, I realize that my experience highlights the need to consider how to meet one's participatory ambitions to a degree that respects the central tenets of the framework without putting undue burden on advisors, participants, or researchers. Further, at the time, my sense of discomfort was so intense that, beyond the 'paralysis' I mentioned above, I frequently questioned whether I should be working in this field at all. Today I understand that far from being a barrier, my discomfort became a motor, a 'methodological tool' even, pushing me beyond what I had been taught (see Chapter 22, this volume). Would I do it the same way if I did it all again? No. Do I regret having done it this way? No: without this—uncomfortable and overwhelming and unwieldy as it often was—my Ph.D. journey would have been lonelier, my thinking more limited, and the result, for what it is worth, poorer.

References

Arnold, L. (2012). Introduction. *Autonomy, the Critical Journal of Interdisciplinary Autism Studies, 1*(1).

Benford, P., & Standen, P. J. (2011). The use of email-facilitated interviewing with higher functioning autistic people participating in a grounded theory study. *International Journal of Social Research Methodology, 14*(5), 353–368.

Bertilsdotter Rosqvist, H., & Jackson-Perry, D. (2021). Not doing it properly? (Re) producing and resisting knowledge through narratives of autistic sexualities. *Sexuality and Disability, 39*(2), 327–344.

Bertilsdotter Rosqvist, H., Kourti, M., Jackson-Perry, D., Brownlow, C., Fletcher, K., Bendelman, D., & O'Dell, L. (2019). Doing it differently: Emancipatory autism studies within a neurodiverse academic space. *Disability and Society, 34*(7–8), 1082–1101.

Brewer, E., Selfe, C. L., & Yergeau, M. (2014). Creating a culture of access in composition studies. *Composition Studies, 42*(2), 151–154.

Byers, E. S., Nichols, S., & Voyer, S. D. (2013). Challenging stereotypes: Sexual functioning of single adults with high functioning autism spectrum disorder. *Journal of Autism and Developmental Disorders, 43*(11), 2617–2627.

CAN (2020). Critical autism network. Available at: http://www.open.ac.uk/health-and-social-care/research/critical-autism-network.

Chown, N., Robinson, J., Beardon, L., Downing, J., Hughes, L., Leatherland, J., Fox, K., Hickman, L., & MacGregor, D. (2017). Improving research about us, with us: A draft framework for inclusive autism research. *Disability & Society, 32*(5), 720–734.

Corbin, J. (2009). Taking an analytic journey. In J. M. Morse, et al. (Eds.) *Developing grounded theory: The second generation*. Left Coast Press.

Crane, L., Adams, F., Harper, G., Welch, J., & Pellicano, E. (2019). 'Something needs to change': Mental health experiences of young autistic adults in England. *Autism, 23*(2), 477–493.

Fletcher-Watson, S., Adams, J., Brook, K., Charman, T., Crane, L., Cusack, J., Leekam, S., Milton, D., Parr, J. R., & Pellicano, E. (2019). Making the future together: Shaping autism research through meaningful participation. *Autism, 23*(4), 943–953.

Hobson, H., Linden, A., Crane, L., & Kalandadze, T. (2023). Towards reproducible and respectful autism research: Combining open and participatory autism research practices. *Research in Autism Spectrum Disorders, 106*, 102196.

Howard, P. L., & Sedgewick, F. (2021). 'Anything but the phone!': Communication mode preferences in the autism community. *Autism, 25*(8), 2265–2278.
Intimate Lives Autism Network. (2018). Intimate lives? Autism, sexuality, gender, and identity. Available at: https://autgensex.wordpress.com/
Jackson-Perry, D., Rosqvist, H. B., Annable, J. L., & Kourti, M. (2020). Sensory strangers. Neurodiversity studies: A new critical paradigm.
Jackson-Perry, D. (2023). Exploring autistic accounts of sexuality, intimacy, and authenticity (Doctoral dissertation, Queen's University Belfast). Available at: https://pure.qub.ac.uk/en/studentTheses/exploring-autistic-accounts-of-sexuality-intimacy-and-authenticit
Jackson-Perry, D. (2024). Exploring autistic accounts of sexuality, intimacy, and authenticity. In *Exploring autistic sexualities, relationality, and genders: living under a double rainbow* (edited by Bertilsdotter Rosqvist, H., Day, A. & Krazinski, M). Routledge
MacLeod, A., Lewis, A., & Robertson, C. (2014). CHARLIE: PLEASE RESPOND! Using a participatory methodology with individuals on the autism spectrum. *International Journal of Research & Method in Education, 37*(4), 07–420.
Martin, J. A. (2015). Research with adults with Asperger's syndrome—participatory or emancipatory research? *Qualitative Social Work: Research and Practice, 14*(2), 209–223.
McAdams, D. P. (2008). *The life story interview*. Northwestern University.
Milton, D. (2014). Autistic expertise: A critical reflection on the production of knowledge in autism studies. *Autism: The International Journal of research and Practice, 18*(7), 794–802.
Milton, D. E., & Green, J. (2024). Theorising autism. *Autism*, 28(4), 795–797. https://doi.org/10.1177/13623613241235786
Milton, D., & Bracher, M. (2013). Autistics speak but are they heard. *Medical Sociology Online, 7*(2), 61–69.
Murray, D., Lesser, M., & Lawson, W. (2005). Attention, monotropism and the diagnostic criteria for autism. *Autism, 9*(2), 139–156.
Nicolaidis, C., Raymaker, D., Kapp, S. K., Baggs, A., Ashkenazy, E., McDonald, K., Weiner, M., Maslak, J., Hunter, M., & Joyce, A. (2019). The AASPIRE practice-based guidelines for the inclusion of autistic adults in research as co-researchers and study participants. *Autism, 23*(8), 2007–2019.
PARC. (2020). The participatory autism research collective. *The Participatory Autism Research Collective*.
Pellicano, E., Dinsmore, A., & Charman, T. (2014). Views on researcher-community engagement in autism research in the United Kingdom: A mixed-methods study. *PLoS ONE, 9*(10), 109946.
Pickard, H., Pellicano, E., den Houting, J., & Crane, L. (2022). Participatory autism research: Early career and established researchers' views and experiences. *Autism, 26*(1), 75–87.
Raymaker, D., & Nicolaidis, C. (2013). Participatory research with autistic communities: Shifting the system. In *Worlds of autism: Across the spectrum of neurological difference* (pp. 169–188). University of Minnesota Press.
Raymaker, D. M. (2020). Shifting the system: AASPIRE and the loom of science and activism. Autistic community and the neurodiversity movement: Stories from the frontline, 133–145.

Sinclair, N. (2005). History of ANI. Autism Network International. https://www.autismnetworkinternational.org/History_of_ANI.html

Woods, R., & Waltz, M. (2019). The strength of autistic expertise and its implications for autism knowledge production: A response to Damian Milton. *Autonomy, the Critical Journal of Interdisciplinary Autism Studies*, *1*(6).

Yergeau, M. (2013). Clinically significant disturbance: On theorists who theorize theory of mind. *Disability Studies Quarterly*, *33*(4).

Zarb, G. (1992). On the road to Damascus: First steps towards changing the relations of disability research production. *Disability, Handicap and Society*, *7*(2), 125–138.

CHAPTER 21

Using Email Interviews to Understand Autistic Experiences

Claire M. Chapman

Introduction

This chapter draws on insights from postgraduate criminological fieldwork (Chapman, 2022) to discuss an email interview method that was purposefully used to create an inclusive research experience with autistic women. Digital environments have increasingly become a place of social engagement and interaction on a larger and more dispersed scale, almost entirely negating both time and space (Gibson, 2017; Hawkins, 2018). With this, there has been an interest from qualitative researchers in the social sciences to explore how the internet can best be used for undertaking research with people and generating rich data (Dahlin, 2021; Gibson, 2017; Hawkins, 2018; James, 2007).

In recent years, autism research has become a more prevalent field of study, with growing interest in understanding more about who autistic people are and how they experience the world (Poulsen et al., 2022; Rice & Lee, 2017). However, a vast amount of autism research is still focused on being about autistic people without necessarily doing research *with* them (Chown et al., 2017), and it is important that research intentionally involves them. Moreover, autism research as a whole also lacks the representation and perspectives of autistic women (Cascio et al., 2021), particularly from a qualitative standpoint (Sohn, 2019).

C. M. Chapman (✉)
University of Glasgow, Glasgow, UK
e-mail: c.chapman.3@research.gla.ac.uk

H. Bertilsdotter Rosqvist and D. Jackson-Perry (eds.), *The Palgrave Handbook of Research Methods and Ethics in Neurodiversity Studies*,
https://doi.org/10.1007/978-3-031-66127-3_21

In the postgraduate study (Chapman, 2022), email interviews were used to try and improve the experiences of autistic women in research, giving them the agency and environment to participate more meaningfully. The method sought to recognise the individuality and uniqueness of all participants and the researcher, and their communication preferences. Guided by Chown et al. (2017) and their draft framework for inclusive autism research, this research was led by an autistic researcher, had the social model of disability (Oliver, 1990) at the core, and the outcomes of the research aimed to better the lives of autistic people. It aimed to challenge mainstream assumptions and stereotypes and reclaim dismissed or misrepresented stories intersecting disability and gender (Garland-Thomson, 2005) whilst addressing the paucity of qualitative research in understanding autistic women, and the gap in perspectives within criminological research. It also sought to directly address key research priorities identified by the autistic community in Scotland, notably: support services; issues impacting autistic women; and victimisation (Cage et al., 2022) (See also Chapters 4, 10, 13,14, 20, 22, this volume).

This chapter will present an overview of the research, reflecting on the email interview method and how it might contribute to autism research or research involving sensitive subjects. Firstly, it will provide a background to the research in question, describe the method, and provide the justifications behind it. Secondly, it will present an overview of how the email interviews were conducted in practice. Lastly, it will provide methodological reflections of the fieldwork, relating to time and control, depth and authenticity, and suggestions to help inform future research practice.

Overview of the Research

The postgraduate study discussed in this chapter (Chapman, 2022) aimed to address key gaps in academic literature in relation to autistic women, victimology, and associated support services. The initial, in-depth literature review of the study found that as a group, autistic women are underrepresented in research (Cascio et al., 2021; Saxe, 2017; Sohn, 2019). With this deficit in representation, an area of questioning has opened up of how they differ and how their experiences are distinct, with increased interest in engaging with them to answer these questions (Milner et al., 2019; Sohn, 2019).

The research available suggests that autistic women encounter barriers in the social world based on their gender and perceived differences (Bargiela et al., 2016; Saxe, 2017). As women, they have marked differences in both their autism and their position in society, often facing unfair and differential treatment in a world that is inaccessible to them (Davidson, 2007; Hull et al., 2020). The literature review found that disabled women are disproportionately victimised (Ballan & Freyer, 2012). More specifically, autistic women can be at risk of greater and repeat victimisation, especially interpersonal violence or sexual abuse (Cazalis et al., 2022; Gibbs & Pellicano, 2023; Gibbs et al., 2023; Hull et al., 2020; Paul et al., 2018; Pecora et al., 2020). Despite this, there has been little research within feminist criminology exploring the significance of disability[1] and the experiences of those who have been victimised (Macdonald et al., 2023; Shaw et al., 2012).

Autistic women deal with diverse and complex challenges such as masking[2] and internalising emotion (Baldwin & Costley, 2016; Bargiela et al., 2016; Greenlee et al., 2020), and in response to victimisation, there is a greater need for both practical and emotional support services that are substantial, suitable, and tailored to these experiences (Babb et al., 2021; Ballan & Freyer, 2012). This begged the question of what their experiences were of accessing support services following victimisation and how these could be improved. Therefore, the aim of my research was to understand the experiences of accessing support for autistic women who have been victims of crime.[3] This involved exploring participants' perspectives of what good practice looked like, as well as the barriers to support provision and where the needs of autistic women were not being met. Specifically, the research looked to answer the following questions:

1. What barriers do autistic women who have been victims of crime face in accessing support?
2. To what extent are the needs of autistic women being met in this support provision and what does good practice look like for them?
3. How could support provision for autistic women who have been victims of crime be improved?

The research design was rooted in a constructivist approach to grounded theory, adopting a focus on flexibility, reflexivity, and a ground-up, iterative approach to inquiry (Charmaz, 2007). Therefore, to address the aims and objectives of this research, I used qualitative email interviewing to give an alternative platform within research for the voices of autistic women and develop a deeper understanding of their engagement with support services (Bryce et al., 2016). The objective of this was to gather insight into their experiences and formulate recommendations that could be practically implemented, as well as provide theoretical contributions to build upon existing academic work. This emphasised the importance of autistic women's own experiences as individuals set within their subjective realities and contexts, and in doing so aimed to offer email interviewing for them to share their personal stories. Overall, the fieldwork lasted six months (April–September 2022), with 11 email interviews undertaken with autistic women during this time.

[1] It is worth noting that there is more than one school of thought regarding the construction of autism as a disability; this research allowed participants to take the lead in their self-identification.

[2] Masking: Concealing traits, emotions, or true responses to meet perceived expectations in social situations and appear 'less' autistic (Bargiela et al., 2016; Hull et al., 2020).

[3] To minimise exclusions, the type of victimisation/crime was not specified and is open to the personal interpretation of each participant.

Email Interviewing

Qualitative research can provide in-depth, rich data, adding value and meaning in ways that cannot be obtained from quantitative research alone (Lichtman, 2014; Silverman, 2016). As a key qualitative research method, interviewing is 'one of the most powerful tools available to the qualitative researcher' (Lichtman, 2014, p. 241). Interviews were the primary method of generating data in my research and were used to explore the experiences of autistic women in relation to support services by emphasising the realities and contexts of their personal experiences, offering a space for me to listen, understand, and interpret these. Specifically, I used semi-structured interviews via email as a central component in promoting inclusivity, aiming to ensure that the views of my participants were heard.

Email interviews are a form of asynchronous communication between a researcher and participant where text-based information is repeatedly exchanged out with strict temporal and spatial limitations (Gibson, 2017; James & Busher, 2006; Ratislavová & Ratislav, 2014). All communication is undertaken through email which allows participants to respond in their own time, and interview questions can be sent one or several at a time (Gibson, 2017). The use of email interviewing in qualitative research has been increasing (Meho, 2006), although the method is still somewhat underused, possibly due to the lack of awareness and understanding of it (Hawkins, 2018). Email interviews can remove some of the limitations of traditional interview methods, such as time, cost, and accessing participants, and have been found to be a reliable method, offering a positive experience for both participant and researcher and generating a wealth of rich data (Benford & Standen, 2011; Dahlin, 2021; Fritz & Vandermause, 2018; Gibson, 2017; Hawkins, 2018; Meho, 2006). In a practical sense, email interviewing can break down barriers to engaging in research and facilitate involvement for those who may otherwise have been unable to take part (Benford & Standen, 2011; Hawkins, 2018; McCoyd & Kerson, 2006; Meho, 2006; Ratislavová & Ratislav, 2014). It does this by removing space, time, and financial constraints; addressing challenges with language and communication differences; and has been considered a particularly useful method to engage with isolated or stigmatised groups who may get overlooked in research, such as disabled people, or those who may better express themselves in writing (Benford & Standen, 2011; Bowker & Tuffin, 2004; Hawkins, 2018; McCoyd & Kerson, 2006; Meho, 2006; Ratislavová & Ratislav, 2014).

Why this Method?

Studies that have previously used email interviews have done so due to the potential of the method to be more inclusive, particularly when working with sensitive topics or 'vulnerable' participants. For example, Ratislavová and Ratislav (2014), McCoyd and Kerson (2006) and Beck (2005) all used this method when doing sensitive research with women who had dealt with birth trauma or bereavement, and Egan et al. (2006) used it with traumatic brain injury survivors. When given the choice in these studies, participants overwhelmingly chose email as they felt they could communicate more effectively. The method gave them agency over how they participated in research and allowed them to share their experiences more confidently in a safe, private and comfortable manner (Egan et al., 2006; McCoyd & Kerson, 2006; Ratislavová & Ratislav, 2014). As Gibson (2017) commented, the method is especially suitable for topics dealing with past experiences, memories, and recollections, particularly where the focus is on perceptions of these, or of a sensitive nature.

In line with my research, email interviews have also been used to engage with disabled people (Bowker & Tuffin, 2004) as well as autistic people (Benford & Standen, 2011, see also Chapter 20, this volume). Every person will have their own preference for communicating, but Howard and Sedgewick (2021, p. 2265) found that the type of communication mode used can be either 'enabling or disabling for autistic people'. Their research surveying 245 autistic adults found that written communication was consistently preferred; their participants reported that compared to face-to-face or phone calls, it gave them time to think, provided structure and predictability, and reduced both feelings of anxiety and negative sensory experiences, and this was especially true for participating in research (Howard & Sedgewick, 2021). Moreover, Davidson (2008) and Parsons et al. (2020) discuss how online spaces have become invaluable to autistic people, creating opportunities for inclusive and social engagement in ways that may not be as easy in person. Stepping away from the dominant and neurotypical forms of communication, written text can help autistic people feel safe and empowered as they relay information, valuing the time to think without pressure to respond in haste (Davidson, 2008). Extending this, Cascio et al. (2021) touch on how the internet can facilitate inclusion for autistic people, especially those unable to communicate verbally (although they do note concerns about the total inclusivity of written communication for *all* autistic people).

My research acknowledged the importance of autism researchers using tools to study autism that make sense to autistic people for learning about their experiences (Jones, 2022). Email interviews could open up a space for autistic women to discuss sensitive topics and provide deeper reflective written accounts facilitated by the level of anonymity, comfort, and control for participants that may suit many autistic people (Benford & Standen, 2011; James & Busher, 2006; Ratislavová & Ratislav, 2014; Saarijärvi & Bratt, 2021). They could be conducted in a low-pressure environment at a suitable time for the participants, where they feel comfortable and able to use their 'voice'. Additionally, much like traditional interview methods, they are still said to generate rich, high-quality data capturing participant's constructions of reality, their experiences, and the meaning within these (James & Busher, 2006).

Fig. 21.1 Email interview stages

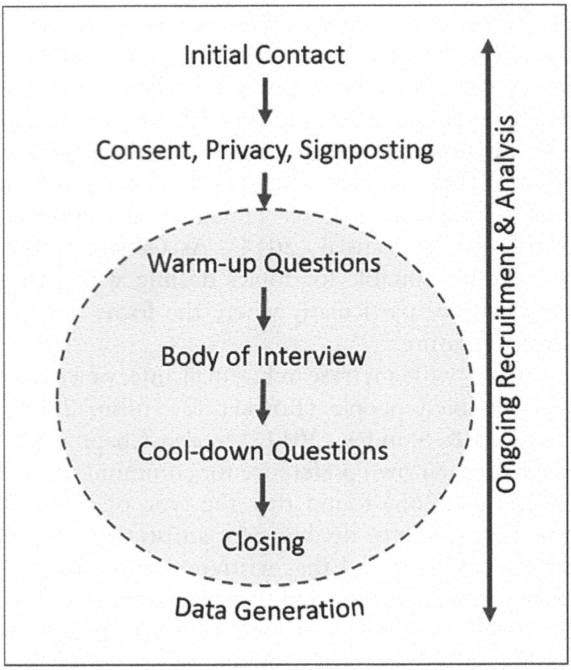

Fieldwork in Action

Through email, I interviewed adults living in Scotland who identified as autistic women and were currently accessing or had previously accessed support services after being victimised. This aligned with constructivist grounded theory principles with co-occurring recruitment, data generation, and analysis—Fig. 21.1 below shows the key stages and how they interacted.

Recruitment and Initial Contact

In recruiting participants, I identified and contacted gatekeeper organisations and online support groups that represented autistic people and/or victims, including those aimed specifically at women. I reached out through email to see if they were willing and able to support the research by sharing a recruitment request (see Fig. 21.2) via their own networks to encourage people to take part in the research. The recruitment request was shared through seven support networks in total.

Potential participants would contact me via email to notify their interest in the research, at which point I would share the information sheet, privacy notice, and consent form for them to read and complete as they felt fit. I also gave them a directory that signposted them to relevant support services and reminded them that their wellbeing was important, and that the interview

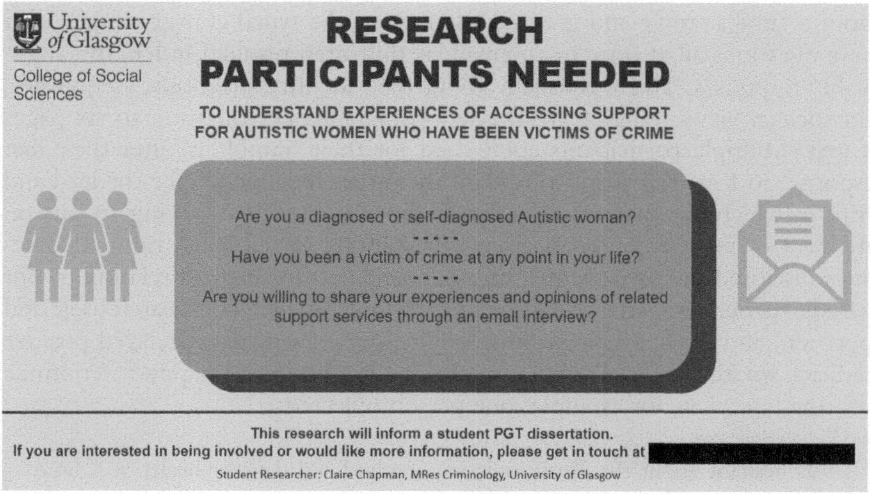

Fig. 21.2 Recruitment request infographic for online dissemination

should work around them (c.f. Chapters 11, 13, 20, this volume). This was also an opportunity for them to ask any initial questions before beginning to alleviate any doubts and provide clarity. Overall, I received interest from 24 individuals, ultimately undertaking interviews with 11 individuals. This was to ensure that it was feasible to conduct the research within the given timeframe and that I was able to provide the necessary attention to each participant, respecting their commitment to the research, and giving sufficient time for analysis. Interviews were with women in their late 20s to late 40s, all of whom identified as autistic.

Data Generation

I followed a general topic guide which was open to change based on responses, however, each interview typically went through four key stages: warm-up, body, cool-down, and closing. To warm-up, I found it helpful to ask questions to understand a little bit about the participants and their autism to learn more about how they viewed the world and themselves in it, for example:

> Could you tell me how you came about understanding that you were autistic? For example, when was this and what prompted this?
>
> What are your perceptions of autism and how you view yourself? For example, negative/positive feelings you may have, how this might have changed over time?

These warm-up questions felt like a good way to initiate the interview, particularly using a relatively uncommon method, as they were able to instantly talk about a familiar topic and get into the flow of the typed conversation. I sent these questions all at once in an email, as this often resulted in lengthy, high-quality responses. This was usually about their identity, diagnosis, life journey, and societal views, ranging from a few paragraphs per question to six pages of text. Multiple participants apologised for their 'rambling' after their first response, so I made a point to remind them that they could take the lead and write in whichever manner they felt suited them, without trying to put on any front or meet any preconceived expectations. When it felt right, I sometimes disclosed my own identity as an autistic person through relating to our shared experiences; participants were unsurprised due to the research topic and approach, and I felt it helped build our rapport. Participants relayed positive feedback for this stage, feeling content at the outset and happy to continue with the interview, for example, one person reflected:

> I was hesitant about the email interview method as I'd never heard of it used before and was a little unsure of what to expect or how it would actually work, but as the interview got underway I felt very relaxed with the process and realised it was perfect for me…It was explained simply at the start, which alleviated some of my hesitancy, and there was no need for protracted social chit-chat which I also appreciated as this put me at ease.

This warm-up stage also provided a useful backdrop into their experiences and opened opportunity to explore further in relation to support services, leading to the main topic of my research. At times, the response to the initial question naturally transitioned into the body of the interview, or I would help move it along by querying about support they may have had, for example:

> Did you reach out to access, or get offered any kind of support? If not, what were the reasons you didn't? If so, who was this with, and how did you come to get in contact with them?

This part of the interview involved back-and-forth communication, where I would send follow-up questions to their answers to dig deeper into the research topic, according to their experiences. Each time, I made sure the interview was tailored to the individual. When I received a new email, I would look back to remind myself of the conversation and ensure that I was not responding generically, but to their distinct experiences. I would also write up key field notes throughout on what we were talking about so that I could easily refer to these when I revisited the interview.

After I felt I had reached the saturation stage of the interview, I asked some more general cool-down questions to ease out of the exchange. This allowed me to understand how their experiences affected their wider perspectives of autistic women and support services, and gradually took some of the more personal aspects out of the conversation. For example:

Thinking more generally about autistic women who have had similar experiences, how do you think they could be better supported?

On closing, I asked participants for feedback on how they felt about email interviewing, as I had been pondering the positives and negatives about using the method. This was also a useful addition to the cool-down approach to help end the interview after discussing sensitive topics and finish the conversation on something lighter. I then asked if they were interested in receiving the research findings before ending the interview.

Co-occurring Analysis

Generally, I interviewed multiple participants at once as there were no set timeframes to complete an interview and my recruitment requests were shared at various stages. For example, by the end of July, I had completed three interviews with two underway whilst still receiving emails of interest to participate. Taking a co-occurring approach, common with constructivist grounded theory (Charmaz, 2007), I also began data analysis as interviews were still ongoing. I found it a particularly useful analytical approach where data generation and analysis worked simultaneously in an inductive, comparative, and iterative process. I was thinking deeply about what the data was saying, whilst using this to help shape ongoing interviews; something which would not be as easily implemented in more traditional, face-to-face interviewing. This was useful as it meant I could use approaches or insights I had learnt to apply to the most recent interviews, such as question wording, or new knowledge prompting additional questions to others. For example, one participant put forward the idea of autistic people leading in training or resource development, and the email interview method allowed me to extend this question to my ongoing interviews.

Whilst data was continually being created, I undertook key analytical steps with the completed interviews. I collated email threads into word documents (already transcribed as the interview occurred) and the data was anonymised. This was followed by an initial coding stage, analysing and naming smaller fragments of data, followed by a focused coding stage to further categorise chunks in the text (Charmaz, 2007). Memo writing and spider diagrams were then used to write about and reflect on the data to visualise and connect it and begin the path towards theoretical categories, developing this until nothing new was found and all interviews were completed (Charmaz, 2007).

Ethical Considerations

As my research involved real-time, online data generation from participants and related to a sensitive research topic, ethical considerations were a critical component, going beyond just informed consent and confidentiality. I reviewed these elements in great detail, and they remained at the forefront throughout the research process.

The digital and online nature added an additional layer of risk surrounding data security and protecting participant information (Gibson, 2017; Sparks et al., 2016). I had certain digital protections and procedures in place to manage and reduce these risks, guided by McCoyd and Kerson (2006), Gibson (2017) and Meho's (2006) research. The university IT systems I used had virus detection software and firewalls and the email account was on a secure platform with strong password protection. Personal data was anonymised as efficiently as possible and saved on secure, password protected file systems and emails were deleted soon after.[4] With the sensitivity of the research topic, there was also the risk of harm for participants, but with the lack of visual and verbal cues in email interviews, detecting harm was a concern (McCoyd & Kerson, 2006; Ratislavová & Ratislav, 2014). I extended support to participants throughout our interactions, including ensuring their full understanding of the process; providing reassurance; regularly checking-in on their wellbeing; giving clear thought to wording to minimise confusion; and sending a list of relevant support services (Benford & Standen, 2011; Egan et al., 2006; McCoyd & Kerson, 2006; Ratislavová & Ratislav, 2014).

Time and Control

A prior concern to email interviewing was that it was time consuming, which could create difficulties sustaining interest and the risk of drop out, as well as raising ethical questions around the commitment required for participants (Benford & Standen, 2011; Gibson, 2017; Hawkins, 2018, see also Chapter 20, this volume). I avoided drawing out the interview period excessively and was conscious of swaying from the research focus where possible to help minimise the effects of time and maximise engagement for myself and the participants. However, from the outset, this was much more challenging in practice, and conducting the interviews did ultimately feel labour and time intensive, at least from my perspective.

Although it was an important stage and helped me to start building rapport with my participants, the time between initial contact and beginning of the interview questions felt lengthy and was not anticipated. There was a somewhat significant number of back-and-forth emails for sharing participant information sheets, ensuring understanding, and discussing consent which could feel onerous, albeit important. On commencing the interview, I had a rough rule of no more than three days between receiving a participant email and responding to it, as I wanted to be able to give it the thought and time

[4] Since this research was undertaken, Pellicano et al. (2023) have uncovered threats to online methods in autism research where scammers disguise themselves as authentic participants for financial gain—this should be an additional consideration when planning online research.

each participant deserved without them waiting too long. Whilst participants reported that they enjoyed the process, they suggested it would be helpful to know when they would expect a response, how far they were from the end, or to receive an outline of the predicted interview timeline from the outset. This could set their expectations and reduce any anxieties around communications:

> I liked doing it via email... Sometimes I got anxious not knowing how long a response would take. Maybe each email could say 'you will hear back from me in x hours' or something?

However, as a researcher I also had little direction over the response lag from participants and when I would receive replies. This meant I had to chase some responses, but also that time to complete the interviews ranged between two and six weeks; slightly longer than anticipated for my research, but not entirely unexpected (Fritz & Vandermause, 2018; Gibson, 2017). It became apparent that some participants found it harder to manage their responses to emails, and tended to have a greater need for reminders throughout the process, and this appeared more prevalent for participants who were also ADHD or dyslexic.

> I think it was good in that I could do it in my own time and I could think a bit on the questions but bad in that it took up a lot more of my time (takes longer to type [than] talk) and I had to remember to reply vs a booked slot for a call.

In general, I found that interviews over a shorter timeframe tended to maintain a better, more natural flow of conversation, so there was the risk that extended delays to the interview process could negatively impact on the quality of the interview and researcher-participant relationship. However, it was important to me as a researcher to tailor the email interview process to each individual, and so I was willing to send reminders to support their participation, without badgering them.

Despite the challenges of time demands in email interviewing, I found that this was still offset by the benefits that came from using this method. Like myself, participants stated that they really enjoyed the minimal interaction and agency that came with this approach, and having control over the time and space they participated in, and how they used this, worked well:

> For me, this was the most suitable method as I could take it at my own pace and could take part from the comfort of my own home (where I'm most relaxed and able to think clearly without distractions) at any hour of the day or night that suited my erratic schedule. I didn't have to worry about appointment times or meeting new people (even online) or travelling somewhere unfamiliar.
>
> I could also respond to the email at a time that was suitable to me and I could take a break from writing if needed, without the awkwardness feeling you might get from asking for frequent breaks if had been a face to face interview.

Similar to what Gibson (2010), Egan et al (2006) and Bowker and Tuffin (2004) found in their research, participants could view, read, consider, and respond to interview questions without haste. The pace was slower, and they could participate whenever and wherever they chose, suiting their schedule and preferences. Both the participants and I responded to everything in our own time, when we felt ready and without pressure to do so quickly or with rash replies. As research has discussed relating to online communication for autistic people, these written accounts take away the pressure of the neurotypical ping-pong and faster-paced conversations, and allow for 'a delay in response that is almost never allowed in real life' (Brosnan & Gavin, 2015, p. 255; Davidson, 2008). In doing so, participants had more control over the direction of the conversation and how they told their story, giving them more time to formulate their responses in a manner that suited them, which may not have happened in a more traditional interview.

With this, participants also took the lead in their writing and approach to sending emails. Some sent very lengthy, detailed responses, whilst others were very concise and to the point; some were grammatically correct and organised, whilst others were somewhat less focused and filled with misspellings, and expressive punctuation, such as capital letters, exclamation marks, ellipses and emojis. They were able to choose how to express their personalities through their writing, and the time and effort they put into this. In my role as the researcher, I was guided by their approach and my way of writing adapted to match theirs, to aid in their comfort levels throughout the conversation and their ability to relay their experiences to me. This process of mirroring participants and responding to their language can help strengthen trust and support the researcher-participant relationship (Fritz & Vandermause, 2018) and allow for thoughtful and somewhat personal conversations to be shared.

Depth and Authenticity

Researchers have shared concerns over the misinterpretation of information that can occur through email interviews and questioned the authenticity of written word in expressing experiences (Benford & Standen, 2011; Bowker & Tuffin, 2004; Gibson, 2017; James, 2007; Meho, 2006). These are based on the fact that responses are solely textual-based data; the extent to which words can be rewritten and edited online; and that it is 'unconversational' interviewing, without 'natural' social interactions (Benford & Standen, 2011; Bowker & Tuffin, 2004; Gibson, 2017; James, 2007; Meho, 2006). This could result in challenges interpreting meaning or the emotional states of participants, resulting in a lack of rich data, and less opportunities to build rapport. However, I intended to help autistic participants feel more comfortable to take part in research, and, with certain adjustments, email interviewing had the potential to provide a platform for them to give a more 'authentic' representation of themselves. Written communication may be less natural to neurotypical people but could be fitting for many autistic people.

During the research, it was important for me to work on building a trusting relationship and provide a space where my participants could feel at ease and able to express their genuine self as much as possible whilst sharing their experiences. I endeavoured to acknowledge each participant as an individual, taking time to respond thoughtfully to every email. I was unable to use typical verbal interjections such as 'mhmm' or nodding to express interest, reactions, or encouragements to their email responses. Although arguably, these may be considered neurotypical encouragements, and not using them may have actually been encouraging for my participant's engagement. Nonetheless, I felt it was important that for each reply I commented on or acknowledged a point they made in their previous response. This was to ensure they felt heard and alleviate any doubts they had about answering 'correctly', which appeared to be a common concern. I found that I got better quality responses and the discussions flowed better and were filled with depth by tailoring each interaction. Similar to Ratislavová and Ratislav (2014), personalisation and considered communication for each individual allowed me to get to know how the participant 'spoke' and brought out emotive elements from their text-based responses, helping to understand their story better and how they were feeling telling it. Feedback from some of this approach was also praised by participants:

> You gave me just enough feedback at every stage of the interview to reassure me that I had answered each question sufficiently but not gushingly so which would have made me feel uncomfortably patronised. I enjoyed it!

Where there may have been a delay between emails sent and received, at times this was awaiting a long, well-crafted response from a participant that powerfully portrayed their experiences; it was worth the wait. As James and Busher (2011) pointed out, through online research participants can be chatting away, you just can't see or hear it yet. Whilst the changing nature of data through editing, rewriting and low spontaneity could be a concern, it allowed more time for introspective thinking and reflective responses (Gibson, 2017; James & Busher, 2011; Meho, 2006). Writing may take away the complexity and stressors for autistic people to follow social expectations and could be considered a more authentic and open representation of their experience and reality (Davidson, 2008; Howard & Sedgewick, 2021). My participants could remove the 'mask' and express themselves more easily, particularly through having additional time to think without pressure, contributing to a truer reflection of their reality. One participant commented favourably on this aspect of the method:

> I found the method of emailing really good. Having the questions sent via email was more accessible because I had time to think about my responses without pressure. It meant I was able to answer exactly what I wanted to say rather than

> give a quick answer on the spot… Another positive was that I didn't have the added pressures to socialise or communicate verbally.

My participants, and those in previous research (Beck, 2005; Ratislavová & Ratislav, 2014), used the terms '*healing*' and '*therapeutic*' in response to email interviewing as they wrote out their experiences in a journal-like manner, enjoying and embracing the time to think and reflect. Whilst writing can be edited, I believe this connection to journalling supports the authenticity of participants' voice through email interviewing to share their experiences. However, guidance in undertaking this method should provide clear instructions and expectations so that participants have a clear understanding of what it is and that they do not treat it like a diary. My participants also found it a much more accessible way to engage more deeply in research where they otherwise could not, bringing a sense of freedom, as well as convenience with how it aligned with their life, even over the extended timeframes:

> I really enjoyed the process of an email interview. I found it to be convenient for me, and I was able to put some thought into my answers in a way that I might not have been able to if the interview was done in person or via zoom (I can get very nervous in spoken conversations and say things I don't mean, also they take a lot out of me!).
>
> I felt much more able to stay on topic and discuss some very personal things in a more descriptive, factual way than I would have in person, on video call, or by phone… I actually really enjoyed it! It was therapeutic to type everything out in such an organised way … it already made sense in my head but I'd never actually put it down on 'paper' like that before. I don't think it would have been so therapeutic if I'd been trying to do it via video call or any other way because I'd have been less focused on accuracy as I worried about interpreting facial expressions and tones of voice etc. and I'd have felt more emotionally and socially drained by the process.

However, it is worth mentioning that some of my participants found the open-ended aspects of the interview questions more difficult, which was evident through their comparably smaller responses. Although according to participants, this was unrelated to the email method and more about the interview itself and would have been the case if it were a different technique. On learning this, I provided clear and more direct questions with examples to help guide them and alleviate these concerns—although it was a delicate balance between this and not influencing the research too much. On top of this, the value of email interviews is that any vagueness, misunderstanding or misinterpretation could be considered and clarified in follow-up emails from either party if necessary (Benford & Standen, 2011; Gibson, 2017; Meho, 2006). There is more time to pause and consider wording, phrases, and meaning and ensure they are being interpreted as intended; clarifying, and expanding on any responses as required whilst remaining in the context of the interview.

A Flexible Approach

In doing sensitive research with autistic women, my asynchronous email interview method was received favourably. However, there were opportunities for improvement in how the method played out in practice which could have helped it function more conveniently for participants and streamline the method for me as a researcher.

At the research design stage, it would have been useful to have a more efficient process of commencing the interview to reduce time spent with back-and-forth emails for administrative tasks. All the required information could have been more concisely collated into one email at the start. I would have benefited from taking more time to work out how I was going to pose questions to participants in the emails, such as my structure of doing this, and whether I was taking an approach of one question at a time or collating questions. Secondly, providing guidance and clearer expectations at the outset and throughout the email interview timeline, or at key stages, would have been beneficial for participants, such as the amount of time or number of exchanges remaining, or receiving questions in advance. This would have been useful to alleviate any unease about length and ongoing commitment and allow them to manage their own time and any individual preferences could have been discussed and pre-prepared.

Whilst most participants enjoyed and gained a lot from the email interview, a few also acknowledged that they might prefer doing the interview through alternative means:

> It was good so I knew the questions to have a think about them I would of preferred doing it via online video link as I find it hard to type too much and I could of included better answers giving them verbally.
>
> I'm fine doing video calls and phone calls most of the time. I know some autistics aren't.

Considering this, it may be beneficial to adjust the email interview approach slightly to suit different ways of thinking, working, and communicating, without losing the value uncovered from the method (see also discussion in Chapter 7, this volume). It could be more flexible, giving thought to other identities or intersections within the autistic community. For example, the interview could be primarily by email, maximising the benefits of an online, written, asynchronous approach. Alternative options (e.g., video calls) could then be offered for those who may prefer it—online or offline, verbal or written, to suit a mixture of communication styles. Although it is worth noting that this may have some limitations, such as coordinating the different techniques through interviewing and analysis, and the organisational risk that comes with this. There may be challenges in different platforms producing different types of responses and the impact this would have, particularly when the questions and prompts are framed differently as this may affect the consistency across interviews, and the quality and depth of exchanges. Despite this, a flexible approach would address some of the limitations of the email interview method, embracing different preferences and prioritising accessibility and comfort of autistic participants. Whilst this was not appropriate for the research in question due to the parameters of the study and timescales, this is something I intend to take on in future research.

Conclusions

In my research to understand autistic women's experiences of support services, I took steps within the research design and fieldwork to use a method that was accessible, comfortable, and enjoyable whilst generating the information required to inform the overall research project. Interviews were conducted via email to explore the applicability of this method to inclusively engage autistic women in qualitative research.

I found it to be an extremely worthwhile method for capturing reflective and vivid personal experiences as written from an individual's point of view. Flexible in time and space, it passed more control to participants and gave in-depth and rich responses about their life experiences relating to sensitive topics of victimisation and support services. The feedback on the method was very positive, and participants enjoyed it and felt included in research where they have often felt unable to participate in. However, future use of email interviewing should provide participants with clear guidance on how the method works in practice and expectations for them as a participant to help manage any anxieties for engaging in research.

Future autism research should consider making use of email interviews through a flexible approach, giving participants options for various communication modes where required and providing greater agency for the participants involved. To inclusively engage autistic people, research should incorporate methods to meet the requirements of participants, not expect them to adopt the requirements of a method which may not be suitable to them. Knowing this, reworking or introducing new methods in qualitative research in a manner that is flexible, considerate, and comfortable should be a priority for autism, broader neurodiversity studies, and criminological research.

Ethical Approval

Full approval was granted by the University of Glasgow School of Social Sciences Ethics Forum (PGT/SPS/2022/136/SOC) and informed consent was obtained from individual participants.

References

Babb, C., Brede, J., Jones, C. R. G., Elliot, M., Zanker, C., Tchanturia, K., Serpell, L., Mandy, W., & Fox, J. R. E. (2021). 'It's not that they don't want to access the support ... it's the impact of the autism': The experience of eating disorder services from the perspective of autistic women, parents and healthcare professionals. *Autism, 25*(5), 1409–1421. https://doi.org/10.1177/1362361321991257

Baldwin, S., & Costley, D. (2016). The experiences and needs of female adults with high-functioning autism spectrum disorder. *Autism, 20*(4), 483–495. https://doi.org/10.1177/1362361315590805

Ballan, M. S., & Freyer, M. B. (2012). Self-defense among women with disabilities: An unexplored domain in domestic violence cases. *Violence against Women, 18*(9), 1083–1107. https://doi.org/10.1177/1077801212461430

Bargiela, S., Steward, R., & Mandy, W. (2016). The Experiences of late-diagnosed women with autism spectrum conditions: An investigation of the female autism phenotype. *Journal of Autism and Developmental Disorders, 46*(10), 3281–3294. https://doi.org/10.1007/s10803-016-2872-8

Beck, C. T. (2005). Benefits of participating in internet interviews: Women helping women. *Qualitative Health Research, 15*(3), 411–422. https://doi.org/10.1177/1049732304270837

Benford, P., & Standen, P. J. (2011). The use of email-facilitated interviewing with higher functioning autistic people participating in a grounded theory study. *International Journal of Social Research Methodology, 14*(5), 353–368. https://doi.org/10.1080/13645579.2010.534654

Bowker, N., & Tuffin, K. (2004). Using the online medium for discursive research about people with disabilities. *Social Science Computer Review, 22*(2), 228–241. https://doi.org/10.1177/0894439303262561

Brosnan, M., & Gavin, J. (2015). Are "Friends" electric? In *The Wiley Handbook of psychology, technology, and society* (pp. 250–270). Wiley. https://doi.org/10.1002/9781118771952.ch14

Bryce, J., Brooks, M., Robinson, P., Stokes, R., Irving, M., Graham-Kevan, N., Willan, V., Khan, R., Karwacka, M., & Lowe, M. (2016). A qualitative examination of engagement with support services by victims of violent crime. *International Review of Victimology, 22*(3), 239–255. https://doi.org/10.1177/0269758016649050

Cage, E., Crompton, C. J., Dantas, S., Strachan, K., Birch, R., Robinson, M., Morgan-Appel, S., MacKenzie-Nash, C., Gallagher, A., & Botha, M. (2022). What are the autism research priorities of autistic adults in Scotland? *PsyArXiv*. https://doi.org/10.31234/osf.io/nbq8h

Cascio, M. A., Weiss, J. A., & Racine, E. (2021). Making autism research inclusive by attending to intersectionality: A review of the research ethics literature. *Review Journal of Autism and Developmental Disorders, 8*(1), 22–36. https://doi.org/10.1007/s40489-020-00204-z

Cazalis, F., Reyes, E., Leduc, S., & Gourion, D. (2022). Evidence that nine autistic women out of ten have been victims of sexual violence. *Frontiers in Behavioral Neuroscience, 16*. https://doi.org/10.3389/fnbeh.2022.852203

Chapman, C. M. (2022). *Supporting victimised autistic women: An inclusive approach to understanding their experiences*. University of Glasgow. Unpublished Assignment (Dissertation).

Charmaz, K. (2007). *Constructing grounded theory: A practical guide through qualitative analysis*. SAGE Publications.

Chown, N., Robinson, J., Beardon, L., Downing, J., Hughes, L., Leatherland, J., Fox, K., Hickman, L., & MacGregor, D. (2017). Improving research about us, with us: A draft framework for inclusive autism research. *Disability & Society, 32*(5), 720–734. https://doi.org/10.1080/09687599.2017.1320273

Dahlin, E. (2021). Email interviews: A guide to research design and implementation. *International Journal of Qualitative Methods, 20*, 16094069211025452. https://doi.org/10.1177/16094069211025453

Davidson, J. (2007). 'In a world of her own': Re-presenting alienation and emotion in the lives and writings of women with autism. *Gender, Place and Culture : A Journal of Feminist Geography*, *14*(6), 659–677. https://doi.org/10.1080/09663690701659135

Davidson, J. (2008). Autistic culture online: Virtual communication and cultural expression on the spectrum. *Social and Cultural Geography, 9*(7), 791–806. https://doi.org/10.1080/14649360802382586

Egan, J., Chenoweth, L., & McAuliffe, D. (2006). Email-facilitated qualitative interviews with traumatic brain injury survivors: A new and accessible method. *Brain Injury, 20*(12), 1283–1294. https://doi.org/10.1080/02699050601049692

Fritz, R. L., & Vandermause, R. (2018). Data collection via in-depth email interviewing: Lessons from the field. *Qualitative Health Research, 28*(10), 1640–1649. https://doi.org/10.1177/1049732316689067

Garland-Thomson, R. (2005). Feminist disability studies. *Signs, 30*(2), 1557–1587. https://doi.org/10.1086/423352

Gibbs, V., Hudson, J., & Pellicano, E. (2023). The extent and nature of autistic people's violence experiences during adulthood: A cross-sectional study of victimisation. *Journal of Autism and Developmental Disorders, 53*(9), 3509–3524. https://doi.org/10.1007/s10803-022-05647-3

Gibbs, V., & Pellicano, E. (2023). 'Maybe we just seem like easy targets': A qualitative analysis of autistic adults' experiences of interpersonal violence. *Autism, 27*(7), 2021–2034. https://doi.org/10.1177/13623613221150375

Gibson, L. (2010). Using email interviews. *Realities Toolkit*, *09*.

Gibson, L. (2017). 'Type me your answer': Generating interview data via email. In V. Braun, V. Clarke, & D. Gray (Eds.), *Collecting qualitative data* (1st ed.). Cambridge University Press. https://doi.org/10.1017/9781107295094.011

Greenlee, J. L., Winter, M. A., & Marcovici, I. A. (2020). Brief report: Gender differences in experiences of peer victimization among adolescents with autism spectrum disorder. *Journal of Autism and Developmental Disorders, 50*(10), 3790–3800. https://doi.org/10.1007/s10803-020-04437-z

Hawkins, J. (2018). The practical utility and suitability of email interviews in qualitative research. *The Qualitative Report*. https://doi.org/10.46743/2160-3715/2018.3266

Howard, P. L., & Sedgewick, F. (2021). 'Anything but the phone!': Communication mode preferences in the autism community. *Autism, 25*(8), 2265–2278. https://doi.org/10.1177/13623613211014995

Hull, L., Petrides, K. V., & Mandy, W. (2020). The female autism phenotype and camouflaging: A narrative review. *Review Journal of Autism and Developmental Disorders, 7*(4), 306–317. https://doi.org/10.1007/s40489-020-00197-9

James, N., & Busher, H. (2011). Engaging with research participants online. In *Online Interviewing* (pp. 19–29). SAGE Publications Ltd. https://doi.org/10.4135/9780857024503

James, N. (2007). The use of email interviewing as a qualitative method of inquiry in educational research. *British Educational Research Journal, 33*(6), 963–976. https://doi.org/10.1080/01411920701657074

James, N., & Busher, H. (2006). Credibility, authenticity and voice: Dilemmas in online interviewing. *Qualitative Research, 6*(3), 403–420.

Jones, S. C. (2022). Measuring the wrong thing the right way? Time to rethink autism research tools. *Autism in Adulthood: Challenges and Management, 4*(2), 104–109. https://doi.org/10.1089/aut.2021.0050

Lichtman, M. (2014). *Qualitative research for the social sciences*. SAGE Publications, Inc. https://doi.org/10.4135/9781544307756

Macdonald, S. J., Donovan, C., & Clayton, J. (2023). 'I may be left with no choice but to end my torment': Disability and intersectionalities of hate crime. *Disability and Society, 38*(1), 127–147. https://doi.org/10.1080/09687599.2021.1928480

McCoyd, J. L. M., & Kerson, T. S. (2006). Conducting intensive interviews using email: A Serendipitous Comparative Opportunity. *Qualitative Social Work, 5*(3), 389–406. https://doi.org/10.1177/1473325006067367

Meho, L. I. (2006). E-mail interviewing in qualitative research: A methodological discussion. *Journal of the American Society for Information Science and Technology, 57*(10), 1284–1295. https://doi.org/10.1002/asi.20416

Milner, V., McIntosh, H., Colvert, E., & Happe, F. (2019). A qualitative exploration of the female experience of autism spectrum disorder (ASD). *Journal of Autism and Developmental Disorders, 49*(6), 2389–2403. https://doi.org/10.1007/s10803-019-03906-4

Oliver, M. (1990). *The politics of disablement*. Macmillan Education.

Parsons, S., Yuill, N., Good, J., & Brosnan, M. (2020). 'Whose agenda? Who knows best? Whose voice?' Co-creating a technology research roadmap with autism stakeholders. *Disability & Society, 35*(2), 201–234. https://doi.org/10.1080/09687599.2019.1624152

Paul, A., Gallot, C., Lelouche, C., Bouvard, M. P., & Amestoy, A. (2018). Victimisation in a French population of children and youths with autism spectrum disorder: A case control study. *Child and Adolescent Psychiatry and Mental Health, 12*, 48. https://doi.org/10.1186/s13034-018-0256-x

Pecora, L. A., Hancock, G. I., Hooley, M., Demmer, D. H., Attwood, T., Mesibov, G. B., & Stokes, M. A. (2020). Gender identity, sexual orientation and adverse sexual experiences in autistic females. *Molecular Autism, 11*, 1–16. https://doi.org/10.1186/s13229-020-00363-0

Poulsen, R., Brownlow, C., Lawson, W., & Pellicano, E. (2022). Meaningful research for autistic people? Ask autistics! *Autism, 26*(1), 3–5. https://doi.org/10.1177/13623613211064421

Ratislavová, K., & Ratislav, J. (2014). Asynchronous email interview as a qualitative research method in the humanities. *Human Affairs, 24*(4), 452–460. https://doi.org/10.2478/s13374-014-0240-y

Rice, C. E., & Lee, L.-C. (2017). Expanding the global reach of research in autism. *Autism, 21*(5), 515–517. https://doi.org/10.1177/1362361317704603

Saarijärvi, M., & Bratt, E.-L. (2021). When face-to-face interviews are not possible: Tips and tricks for video, telephone, online chat, and email interviews in qualitative research. *European Journal of Cardiovascular Nursing, 20*(4), 392–396. https://doi.org/10.1093/eurjcn/zvab038

Saxe, A. (2017). The theory of intersectionality: A new lens for understanding the barriers faced by autistic women. *Canadian Journal of Disability Studies, 6*(4), Article 4. https://doi.org/10.15353/cjds.v6i4.386

Shaw, L. R., Chan, F., & McMahon, B. T. (2012). Intersectionality and disability harassment: The interactive effects of disability, race, age, and gender. *Rehabilitation Counseling Bulletin, 55*(2), 82–91. https://doi.org/10.1177/0034355211431167

Silverman, D. (2016). *Qualitative research* (5th ed.). SAGE Publications.

Sohn, E. (2019). *Righting the gender imbalance in autism studies.* Spectrum | Autism Research News. https://www.spectrumnews.org/features/deep-dive/righting-gender-imbalance-autism-studies/

Sparks, H., Collins, F. L., & Kearns, R. (2016). Reflecting on the risks and ethical dilemmas of digital research. *Geoforum, 77*, 40–46. https://doi.org/10.1016/j.geoforum.2016.09.019

CHAPTER 22

Unknowing as a Methodological Tool: Autism, Authenticity, and Epistemic Injustice

David Jackson-Perry

Introduction

It is "increasingly acknowledged" that only by shaping research agendas "together by researchers and community members" can "autism research adequately address the issues facing autistic people and their allies" (Pellicano et al., 2018, p. 1). This togetherness is often sought through participatory research (PR), that is, "incorporating the views of autistic people and their allies about what research gets done, how it is done and how it is implemented" (Fletcher-Watson et al., 2019). However, despite being myself the product of autism research circles that hold that the path to ethically and epistemologically robust research largely depends on implementing PR, I no longer feel that it is the fix-all that we may have hoped. Indeed, as things stand, I worry, with Orsini (2022, p. 5), that:

> bringing autistic voices or perspectives to the table may reproduce what many have decried in the past: participation or engagement for the sake of crafting a thin veneer of legitimacy.

Before going any further, a few words about what I bring to this chapter, personally and professionally. In 2023, I completed my PhD thesis[1] on

[1] On which I have drawn extensively for parts of this chapter.

D. Jackson-Perry (✉)
Queen's University, Belfast, UK
e-mail: djacksonperry01@qub.ac.uk

H. Bertilsdotter Rosqvist and D. Jackson-Perry (eds.), *The Palgrave Handbook of Research Methods and Ethics in Neurodiversity Studies*,
https://doi.org/10.1007/978-3-031-66127-3_22

autistic experiences of sexuality, intimacy, and authenticity, using a participatory methodology and a critical autism studies framework (Jackson-Perry, 2023). When I started that study six years earlier, I saw and presented myself as a neurotypical researcher: today I have a diagnosis of ADHD. I am also HIV project manager at Lausanne University Hospital, in Switzerland, where much of my work involves contributing to and implementing participatory processes in HIV research. 'Contributing to' because I also live with HIV. In all these roles—living and working with HIV, carrying out autism and neurodiversity research as a non-autistic, neurodivergent sociologist—I promote the integration of lived experience into research processes. However, as things stand, far from being the solution, I increasingly feel that PR itself presents a set of ethical and methodological challenges, similar in both HIV and autism research.

These challenges are particularly pertinent in the current climate where there is a certain pressure on researchers in both fields to be inclusive. Funding and publication may depend on it: researchers, more or (generally) less experienced in PR, more or (sometimes) less in agreement with PR principles (den Houting et al., 2021), are encouraged to carry out participatory research. This may not always be because they believe that it adds value to their research, but because it is a requirement, a box to tick.

This can be problematic on several fronts. Who is chosen to participate, and by whom, is rarely made clear. In my experience, it is often the same people who are called upon, those who are comfortable for researchers to work with, convenient to contact, known quantities. At which part of the research process they are involved is also largely left to researchers to decide. This may often come toward the middle of research processes (den Houting et al., 2021), the check-the-research-tools-stage ("what do you think about this questionnaire?"). Early collaboration or even consultation remains rare (for a counter example see Chapter 4, this volume), meaning that research priorities are generally fixed before seeking community input: unsurprisingly then, the mismatch between autistic people's research priorities and the research that is funded and carried out remains significant (Cage et al., 2024; Cervantes et al., 2021; Pukki et al., 2022).

In HIV research, people living with HIV are trotted out to explain to clinicians, researchers, and funders how and why we should be involved in research processes, so that they will continue to fund and carry out PR. I used to think this was a good thing and consider myself lucky to be called on. I now feel that it is closer to "epistemic exploitation" (Berenstain, 2016, p.569), an "additional labor created by the default skepticism of the privileged." Particularly when PR is carried out for pragmatic reasons rather than belief in its importance, autistic folk are also subject to epistemic exploitation, with, for example, funding mechanisms for payment often complex or inexistent (den Houting et al., 2021; also see Chapters 6,19,20, this volume). Leaving aside financial compensation, the risk of tokenism is never far away, for lay-people or researchers. “Autistic academics” say Thom-Jones and Lowe (Chapter 10, this volume), in a quote that echoes my own HIV experience:

> are all too familiar with the proposition of an informal zoom meeting to discuss autistic perspectives about a study that is already underway, being later positioned as a participatory approach or autistic community consultation.

None of this is to say that I think PR is a bad thing, or that it cannot be meaningful or useful: there are examples that clearly demonstrate the contrary (see, for example, Nicolaidis et al., 2019). However, we largely look to PR to fix all our problems, to move from deficit to difference, from questions important to researchers to questions important to autistic communities, from individualist medical narratives to more social leaning models. And yet, to borrow from the title (and content) of Botha and Cage’s (2022) article, “autism research is in crisis”: high proportions of autism researchers endorse ableist or dehumanizing language, lean on medical narratives over social, and associate autistic people with outdated and prejudicial stereotypes. It is difficult to see how the inclusion of people who are perceived in this way could bring about much change. Participatory research alone is not a panacea that can right the epistemic and ethical wrongs of autism—or indeed HIV—research.

In this chapter, I suggest a step that might be useful in this direction. That step—whatever one’s neurotype—is to unknow what we have been led to believe is true about a condition constructed around outdated and fallacious assumptions of deficit tinkered for years by medical, psychiatric, and aligned disciplines. I want to think about unknowing as a tool that has the potential to get to the root of epistemic injustice and ignorance.

To do so, I first draw on my PhD thesis, in which authenticity emerged as crucial to participants as they navigated their social and intimate lives (Jackson-Perry, 2023). Despite multiple elements pointing to authenticity as a potentially central concept in autistic subjectivity, there is little sustained discussion of it in autism research.[2] I use this absence as a case study to take a critical look at epistemic injustice as it relates to both autism and authenticity, although it is also pertinent for Neurodiversity Studies more broadly. I conclude with some practical suggestions in the form of an unknower’s toolbox.

The Case of Autistic Authenticity

Authenticity is often used in the social sciences “to refer to such different things as sincerity, truthfulness, originality, and the feeling and practice of being true to one’s self or others” (Vannini & Franzese, 2008, p. 1621). While over-simplifying a contentious and complex concept, I use a similarly simple

[2] A notable exception here is regarding the sexual and gender behaviours and identities of autistic people: these are generally read as constituting ‘symptoms of autism’ and so are, necessarily, *in*authentic (Jackson-Perry, 2020).

definition here for authenticity and its proxies: truth-telling, straightforwardness, and an understanding of self, others, and concepts that are less likely to be swayed by context or social bias (see Jackson-Perry, 2023, pp. 60–74 for detailed discussion).

Much thinking on authenticity draws on George Herbert Mead. He distinguished between the 'I' and the 'me': the subjective impulse-oriented part of the self (the 'I') coexists with the other-oriented objective component of the self (the 'me'), allowing individuals to think of themselves as both subjects and objects, to take into consideration their own self-presentation and, by extension, to access authenticity. Mead remains highly influential, and symbolic interactionist Franks (2003, p. 621), for example, leans on his theory and that of a deficit in Theory of Mind[3] to discuss a report in which the authors "study cases in which the self is diminished or non-existent, as in autism."

This categorical leap to a diminished or non-existent self, a person lacking "the capacity to construct persons and subjectivities beyond observable bodies," *stated as fact*, is made possible here by drawing on Mead. Mead's self develops through early stages of childhood socialization, notably "play," whereby the "normally developing" child "begins to take the 'view of the other', imagining situations from another person's perspective" (Scott, 2015, p. 6). This, Scott goes on to say, "echoes Theory of Mind in psychology." And so, leaning on two speculative and contestable theories—Mead on the reflexive self and Baron-Cohen on deficit of Theory of Mind—autistic authenticity becomes unthinkable. If I do not have a self, or if that self is diminished, then what are my possibilities for authenticity, for "a commitment to self-values" (Erickson, 1995, p. 121) for which self-knowledge is a pre-requisite?

And yet, a great deal of empirical evidence points to high levels of authenticity and its proxies among autistic people. They may speak their minds "irrespective of social context or adherence to personal beliefs" (Kirchner et al., 2016, p. 3330). Authenticity is ranked significantly higher as a signature strength by autistic than non-autistic research participants both when self-scored (Kirchner et al., 2016, p. 3333) and in personality traits measurement (Strunz et al., 2015): "honesty," the latter authors conclude (p.4035), "seems to be a primary strength in individuals with ASD." On the other hand, "lying…is not a common phenomenon" among autistic people (Jaarsma et al., 2012, p. 272). Indeed, there is "empirical evidence that children with ASD use an effortful route via their working memory—when they engage in the sociocognitive process of deception," leaving them eight times less likely to tell even a self-protective lie than non-autistic peers (Ma et al., 2019, p. 3374).

Cognitive neuroscientist Reubs Walsh and I (2021) suggested that autistic trans and non-binary people may have heightened access to authentic gender

[3] This highly pervasive theory, which has been robustly and repeatedly refuted elsewhere (Fletcher-Watson and Bird, 2020; McGrath, Peterson and Pennington, 2020; Williams, 2021), seeks to both demonstrate and explain an assumed autistic inability to understand or take into consideration what someone else is thinking or feeling.

identity and expression: they may give a higher priority to how a person *experiences* their own gender, and a lower priority to social representations of what is considered possible or desirable. We drew on three cognitive theories to explore this: imperfect systems (Kristensen & Broome, 2015), bottom-up or unbiased processing (Goldman & Kernis, 2002; Mottron et al., 2006), and Bayesian theories of flattened priors (Pellicano & Burr, 2012). Broadly, these theories suggest that autistic people may be neurocognitively inclined to give less consideration to context, or to pre-existing experiences or beliefs, and more to their immediate experience of a phenomenon. We posited that this might leave autistic people less inclined to conform to rigid, pre-existing categories, preferring (and able) to evaluate their own experience authentically, even if that experience falls outside what is generally considered possible or desirable (see also Wattel et al., 2022). This appears to be supported in qualitative discussions of gender (see, for example, Kourti & MacLeod, 2018), with Späth and Jongsma (2020, p. 76) noting that autistic people appear "less prone to be influenced by habits or options others embrace, which makes them authentic and better 'guardians of their own interests' than non-autists."

Almost all the autistic participants in my doctoral research recurrently referenced honesty, straightforwardness, and authenticity as central to their intimate lives: desired in a partner, facilitating satisfying and trusting intimate relationships, and enabling them to resist heteronormative pressure to arrive at an authentic understanding of their own sexuality or gender. Conversely, participants considered inauthenticity, conformity, a lack of honesty, lying, and so on, to be typically neurotypical traits. This echoes parodic autistic community writing, with Main (2003), for example, detailing a set of diagnostic criteria for "allism" ('non-autism'). The author describes the "underlying trait" of allistic people as lacking:

> the capacity to independently experience emotion...(their) emotional state is not determined by their own thought processes but instead is borrowed from other people that are expressing emotion nearby.

Allistic people, Main goes on, have:

> difficulty with the difference between truth and falsity. In many circumstances, they find it difficult to say accurate things, and will instead say inaccurate things that they find more pleasing.

The Institute for the Study of the Neurologically Typical (ISNT, no date) similarly parodies diagnostic criteria for neurotypicals, who, as well as having an "obsession with conformity," says ISNT, also "find it difficult to communicate directly, and have a much higher incidence of lying as compared to persons on the autistic spectrum."

To summarize, the concept of authenticity and its proxies appear to be "especially pertinent to autistic individuals" (Stark and colleagues, 2021,

p. 195). There are intriguing indications that they may be cognitively predisposed to a tendency for truth-telling, open-ness, honesty, straightforwardness, and attachment to a sense of authentic experience over imposed beliefs. Socially, those same attributes form part of how some autistic people and communities narrate themselves, in contrast to non-autistics who are perceived as having deficits in these same areas (see also Brownlow, 2010). And yet, despite all indications that authenticity might be central to autistic experience, sustained discussion of this overwhelmingly positively perceived attribute (see, for example, Schlegel & Hicks, 2011; Stark et al., 2021; Bottema-Beutel et al., 2018) is missing from the literature. I noted at the beginning of this section that theories formulated by Mead had in a sense definitionally written out any possibilities of autistic authenticity. I now turn to a discussion of this and other forms of epistemic injustice relating to authenticity and autism research more broadly.

Epistemic Injustice: "Very Easy to Commit…Extraordinarily Difficult to Avoid"?

> The reality is that epistemic injustice is very easy to commit. In fact, it is extraordinarily difficult to avoid it (Pohlhaus, 2012, p. 717).

Miranda Fricker (2007, p. 4) proposes that "shared social-imaginative conceptions of the social identities" of "social types" (here, autistic people), produce "testimonial injustice," reducing the credibility given to a "speaker" from a group that the "hearer" is prejudiced toward. This prejudice "typically enter(s) into a hearer's credibility judgement by way of the social imagination, in the form of a prejudicial stereotype—a distorted image of the social type in question" (p.4). A second form of epistemic injustice suggests (a) that there is no existing framework through which a person's experience can be read, and (b) that members of some social groups are marginalized from the creation of knowledge about themselves. Fricker calls this "hermeneutical injustice," whereby "some significant area of one's social experience (is) obscured from collective understanding" (p. 158).

Autistic people are subjected to both types of epistemic injustice. They are disadvantaged testimonially, as Yergeau (2018, p.8) reminds us:

> autism's essence…has been clinically defined as a disorder that prevents individuals from exercising free will and precludes them from accessing self-knowledge and knowledge of human others.

They are also disadvantaged hermeneutically when making sense of their own experiences. When knowledge about autism is overwhelmingly (re)produced by medical and psychological authorities based on assumptions of deficit, it may be complex, for example, "for autistic people to imagine themselves and their intimate experience other than through deficit" (Bertilsdotter Rosqvist & Jackson-Perry, 2021, p.340), or doing so will require a certain level of labor: will require, in fact, *unknowing*. This labor of unknowing may involve participating in autistic spaces in which to "unlearn everything your parents said was wrong with you" (Belek, 2022, p. 2). It may involve meeting neurodivergent and/or professional allies, interacting with, reclaiming, and renaming diagnostic criteria, navigating definitions of what it means to be autistic (Jackson-Perry et al., 2020).

In an editorial of the journal Autism, *A new era for autism research, and for our journal*, autistic participation in research processes is considered crucial (Pellicano et al., 2018). PR is, of course, an element in creating research that on the one hand directly and positively impacts issues of relevance to autistic people's needs, and on the other potentially positions autistic people as holders of knowledge, thus theoretically perhaps going some way to righting the testimonial injustice of which Fricker speaks.

However, in an article published a few years earlier in the same journal, Hobson (2010, p. 398), discusses, among other things, the autistic:

> difficulty in shifting among person-centered perspectives (which) undermines both their grasp of what it means to hold a perspective and, beyond this, what it means to claim that any given perspective is true of that which transcends individuals' perspectives, namely reality.

While this is not a recent publication, nor is it ancient, and variations of statements of this kind litter autism literature to this day. There is, for example, considerable, robust, and long-standing body of literature that refutes the notions that Theory of Mind tests are rigorous, that autistic people are necessarily lacking in Theory of Mind, or indeed that non-autistic people have mind-reading super-powers (see, for example, Fletcher-Watson & Bird, 2020; McGrath, Peterson & Pennington, 2020; Williams, 2021). However, apparently, we 'know' much about autistic people: that "individuals with ASD have a specific deficit in taking into account their reputation in the eyes of others" (Izuma et al., 2011, p. 17,305); that "(A) hallmark of autism is difficulty with social-emotional reciprocity, and...(impairments in) the ability to reflect on one's own mental state, (and) infer others' mental states" (Sala et al., 2020, p. 4135); that in autistic people "the self is diminished or non-existent" (Franks, 2003, p. 621). Pages are filled with examples of what we 'know' about autism and autistic people, most of it along similar lines. Autistics have been constructed as imperfect versions of non-autistics, as representing "the disordered and damaged other" (Milton & Sims, 2016, p. 524), and, perhaps most pertinently to discussion of authenticity, and most damagingly given the centrality of these qualities to what it means to be human, as lacking capacity "to be volitional, to be social, and to be selves" (Yergeau, 2018, p. 12). In various forms, this line of thinking has become so pervasive as to be routinely stated as fact both within and beyond medical or psychological frameworks.

Despite being "qualified knowers with deep, situated knowledge of how they engage with the world and with ableist structures" their knowledge risks being considered, at best, as an "adjunct" to existing research (Orsini, 2022, p. 10), epistemic injustice *oblige*. At worst, their knowledge might simply be discounted: as Yergeau (2013, p. 10, original italics) notes, "I am writing this essay, but anyone with a Theory of Mind can refute it, can refute *me.*" Can change be expected from autistic input into knowledge production in the context of a psychopathological reading of autism considering autistic people to stand outside the possibility of grasping reality?

This could constitute what Fricker (2007, p. 152) refers to as "circumstantial epistemic bad luck" both for researchers and for autistic people: that rather than being a purposeful injustice perpetuated by one person on another, the injustice done is harmful for the 'accidental' perpetrator as well as the 'victim' and is reliant on historical contingencies. Fricker (p. 152) gives the example of what could describe autism when she says:

> If, for instance, someone has a medical condition affecting their social behavior at a historical moment at which that condition is still misunderstood and largely undiagnosed, then they may suffer a hermeneutical disadvantage that is, while collective, especially damaging to them in particular...But they are not subject to hermeneutical injustice; rather, theirs is a poignant case of circumstantial epistemic bad luck.

In this situation, Fricker says (p. 33), "it is simply too much to expect the (possessor of this "bad luck") to achieve awareness that a certain prejudice is structuring his social consciousness."

However, the generous reading that those clinging to assumptions of deficit are victims of epistemic bad luck is leaky. "Ignorance," counters Pohlhaus (2012, p. 731), "is not something to which one is doomed because of social position, but rather something one chooses to maintain." There is so much literature that refutes, for example, deficits in Theory of Mind as being a hallmark of autism, that to ignore it looks less like bad luck and more like bad science, the willful epistemic ignorance of Pohlhaus' title. This bad science in turn demonstrates the extent to which autistic people and their accounts are "hermeneutically marginalized," subordinated to and excluded from "a practice that would have value for the participant" (Fricker, 2007, p. 153).

Unknowing

> The traditional systems of epistemology may be said to result from yes-answers or no-answers to the questions about the sources of knowledge. *They never challenge these sources, or dispute their legitimacy*: the questions are taken as perfectly natural, and nobody seems to see any harm in them (Popper, 1960, p.65, original emphasis).

Sweeping statements of what we know about autism like those I mentioned above not only do not challenge the sources or dispute their legitimacy: they do not even consider, are in blissful ignorance of, research that ***does*** this work. This results in circular logic, re-demonstrating and confirming what is already considered, uncritically, to be known. As Kourany (2023, p.178) points out, "(I)gnorance as well as knowledge is produced by science…such as by framing research problems to foreground certain issues rather than others."

The problem, then, is not what we ***don't*** know about autism and autistic people, but what we think we ***do*** know, or what we are unwilling, through arrogance or laziness, to ***unknow***. Medina (2013, p.33) describes "a socially produced and carefully orchestrated lack of curiosity" that he calls the "epistemic vice of laziness" that compromises "one's learning potential and contributions to knowledge" (p.34). Like other forms of epistemic vice and injustice, "epistemic laziness" negatively impacts and limits both the "subject" (here, the autism researcher) and the "cognitive perspectives of those around the subject, and on the social knowledge that becomes available or unavailable to the relevant communities" (p.34).

In a Foucauldian sense, of course, the oppressive discourses in much autism research that produce autistic people as just so many examples of deficit also produce the possibility for "reverse discourse," whereby autistic people form communities and create knowledge that resists that oppression (Foucault, 1990, p.101, concerning sexuality). However, the processes supporting and perpetuating epistemic injustice in autism research are intricate, long-standing, deeply entrenched, and powerful. As things stand, including autistic voices in research cannot alone constitute a serious challenge to existing knowledge systems, does not have the weight to undo stereotypes and prejudice so heavily fixed in the popular and scientific imagination.

Much autism research goes in circles, sticking to safe paths, it "confirms what is already known according to approved methods of knowing" within disciplinary silos (Halberstam, 2011, p. 6). It is not for nothing that critical autism studies (see Chapter 2, this volume), and Neurodiversity Studies in its turn, put considerable value on interdisciplinarity, for, as Gross and McGoey (2023, p.3) point out, "disciplinary silos can compound ignorance." Disciplinary assumptions "signal a form of training and learning that confirms what is already known according to approved methods of knowing, but they do not allow for visionary insights or flights of fancy" (Halberstam, 2011, p. 6). In succeeding scientifically, in confirming what we/they know about autism, researchers are also failing. This is not a creative failure allowing for visionary insights or flights of fancy, or for the creation of new and unexpected knowledge. It is a stultifying failure that allows us only to see what is already (assumed to be) known.

We need to go back to the sources, to reread the corpus, in order to "detect(ed) and correct(ed) for" identity prejudice (Fricker, 2007, p. 6). As Murray et al., (2005, p.140) put it, rather than "attempting to establish new facts about autism" we might already start by "trying to interpret what is already known." Going back to sources, unknowing the assumptions on which their hypotheses rest and coming with a new eye to what results might suggest if we did not 'know' about, for example, deficit in Theory of Mind (which in turn shuts down knowledge creation around authenticity) opens up possibilities for 'reknowing' that are otherwise left disregarded.

In this regard, Hoeyer and Winthereik (2022, p. 228) encourage the use of "playfulness," by which they mean:

> to dare to leave the path you know—as well as to dare to return to 'old' insights, those that are no longer in fashion, when they can help us approach a problem from a new, or a forgotten, angle.

Unknowing of the type I am suggesting, then, is not a lack of knowledge, but rather an intentional, creative, curious, and playful act rejecting a priori premises (notably of deficit) with the aim of coming to new understandings rather than confirming and contributing to existing assumptions. This process, that could also be called unlearning and then rereading, "demands no less than immersion in acts and practices of unknowing, a vigorous movement to flood the institutions that govern autism with the many ways in which autism must be unthought" (Orsini, 2022, p. 12). In the unknower's toolbox below I share, some tentative tools that I find useful in contributing to this movement toward creative unknowing.

Conclusion

In this chapter, I have shown how autistic people have largely been written out of authentic possibility through deficit-driven theories positioning them as impaired in reference both to self and to (non-autistic) others. Research recurrently points to enhanced truthfulness, lack of guile, straightforwardness, and originality on the part of autistic people—all of which direct us to the concept of authenticity (Vannini & Franzese, 2008). Neurocognitive theories, albeit tentatively, suggest that autistic people may have facilitated access to those same proxies of authenticity. The concept of authenticity also circulates through autistic communities and writing, to function as a sort of community marker.

However, I am aware that this line of thinking comes with certain risks. I agree with Burks-Abbot (2022, p.3) when he says that:

> As it stands now, autism research tends to divide the world into two distinct groups—autistics and non-autistic—and treat them as if they have little in common; this approach only serves to further marginalize autistics by positioning them as an 'Other' that no 'normal' person could ever really fathom.

So, does authenticity as a potential autistic specificity—particularly when looking to neurocognition as one of the pathways to this theory—risk essentialising and reifying difference? Does it reinforce the positioning of autistic people as unfathomable others?

Just as, in considering ToM, "the understanding of others' minds is not all or nothing in autism" (Davis et al., 2022, p. 1), neither is authenticity all or nothing, whether 'in' or 'elsewhere than' autism. Autistic people—contrary to how much scientific literature portrays them—are "as different from each other as 'neuro-typical' people are" (Schneid & Raz, 2020, p. 7). However, while Robert Chapman (2020, p. 424) notes that no core traits "have been found to be shared by all or most autistic people," he also considers that "there may be very general statistical tendencies toward certain cognitive and neurological characteristics." This is echoed by Dinah Murray (2020, p. 24), who suggests that "there is a common thread that unites autistic dispositions and perspectives." Murray[4] proposes an important thread as being 'a "tendency to give more to one's current self-generated, authentic, interest...and commensurately less to all other processing needs."

A consideration of authenticity and its proxies in what being autistic means for some autistic people may constitute a general statistical tendency, an understudied common thread. Notwithstanding the risks of reification, research in this direction holds both epistemological and ethical potential. The former has been demonstrated here: I have briefly illustrated how autistic people are largely written out of authentic possibility. I hope also to have shown how—if we unknow deficit as being the whole story—various disciplines point to interesting paths of exploration (see, for example, Chapter 14, this volume), including that of neurological predispositions to authenticity and its proxies.

Coghlan (2008, p. 360) defines authenticity as being "at the heart of being human," which "means to be experiencing, understanding, judging and deciding/acting." Autistic people, through metaphor, through identity prejudice, through medical certainty, have been constructed as being on the very edge of humanity (Botha et al., 2022; Danforth & Naraian, 2007). The question of authenticity repositions autistic people as holders of knowledge, as reflexive, autonomous beings: as ***human*** beings. The epistemic and the ethical are here inextricably linked, for as Pohlhaus (2012, p. 733) points out:

> the more unjust one is with regard to the development and maintenance of epistemic resources, the less accurate and robust will be one's account of the world we inhabit together.

Correcting the epistemic injustice implicit in much autism research is not just a question of including as many neurodivergent voices as possible. It requires—whatever our neurotype—unlearning what we think to be true, creatively

[4] Murray was discussing focus, monotropism, also pointing to authenticity in the context of passionate interests that may not follow social expectations of what constitutes a credible or acceptable 'hobby.'.

and critically engaging with and rereading existing literature, and intensifying epistemic pushback as to how we are able to know—or unknow, unlearn, re-imagine—what it means to be autistic.

An Unknower's Toolbox

Be alert to—and suspicious of—sweeping statements of certainty

Phrases like "it is well-known that...," "A hallmark of autism is...", "individuals with ASD have a specific deficit in...," etc., should send alarms ringing! Articles starting like this may well conclude with "our findings confirm that...." Allow yourself a wry smile as you see research going about business as usual.

Go back to the sources

Related to the previous point: once you have had your wry smile, wonder, could those results have been interpreted in any other way if we did not already "know that..."? For example, an article might start "it is well-known that autistic people have a deficit in managing their social reputations." The authors might then find that autistic people give more to charity when they are *not* observed than when they are observed, whereas non-autistic people give more when they *are* observed. The authors might conclude by stating "this confirms that autistic people have a deficit in managing their social reputations" (I am going to ignore one of my following tools, citational policies, and say that Izuma et al., 2011 is a great example of this, but plenty of others operate similarly). Had the authors *not* gone into their study knowing what they did, how might they have hypothesized their findings differently? Heightened altruism or authentic gifting processes in autistic folk? Heightened social conformity in non-autistics? Unknow the assumptions on which their hypotheses rest and come with a new eye to what results might then suggest.

Your discomfort is your friend

As an ADHD academic working largely in autism research, I have often felt an extreme sense of discomfort, a lack of legitimacy given the attention paid in certain UK contexts to the importance of autistic voices speaking about autistic experience (see Chapter 20, this volume). I now realize that this discomfort, however, well, uncomfortable, at the time, pushed me to engage as intensely as possible with autistic communities, lay-people, and colleagues, to listen and to collaborate. Instead of fighting the discomfort, stay with it, use it as a motor: it may just turn out to be your best friend and guide.

Be aware of who you cite and why

Feminist theorist Sarah Ahmed (2017, p. 270) has a "blunt citational policy" of not citing those she sees as "part of the institutional apparatus of white men." This allows her to follow what she calls "desire lines," paths of thinking other than those we are expected to follow, or that simply reproduce what has come before. This echoes Halberstam (2011, p. 6) warning us against "learning that confirms what is already known according to approved methods of knowing." Citing Baron-Cohen on Theory of Mind, for example, adds to the weight of his words. It is difficult to ignore Theory of Mind, if only to refute it, but how about citing research that critiques it, rather than that propagates it? I am a beginner here, and unable to be as blunt or radical as Ahmed in this chapter, but I have tried to be aware of who I am citing and why. A choice of citation is also a choice of reading: are there any articles or books or blogs written by neurodivergent folk and their allies on the subject I am interested in? If yes, read them, cite them, and give them space in your mind and on paper. Don't be afraid of engaging with (and citing!) non-academic neurodivergent sources (see Chapter 19, this volume; for an annotated list of resources, see Zisk, 2023).

Be playful!

Playful is perhaps not the first word that comes to mind in a toolbox of epistemological unknowing, but, following Hoeyer and Winthereik (2022), it is perhaps the most important tool here. This could mean creating new research paths or revisiting old ones to see where they might take us if we approach them differently. Another way of getting off the epistemic hamster wheel confirming what we already (think we) know, might be to mix up disciplines. Approach someone in another area to work with, as Reubs Walsh and I did (2021, p. 50), putting cognitive neuroscience (Reubs' area) in play with sociology to critically review the literature that "attempts to describe and explain (away) people who may be both trans/non-binary and autistic." It is not easy, it takes time to get to grips with another discipline's language, but we both learnt from it and had some ideas that I do not think we would have had otherwise.

Or you might have a new idea but are worried that it might be unpopular or has not been identified in the literature. Does everything we say need to have been said before, do we always need to stand on someone else's shoulders? Unfortunately, the answer to that is often yes: but is not the point of this chapter, of this book, to think about how to do things differently?

References

Ahmed, S. (2017). Living a feminist life. Duke University Press; Berenstain, N. (2016). Epistemic exploitation. *Ergo*, *3*, 569–590.

Belek, B. (2023). 'A smaller mask': Freedom and authenticity in autistic space. *Culture, Medicine, and Psychiatry*, *47*(3), 626–646.

Bertilsdotter Rosqvist, H., & Jackson-Perry, D. (2021). Not doing it properly? (Re) producing and resisting knowledge through narratives of autistic sexualities. *Sexuality and Disability, 39*(2), 327–344.

Botha, M. (2021) Academic, activist, or advocate? Angry, entangled, and emerging: A critical reflection on autism knowledge production. *Frontiers in Psychology*, 4196.

Botha, M., & Cage, E. (2022) "Autism research is in crisis": A mixed method study of researcher's constructions of autistic people and autism research, *13*, 7397.

Botha, M., Dibb, B., & Frost, D. M. (2022). Autism is me: An investigation of how autistic individuals make sense of autism and stigma. *Disability and Socciety, 37*(3), 427–453.

Bottema-Beutel, K., Park, H., & Kim, S. Y. (2018). Commentary on social skills training curricula for individuals with ASD: Social interaction, authenticity, and stigma. *Journal of Autism and Developmental Disorders, 48*, 953–964.

Brownlow, C. (2010). Presenting the self: Negotiating a label of autism. *Journal of Intellectual and Developmental Disability, 35*(1), 14–21.

Burks-Abbott, G. (2022). A theory of mine: My perspective on perspective-taking. *Autism in Adulthood, 4*(4), 265–267.

Cage, E., Crompton, C. J., Dantas, S., Strachan, K., Birch, R., Robinson, M., Morgan-Appel S, MacKenzie-Nash, C., Gallagher, A., & Botha, M. (2024). What are the autism research priorities of autistic adults in Scotland? *Autism*, 13623613231222656.

Cervantes, P. E., Matheis, M., Estabillo, J., Seag, D. E., Nelson, K. L., Peth-Pierce, R., Hoagwood, K. E., & Horwitz, S. M. (2021). Trends over a decade in NIH funding for autism spectrum disorder services research. *Journal of Autism and Developmental Disorders, 51*, 2751–2763.

Chapman, R. (2020). The reality of autism: On the metaphysics of disorder and diversity. *Philosophical Psychology*, pp. 1–21.

Coghlan, D. (2008). Authenticity as first-person practice: An exploration based on Bernard Lonergan. *Action Research, 6*(3), 351–366.

Danforth, S., & Naraian, S. (2007). Use of the machine metaphor within autism research. *Journal of Developmental and Physical Disabilities, 19*(3), 273–290. https://doi.org/10.1007/s10882-007-9061-9

Davis, P. E., Slater, J., Marshall, D., & Robins, D. L. (2023). Autistic children who create imaginary companions: Evidence of social benefits. *Autism, 27*(1), 244–252.

den Houting, J., Higgins, J., Isaacs, K., Mahony, J., & Pellicano, E. (2021). 'I'm not just a guinea pig': Academic and community perceptions of participatory autism research. *Autism, 25*(1), 148–163.

Dinishak, J. (2022). The deficit view and its critics. *Disability Studies Quarterly*, *36*(4).

Erickson, R. J. (1995). The importance of authenticity for self and society. *Symbolic Interaction, 18*(2), 121–144.

Fletcher-Watson, S., Adams, J., Brook, K., Charman, T., Crane, L., Cusack, J., Leekam, S., Milton, D., Parr, J. R., & Pellicano, E. (2019). Making the future together: Shaping autism research through meaningful participation. *Autism, 23*(4), 943–953.

Fletcher-Watson, S., & Bird, G. (2020). *Autism and empathy: What are the real links?* SAGE Publications Sage UK.

Foucault, M. (1990). *The history of sexuality volume 1: An introduction*, trans Robert Hurley. Vintage.

Franks, D. D. (2003). Mutual interests, different lenses: Current neuroscience and symbolic interaction. *Symbolic Interaction, 26*(4), 613–630.
Fricker, M. (2007). *Epistemic injustice: Power and the ethics of knowing*. Oxford University Press.
Goldman, B. M., & Kernis, M. H. (2002). The role of authenticity in healthy psychological functioning and subjective well-being. *Annals of the American Psychotherapy Association, 5*(6), 18–20.
Gross, M., & McGoey, L. (2022). Revolutionary epistemology: The promise and peril of ignorance studies. In Routledge International Handbook of Ignorance Studies (pp. 3–14). Routledge.
Halberstam, J. (2011). The queer art of failure. Duke University Press.
Hobson, R. P. (2010). Explaining autism: Ten reasons to focus on the developing self. *Autism, 14*(5), 391–407.
Hoeyer, K., & Winthereik, B. R. (2022). Knowing, unknowing, and re-knowing: Introduction. In The Palgrave Handbook of the Anthropology of Technology (pp. 217–235). Springer Nature Singapore.
ISNT. (no date). Institute for the Study of the Neurologically Typical. Retrieved March 27, 2024, from https://erikengdahl.se/autism/isnt/
Izuma, K., Matsumoto, K., Camerer, C. F., & Adolphs, R. (2011). Insensitivity to social reputation in autism. *Proceedings of the National Academy of Sciences, 108*(42), 17302–17307.
Jaarsma, P., Gelhaus, P., & Welin, S. (2012). Living the categorical imperative: Autistic perspectives on lying and truth telling–between Kant and care ethics. *Medicine, Health Care and Philosophy, 15*, 271–277.
Jackson-Perry, D., Bertilsdotter Rosqvist, H., Annable, J. L., & Kourti, M. (2020). Sensory strangers: Travels in normate sensory worlds. In H. Bertilsdotter Rosqvist, N. Chown, & A. Stenning (Eds.), *Neurodiversity studies: A new critical paradigm* (pp. 125–140). Routledge.
Jackson-Perry, D. (2023). *Exploring autistic accounts of sexuality, intimacy, and authenticity* (Doctoral dissertation, Queen's University Belfast). Available at: https://pure.qub.ac.uk/en/studentTheses/exploring-autistic-accounts-of-sexuality-intimacy-and-authenticit
Jackson-Perry, D. (2020). The autistic art of failure: Unknowing imperfect systems of sexuality and gender. *Scandinavian Journal of Disability Research, 22*(1), 221–229.
Kirchner, J., Ruch, W., & Dziobek, I. (2016). Brief report: Character strengths in adults with autism spectrum disorder without intellectual impairment. *Journal of Autism and Developmental Disorders, 46*(10), 3330–3337.
Kourany, J. A. (2023). Science for better or worse, a source of ignorance as well as knowledge. In Routledge international handbook of ignorance studies (pp. 178–190). Routledge.
Kourti, M., & MacLeod, A. (2019). I don't feel like a gender, I feel like myself: Autistic individuals raised as girls exploring gender identity. *Autism in Adulthood, 1*(1), 52–59; Kristensen, Z. E., & Broome, M. R. (2015). Autistic traits in an internet sample of gender variant UK adults. *International Journal of Transgenderism, 16*(4), 234–245.
Ma, W., et al. (2019). Children with autism spectrum disorder's lying is correlated with their working memory but not theory of mind. *Journal of Autism and Developmental Disorders, 49*(8), 3364–3375.

Main, A. (2003) *Allism: an introduction to a little-known condition*. Available at: http://fysh.org/~zefram/allism/allism_intro.txt
McGrath, L. M., Peterson, R. L., & Pennington, B.F. (2020) 'The Multiple Deficit Model: Progress, Problems, and Prospects', Scientific Studies of Reading, 24(1), pp. 7–13. Available at: https://doi.org/10.1080/10888438.2019.1706180.
McGrath, J. (2017). *Naming adult autism: Culture, science, identity*. Rowman and Littlefield.
Medina, J. (2013). *The epistemology of resistance: Gender and racial oppression, epistemic injustice, and the social imagination*. Oxford University Press.
Milton, D. (2014). Embodied sociality and the conditioned relativism of dispositional diversity. Autonomy: The Journal of Critical Interdisciplinary Autism Studies. Vol. 1 (3).
Milton, D. (2016). Re-thinking autism: Diagnosis, identity and equality. *Disability & Society, 31*(10), 1413–1415.
Milton, D., & Sims, T. (2016). How is a sense of well-being and belonging constructed in the accounts of autistic adults? *Disability & Society, 31*(4), 520–534.
Mottron, L., Dawson, M., Soulières, I., Hubert, B., & Burack, J. (2006). Enhanced perceptual functioning in autism: An update, and eight principles of autistic perception. *Journal of Autism and Developmental Disorders, 36*, 27–43.
Murray, D. (2020) 'Dimension of Difference', in Milton, Damian *et al.* (eds) *The Neurodiversity Reader*.
Murray, D., Lesser, M., & Lawson, W. (2005). Attention, monotropism and the diagnostic criteria for autism. *Autism, 9*(2), 139–156.
Nelson, L.H. (1993). Epistemological communities. In: Alcof, L., Potter, E. (eds.) Feminist Epistemologies. Routledge, New York
Nicolaidis, C., Raymaker, D., Kapp, S. K., Baggs, A., Ashkenazy, E., McDonald, K., Weiner, M., Maslak, J., Hunter, M., & Joyce, A. (2019). The AASPIRE practice-based guidelines for the inclusion of autistic adults in research as co-researchers and study participants. *Autism, 23*, 2007–2019. https://doi.org/10.1177/1362361319830523
Orsini, M. (2022). Who needs to (un) know? On the generative possibilities of ignorance for autistic futures. International Journal of Qualitative Studies in Education, 1–18.
Pellicano, E., & Burr, D. (2012). When the world becomes "too real": A Bayesian explanation of autistic perception. *Trends in Cognitive Sciences, 16*(10), 504–510.
Pellicano, L., Mandy, W., Bölte, S., Stahmer, A., Lounds Taylor, J., & Mandell, D. S. (2018). A new era for autism research, and for our journal. *Autism, 22*(2), 82–83.
Pohlhaus, G. (2012). Relational knowing and epistemic injustice: Toward a theory of willful hermeneutical ignorance. *Hypatia, 27*(4), 715–735.
Popper, K. R. (1960). On the sources of knowledge and ignorance, from Proceedings of the British Academy, Vol. XLVI.
Pukki, H., Bettin, J., Outlaw, A. G., Hennessy, J., Brook, K., Dekker, M., Doherty, M., Shaw, S. C., Bervoets, J., Rudolph, S., Corneloup, T., & Yoon, W. H. (2022). Autistic perspectives on the future of clinical autism research. *Autism in Adulthood, 4*(2), 93–101.
Sala, G., Hooley, M., & Stokes, M. A. (2020). Romantic intimacy in autism: A qualitative analysis. *Journal of Autism and Developmental Disorders, 50*(11), 4133–4147.

Schlegel, R. J., & Hicks, J. A. (2011). The true self and psychological health: Emerging evidence and future directions. *Social and Personality Psychology Compass, 5*(12), 989–1003.

Schneid, I., & Raz, A. E. (2020). The mask of autism: Social camouflaging and impression management as coping/normalization from the perspectives of autistic adults. *Social Science and Medicine, 248*, 112826.

Scott, S. (2015). *Negotiating identity: Symbolic interactionist approaches to social identity*. Polity Press.

Späth, E. M., & Jongsma, K. R. (2020). Autism, autonomy, and authenticity. *Medicine, Health Care and Philosophy, 23*(1), 73–80.

Stark, E., Ali, D., Ayre, A., Schneider, N., Parveen, S., Marais, K., Holmes, N., & Pender, R. (2021). Coproduction with autistic adults: Reflections from the authentistic research collective. *Autism in Adulthood, 3*(2), 195–203.

Strunz, S., et al. (2015). Personality pathology of adults with autism spectrum disorder without accompanying intellectual impairment in comparison to adults with personality disorders. *Journal of Autism and Developmental Disorders, 45*(12), 4026–4038.

Vannini, P., & Franzese, A. (2008). The authenticity of self: Conceptualization, personal experience, and practice. *Sociology Compass, 2*(5), 1621–1637.

Walsh, R., & Jackson-Perry, D. (2021). Autistic cognition and gender identity: Real struggles and imaginary deficits. In *Working with autistic transgender and non-binary people*. Jessica Kingsley Publishers.

Wattel, L. L., Walsh, R. J., & Krabbendam, L. (2022). Theories on the link between autism spectrum conditions and trans gender modality: A systematic review. *Review Journal of Autism and Developmental Disorders*, 1–21.

Williams, G. L. (2021). Theory of autistic mind: A renewed relevance theoretic perspective on so-called autistic pragmatic "impairment." *Journal of Pragmatics, 180*, 121–130. https://doi.org/10.1016/j.pragma.2021.04.032

Yergeau, M. (2013). Clinically significant disturbance: On theorists who theorize theory of mind. *Disability Studies Quarterly*, 33(4).

Zisk, A. H. (2023). Critical autism studies beyond academia: An annotated list. *Ought: The Journal of Autistic Culture*, 5(1), 6.

PART III

Looking Back to the Future

CHAPTER 23

ADHD, Academics, and Communities: Who Are the "ADHD Experts"?

Andrew Ivan Brown

Introduction

Several years ago, when I was still a graduate student, I attended the 2016 annual Canadian ADHD conference hosted by CADDRA (the Canadian ADHD Resource Alliance). I was not a presenter. I only went to observe and learn from the established "ADHD experts." I figured that they must be experts to have been invited to speak at Canada's largest ADHD conference. I did not recognize most of the names of the speakers or panelists, but they all had credentials like "PhD," "PsyD," or "MD" beside their names (graduate students were relegated to poster presentations only). The talks were about the history of ADHD science, emerging genetic research, functional outcomes of adults with ADHD, problems with treatment adherence, parent satisfaction outcomes, teacher strategies, and so on.

As I sat through these various talks and panels, it quickly became clear to me that I was an outsider. Nobody at the conference—not the speakers, panelists, graduate students, or anyone I spoke to—came out as having ADHD (except for me—I outed myself on multiple occasions). Despite researching ADHD for my doctoral studies, all the presentations, posters, and people felt alien to me and my project. They did not seem to me to be studying ADHD at all, at least not the ADHD I knew through my own experiences and the community approaches I was familiar with. They were treating ADHD as an

A. I. Brown (✉)
University of Washington, Seattle, Washington, USA
e-mail: aibrown@uw.edu

H. Bertilsdotter Rosqvist and D. Jackson-Perry (eds.), *The Palgrave Handbook of Research Methods and Ethics in Neurodiversity Studies*,
https://doi.org/10.1007/978-3-031-66127-3_23

object of research, viewed through a largely medical lens. I ultimately left that conference feeling demoralized and skeptical that I would ever be an "ADHD expert," at least in the eyes of these researchers and clinical practitioners who appeared to control the dominant discourse on ADHD (c.f. "credibility deficit" as well as testimonial and hermeneutical injustice discussed by Hultman and Hultman, (Chapter 12, this volume).

Jumping forward a decade to 2024: CADDRA now accepts conference proposals relating to "community approaches" as well as "equity, diversity, and inclusion," and has opened up presenter slots so that anyone—not just established "experts"—can give a talk (assuming their proposal is accepted). In a similar vein, the American-based, and much larger, Annual International Conference on ADHD, hosted by CHADD (Children and Adults with ADHD), ACO (ADHD Coaches Organization), and ADDA (ADHD Association) includes spaces for ADHDers (usually ADHD coaches or social media influencers) to present, innovate, workshop, lead peer support groups, and more. Indeed, I was recently contacted by a well-established ADHD coach who suggested I apply to present at the 2024 International Conference, as he felt my dissertation theory of ADHD (Brown, 2023) could be "big." But even if I am accepted to present as an "expert," something feels off about this individual pursuit of mine. Why do I want so badly to be accepted as an "ADHD expert"? Am I pursuing stardom? Is my intention to write a trade book on ADHD and make money? The problem I am conveying here is that these individual pursuits toward "ADHD expertise" represent a structural issue: ADHD communities and conferences—at least in Canada and the United States—are often structured in a way that relies on and rewards (fame, money, careers) individual "experts," but at what cost to collective organizing and theorizing?

In this chapter, I provide a brief commentary on the structure of "ADHD expertise," drawing on both my own experiences attending various ADHD community groups over the years, and from my decade-long, interdisciplinary research project on ADHD. I suggest that cultivating "expertise" in ADHD communities has traditionally been an individual rather than a collective pursuit, and that future ADHD researchers should instead aim to foster community knowledge production and collective theorizing. I do not take a systematic or methodologically rigorous approach here. Still, in writing this, I hope to give researchers of ADHD a sense of some of the less-discussed, micro-social changes I observed in the ADHD community landscape before and during the popular rise of neurodiversity movements, and to think about the future of "ADHD expertise."

This chapter is divided into three sections. In the first section I ask, who gets counted as an "ADHD expert"? Who gets included or excluded from receiving this status? I provide a brief sketch of the "individual ADHD expert" classifications that I have observed over the years. First, traditional academic researchers of ADHD (who are almost exclusively non-ADHDers). Second, practicing clinicians (who are sometimes ADHDers themselves) specializing in ADHD. Third, ADHD coaches. And fourth, social media influencers who, despite not fitting into any of the aforementioned kinds of "ADHD experts," produce ADHD-related content for their followers. In the second section I ask, how does "ADHD expertise"—even if this specific phrase is not always used—operate in and on some ADHD communities? To answer this, I provide short vignettes of two paradigmatic types of ADHD community groups: those that are facilitated by individual "ADHD experts," and those that are not. In doing so, I describe two operations of "ADHD expertise": what can be called "self-cultivated expertise" (the individual pursuit of learning from the "experts") and "community expertise" (a form of "ADHD expertise" that lends itself to community building). In the third section, I consider how participatory research methodologies discussed in this handbook demonstrate ways that future researchers can challenge individualistic structures of "ADHD expertise." With these examples in mind, I also ask, what mistakes from the past should academic ADHDers avoid repeating? What role might "individual ADHD experts" play in critical ADHD collectives of the future?

Individual "ADHD Experts"

Googling "ADHD expert" returns a link to Russell Barkley's website followed by a link to Edward Hallowell's website (Barkley, 2024a; Hallowell, 2024). They exemplify two kinds of traditional "ADHD experts," similar to the contrast Pearson et al. (Chapter 7, this volume) made between "researcher expertise" and "professional expertise or 'expertise through experience.'" Akin to the "king" and "queen" of autism (Simon Baron-Cohen and Lorna Wing), Barkley, a neuropsychologist, might be considered the de facto king of ADHD, with over 800 presentations, workshops, and public addresses on the subject, and his executive dysfunction theory of ADHD remaining widely influential (Barkley, 2024b). Yet, he does not have ADHD. He is an exemplar of the first traditional kind of "ADHD expert": academic researchers (almost exclusively non-ADHDers) who study ADHD, often in neuropsychology but also in education, genetics, and other various disciplines and fields that cooperate with the medical model of ADHD. These non-ADHDer academic researchers generally do not treat ADHDers as knowers, positioning them instead as passive research subjects. The knowledge that is produced by these researchers is usually disseminated in academic publications, but also in some cases, popular trade books, usually in the self-help genre.

Hallowell and his colleague John Ratey, on the other hand, represent the second traditional kind "ADHD expert": practicing clinicians (psychiatrists, pediatricians, therapists, physicians, and so on) specializing in ADHD, and who are sometimes ADHDers themselves. Hallowell and Ratey are both ADHDers, for example, and their seminal trade book from 1994, *Driven to Distraction*, introduced ADHD self-advocacy to the world (Brown, 2023). What most differentiates this second traditional kind of "expert" from the first, in my view, is that clinicians are far more likely to treat their patients as active knowers. That is, through years of clinical experience listening to ADHD patients (or parents of ADHD children) talk about their struggles, clinicians start to notice patterns between patients, and can create new discourse about ADHD. As Hallowell writes, "all through the 1980s and into the 1990s, John and I explored the topic, comparing notes on patients, speculating on what more went into ADD than was in the [academic] books" (2011: introduction, para. 7).

Still, treating ADHD patients as knowers is not the same thing as including ADHDers as co-collaborators in collective theorizing. *Driven to Distraction* and many of the popular ADHD self-help books that followed still emphasize individual "experts"—the authors—telling stories about ADHDers, rather than, to use Romy Hume's language, having a "dialogue with" ADHDers in a collective storymaking sense (Chapter 13, this volume). For example, *Driven to Distraction* is mostly a collection of (fictional retellings of) case studies of ADHD patients and their specific problems, and how Hallowell and Ratey helped them to figure out solutions to their problems over the course of their therapy sessions. In other words, the kind of knowledge that is produced through clinical practice is still filtered through the therapeutic lens of the "ADHD experts" who write these books. Hultman and Hultman discuss this same "power asymmetry" between patient and clinician in their chapter on ADHD-led research (Chapter 12, this volume; see also Hultman and Hultman, 2023).

Recognizable "ADHD experts" expanded over the years to include ADHD coaches and social media influencers. ADHD coaches are similar to the second kind of traditional "ADHD expert." While usually not practicing clinicians, ADHD coaches can become certified by accredited coaching organizations like ADDCA (ADD Coach Academy), and work with several ADHDers (clients) over the years. They sometimes write trade books based on what they learn through their coaching experience.

Social media influencers, on the other hand, may represent a whole new kind of "ADHD expertise" and I can only offer a cursory analysis here. I am not referring to traditional "ADHD experts" who try their hand at social media and become popular gurus. Traditional "ADHD experts" and coaches have been using various forms of multimedia to spread the word about ADHD since at least the 2000s (TotallyADD.com, for example). Rather, I am thinking of the "influencer-turned-coach" phenomenon (Berkoben et al., 2023: 20). What sets these influencers apart is that they do not need the credentials—academic/clinical degrees or coaching certificates—that would traditionally be required to obtain the status of "ADHD expert." Instead, social media influencers can become "ADHD experts" simply by virtue of having an engaging personality and skill at using a particular platform and developing a large following. From my observations, these kinds of influencers are almost always ADHDers themselves, and sometimes talk about their own experiences. However, most of the knowledge they are presenting about ADHD tends to be re-filtered from traditional "ADHD experts," or, in some cases, a curation of social media posts by other ADHDers. The latter case can lead to a potentially concerning proliferation of ideas of what counts as ADHD, such as reproducing stereotypes or spreading an "epistemic infection" of deficit narratives (Bertilsdotter Rosqvist & Jackson-Perry, 2021: 341).

Since the mid-1990s, the status of "ADHD expert" has increasingly been conferred to ADHDers in various roles. While there are still plenty of non-ADHDers who are considered "ADHD experts," especially in areas of neuropsychology, pediatrics, neuroscience, genetics, education, and even the emerging field of neurotechnology (neurofeedback and deep brain stimulation), there is an ever-growing number of ADHD-specialized therapists, coaches, and social media influencers who are all ADHDers themselves. While this is no doubt a favorable development in the sense of "Nothing About Us Without Us" (Charlton, 2000: 14), the issue I presented at the beginning of this chapter remains: the structure of "ADHD expertise" still appears to be highly individualistic and, in many cases, reliant on non-ADHDer-produced knowledge (e.g., Barkley's executive dysfunction theory of ADHD). Deficit narratives such as these are still pervasive even in community discourse professed by ADHDers as affirmative. Finally, while it is true that ADHDers have collectively played a role in cultivating much of the knowledge on ADHD claimed by the "experts," these community ADHDers often do not consider themselves as "expert knowers" in their own right, as the next section demonstrates. They appear to largely depend on the "experts" to guide them in one way or another, directly or indirectly.

Vignettes of Two Paradigmatic Types of ADHD Community Groups[1]

ADHD community groups are not standardized but are entirely dependent on how the organizers decide to set up the group and the demographic it is tailored for. Nevertheless, for the purpose of exploring how "ADHD expertise" operates in and on these groups, I provide vignettes of two types that I consider paradigmatic: groups that are not facilitated by an "ADHD expert," and groups that are. The first type is common in online community groups, but also sometimes found in student peer support group organization, and, as demonstrated by the first vignette, adult (non-student) ADHD support groups. The second paradigmatic type is, in my experience, more prevalent than the first, and usually facilitated by a psychologist, or by a non-profit ADHD organization that brings in various "experts" on a rotating cycle to lead the group each session. For my second vignette, however, I describe a somewhat peculiar case of a group led by an ADHD coach who operationalizes "community expertise," a form of "ADHD expertise" that lends

[1] The following descriptions are not "qualitative data." I attended these community groups many years ago for personal support, not for research. The descriptions are instead anecdotal, drawing from memory and journal entries I made long ago. The quotations are reconstructions. The individuals described are amalgamations of multiple individuals I remember encountering.

itself to community building, and which may prove useful for critical ADHD collectives of the future, as I discuss below.

Group I. Self-cultivated "Expertise"

There are approximately 20 ADHDers, including myself, sitting in a large circle facing inwards. The floor is open to anyone who wants to speak. One member puts up her hand and begins talking about something called "CAPD" (central auditory processing disorder). It's something about "how you can't hear the beginning of sentences when someone speaks to you because you can't process the low tonal resonances in people's voices until you tune in." She says everyone she's ever talked to with ADHD has CAPD, and that it's a "symptom of ADHD." A man cuts in stating how that's very interesting and that he must have CAPD too. Another member, in rebuttal, then states that "not everyone with ADHD has CAPD." The initial speaker yells at her that she's not right, and everyone with ADHD does have CAPD. The other member then holds up her phone and yells back, "I just looked it up!" A battle of self-cultivated "expertise" ensues.

ADHD physician Gabor Maté, in his trade book on ADHD, writes, "so few doctors or psychologists are familiar with attention deficit disorder that people are forced to become self-cultivated experts" (2000: 5). Indeed, this support group prides itself on being self-driven without the help of professionals. Professional psychologists are not allowed to attend unless invited for a special talk. One problem I observed with this rule, however, is that those members who are loudest and most assertive tend to receive the most speaking time, and so their self-cultivated "expertise" ends up being passed around regardless, no matter how dubious it is or where it came from.

A woman who appears to be in her late 60s suddenly wanders into the room and exclaims to everyone that she is covered in cat hair in case anyone is allergic. She sits down and the group conversation continues. She begins yelling at people to speak up because she can't hear them. A bunch of members start simultaneously telling her that she has CAPD. "CAP-*what?*" she yells back. Everyone replies in unison, "*CAPD!*" A big, burly looking guy tells her he's in the same boat as her, that he can't hear well because he was in the military blowing stuff up when he was nineteen and didn't care about whether it was harmful. She cuts him off—"yeah, yeah, I understand, but people need to speak louder." After the meeting ends, I have a chat with the only other person in their 20s in attendance. He tells me that this group is "scary" and that he won't be coming back. He thinks there are "lots of comorbidities here, lots of negativity, and lots of people talking about issues that he's never experienced or heard about" in his own self-guided research on ADHD. "God, I hope I don't end up like this when I'm older," he concludes.

Group II. Community Expertise

An ADHD coach runs this group. It is free and open to the public. Every month, we show up, take a seat, and wait for the coach to begin his lecture on the chosen topic for the day. He is our community expert. He gets quoted a lot in local newspapers since he is big into ADHD political activism, lobbying the mayor on issues like jail reform for ADHDers, and challenging stereotypes about ADHD in schools. I see him as a role model. He even taught us that ADHD can have its positives. It is clear to everyone in the group who the "expert" is. He is not a psychologist or medical practitioner, but knows ADHD well, and can often recite statistics about ADHD off the top of his head. Most importantly, he is an ADHDer himself. After about 30 minutes of lecturing, he has us get into small groups to discuss specific issues we have been experiencing related to our ADHD (usually in connection to the topic of the lecture). 30 minutes later, we reconvene and tell him what our group talked about. He connects our ideas to ADHD literature, and often provides additional pointers for us relating to the issues we discussed, sometimes on a very individual level based on specific circumstances or contexts.

From Neurodivergent Care Communities to Critical ADHD Collectives

The two vignettes above offer some micro-social glimpses into some of the ways that "ADHD expertise" operates in and on ADHD communities. Self-cultivated "expertise" usually consists of an ADHDer doing their own research, such as reading ADHD trade books, online blogs or articles, following social media influencers specializing in ADHD, watching YouTube videos, listening to podcasts with rotating "experts" as guest speakers, and if inclined, reading academic publications on ADHD. One might even consider therapy a form of self-cultivating one's "ADHD expertise," insofar as one begins to better understand themselves, their struggles, and their differences through a lens of ADHD knowledge conveyed to them with the help of their therapist.

With some exceptions, developing self-cultivated "expertise" in ADHD communities tends to be an individual rather than a collective pursuit, and is mostly dependent on extraneous "ADHD experts" from which the knowledge is obtained or filtered. This individual approach can become antagonistic to developing care communities. I have seen countless attempts of ADHDers telling other ADHDers that they are wrong about something, that they should read this or that article or book, or look up this or that term on the internet, or listen to this or that podcast. This kind of self-cultivated "expertise," in my view, is not a good model for community knowledge production or collective theorizing. We end up treating the "ADHD experts" as the only credible knowers, and in doing so, we don't treat each other as experts in our own right. We don't look up to each other as co-collaborators in ADHD knowledge production. Respect is rather given to those we perceive as having done their own research on ADHD.

A growing number of online neurodivergent care communities emerged in the late 2010s and early 2020s, a few of which I became a part of. These differed significantly from the traditional ADHD communities and groups I had previously attended. For one, they were not exclusive to ADHD, but included autistic members, AuDHDers, and others who broadly identified as neurodivergent. More than that, they emphasized care over sharing knowledge. There was little discussion of strategies, or recommendations of "experts" to follow, watch or read. In other words, these communities were not so much about cultivating expertise, but caring for each other. As crip theorist Logan Smilges writes of such communities, they are more about "holding space for bad feelings in the present than about making promises to fix the future" (2023: 73).

In our introduction chapter to Critical ADHD Studies, myself, Hanna Bertilsdotter Rosqvist, and David Jackson-Perry (Chapter 3, this volume) emphasize the need for phenomenological and community theorizing approaches to ADHD in academic research, some of which has already begun in recent years (Bertilsdotter Rosqvist et al., 2023a, 2023b, 2023c, see also Chapter 12, this volume). These novel approaches position all ADHDers as equal knowers and co-contributors in the realm of ADHD research. They start with the model of collective care communities—such as acknowledging how neurotypical epistemic privilege disables us neurodivergents and make us feel miserable in so many ways (see Chapters 16,17, this volume)—and go from there, drawing from well-established academic methodologies (such as phenomenology). As Pearson et al. put it, "creating safe spaces for neurodivergent scholars [serves] as a vehicle for scholars to work together" (Chapter 7, this volume). It might be said that Critical ADHD Studies is partly an attempt to hold new spaces for collective ADHD expertise to develop outside of the individual forms. These critical ADHD collectives of the future would be grounded in neurodivergent caring for each other in the senses just described, but also use such care as a way of producing new kinds of ADHD knowledge not previously possible.

Indeed, many of the chapters in this book provide examples of methods that may help to push back against the individualistic structures of "ADHD expertise," and make this new kind of knowledge production flourish. Shruti Taneja-Johansson notes that many of her ADHDer research participants take her to be the "expert," and that to help counter this, she downplays her "expert status" and tells them that she sees them as the experts—that this research is about *their* lived experiences (Chapter 11, this volume). Romy Hume makes a similar point, and emphasizes "dialogue *with*" rather than "story *about*" (Chapter 13, this volume). In their chapter exploring the limits and possibilities of ADHD-led research, Lill and Maya Hultman test out a method that future researchers might want to try out. They both take turns as ADHD researcher and ADHD research participant, and try to identify negative consequences that arise with this power asymmetry, in hopes of finding ways to counter it in future research (Chapter 12, this volume).

In his chapter on developing autistic-friendly phenomenological research design, Ned Redmore reminds us that, given how neurodivergent research participants have historically been situated as "voiceless subjects," the "tools with which to conduct research with people from this group are historically underdeveloped" (Chapter 15, this volume). He stresses the need for "dialogue" so that "trusting relationships can begin to form" between participants and researchers; such dialogue includes an "openness to all forms of communication, expression and engagement" (Chapter 15, this volume). From their experiences of working in a neurodiverse group of authors, Pearson et al. suggest a move to "collective storytelling," and facilitating "an ability to be vulnerable with each other, sharing our inner thoughts and feelings about the working process" (Chapter 7, this volume). In the same vein, Hultman et al. describe that forming a "tight-knit group" of community researchers can be helpful (Chapter 8, this volume). In terms of research design, Joke Struyf et al. argue that "co-creation and inclusive research are [...] more than just asking the community for feedback"; they also mean including the community in discussions of how this kind of research should be defined, and how ways of doing research should be changed to be more emancipatory (Chapter 5, this volume). They imagine "involving participants from the beginning," even at the stage of writing the project proposal (Chapter 5, see also Chapter 6, this volume). Sandra Jones and Jennifer Lowe describe these sorts of processes as "adopting an autistic approach to designing autism studies," and building "autistic comradery in research" (Chapter 10, this volume)—something that could surely apply for ADHD research as well.

These various methods—and I have only listed a few of the many examples available throughout this handbook—will help future ADHDer researchers avoid repeating the mistakes made by traditional, non-ADHDer researchers. They also point to the benefit of conducting collective ADHD research in academic institutions: there are rich, methodological tools out there that can help us avoid falling into the "individual expert" trap so prevalent in existing ADHD communities. This trap, of course, is also common in how academic institutions are structured, and future researchers need to be extra careful to avoid falling prey to it. To elaborate, some critics have argued that the term "ADHD" refers not only to the condition itself, but is also symbolic of a "field of force relations" through which specific regimes of knowledge (behavioral psychology, medical sociology, psychiatry, psychoanalysis, neuroscience, genetics, neurotechnology, etc.) "are constantly in flux" as they attempt to gain "temporary dominance over one another," "strategize" for claims over ADHD's truth, try to assert their fields of expertise as the most authoritative, compete for funding, and the like (Rafalovich, 2002: 87–88). Critical ADHD Studies is not meant to be just another addition to the many regimes of knowledge vying for control over ADHD's discourse, using the guise of "new interventions" to build the careers of individual researchers. It is meant to break the individualistic structures of "ADHD expertise" by collectivizing ADHD knowledge production in new ways that are geared toward emancipatory, rather than capitalist, ends.

With that all said, I end this chapter by considering one last question: what role, if any, might individual "ADHD experts" play in critical ADHD collectives of the future? The reason I described "community expertise" in my second vignette was to remind us that individual "ADHD experts" can often be inspirational, affirming, and dedicated to ADHD community building. They can be especially helpful for "new" ADHDers who are seeking to learn more about ADHD. Above all, many "ADHD experts" have already provided us with so much knowledge to build our communities around—for instance, several chapters in this handbook cite Hallowell and Ratey's 2021 trade book, *ADHD 2.0*. At the same time, we must also be wary of much of this pre-produced knowledge on ADHD, because so much of it is still grounded in deficit narratives. As David Jackson-Perry notes in his chapter on autistic research, countering deficit narratives requires a "labor of unknowing" (Chapter 22, this volume). We may have to unlearn much of what the individual "experts" have told us, even if some of those "experts" are ADHDers themselves.

The individualistic structures of "ADHD expertise" will not disappear overnight, and this brings me back to why I want to try to present my dissertation theory of ADHD at the Annual International Conference—if not this year than next year, or the year after. While I cannot completely divorce myself from the seductive rewards that would come with obtaining the status of "ADHD expert," my deeper aim is strategic. The participatory research methods and collective theorizing approaches described throughout this handbook are no doubt central to developing Critical ADHD Studies, but in the meantime, I want to test out an alternative strategic approach: I want to get inside the power structures that currently exist—to become an individual "ADHD expert"—and then use that position of power to push back against deficit narratives. Update: while working on revisions for this chapter, I received an email informing me that my proposal for the ADHD conference was rejected after careful review by a "panel of ADHD experts"—their words. It's been eight years since I attended my first ADHD conference, and I still feel just as much of an outsider to the field as I did back then.

Regardless of the approach future ADHD researchers decide to take, Critical ADHD Studies will have to continue to think deeply about the evolution of "ADHD expertise" I outlined in this chapter, and how to contend with its individual forms still present in the heart of many ADHD communities.

References

Barkley, R. (2024a). Russell A. Barkley, Ph.D. Dedicated to Education and Research on ADHD." Retrieved April 4, 2024 (https://www.russellbarkley.org/).

Barkley, R. (2024b). About Russell A. Barkley, Ph.D. Retrieved April 4, 2024 (https://www.russellbarkley.org/about.html).

Berkoben, Frankie, Dusty Chipura, IngerShaye Colzie, and Ron Capalbo. 2023. "Social Media for ADHD Coaches: Integrity, Influence, and Imposter Syndrome" in *Annual International Conference on ADHD 2023 Program*.

Brown, A. I. (2023). The Univocity of attention: ADHD and the case for a renewed self-advocacy. Dissertation, York University.

Charlton, J. I. (2000). *Nothing about us without us: Disability oppression and empowerment*.

Hallowell, E. (2024). Dr. Hallowell: The Hallowell ADHD centers. Retrieved April 4, 2024 (https://drhallowell.com/).

Hallowell, E. M., & Ratey, J. J. (2011). *Driven to distraction (Revised): Recognizing and coping with attention deficit disorder* (Original). Anchor.

Hultman, L., & Hultman, M. (2023). Believe me, only I know how I feel. An autoethnographic account of experiences of epistemic injustice in mental health care. *Frontiers in Psychiatry*, *14*, 1058422.

Maté, G. (2011). *Scattered minds: The origins and healing of attention deficit disorder*. Vintage Canada.

Rafalovich, A. (2002). Framing the ADHD child: History, discourse and everyday experience. University of British Columbia.

Bertilsdotter Rosqvist, H., & Jackson-Perry, D. (2021). Not doing it properly? (Re)Producing and resisting knowledge through narratives of autistic sexualities. *Sexuality and Disability*, *39*(2):327–44. https://doi.org/10.1007/s11195-020-09624-5.

Rosqvist, B., Hanna, L. H., Wiklund, S. Ö., Nygren, A., Storm, P., & Sandberg, G. (2023a). Intensity and variable attention: Counter narrating ADHD, from ADHD deficits to ADHD difference. *The British Journal of Social Work, 53*(8), 3647–3664. https://doi.org/10.1093/bjsw/bcad138

Bertilsdotter Rosqvist, H., Hultman, L., Wiklund, S. Ö., Nygren, A., Storm, P., & Sandberg, G. (2023b). Naming ourselves, becoming neurodivergent scholars. *Disability and Society*, 1–20. https://doi.org/10.1080/09687599.2023.2271155.

Rosqvist, B., Hanna, L. H., Wiklund, S. Ö., Nygren, A., Storm, P., & Sandberg, G. (2023c). ADHD in higher education and academia. *Canadian Journal of Disability Studies, 12*(3), 1–26.

Smilges, J. L. (2023). *Crip negativity* (1st ed.). University of Minnesota Press.

CHAPTER 24

Power to the People

Jacquiline den Houting, *Abby Sesterka*, *and Elizabeth Pellicano*

The Changing Tide

There is much to be positive about. A mere decade ago, only two published papers had reported community engagement in autism research, both from the ground-breaking Academic Autistic Spectrum Partnership in Research and Education (AASPIRE) (Jivraj et al., 2014; Nicolaidis et al., 2011; Nicolaidis et al., 2013; see also Chapters 4, 19, 25, this volume). Across the intervening ten years, we have seen an exponential increase in participatory autism research, with a recent systematic review (Tan et al., 2024a) identifying 312 autism research papers that reported community engagement. Of these, more than 200 reported involvement by autistic people; many also reported engagement with professionals, family members, and other relevant stakeholders.

During this reasonably short time, we have also witnessed a growing cohort of openly autistic (and otherwise neurodivergent) scholars joining the

J. den Houting (✉)
Olga Tennison Autism Research Centre, La Trobe University, Melbourne, Australia
e-mail: J.denHouting@latrobe.edu.au

A. Sesterka
Flinders University, Adelaide, Australia

E. Pellicano
University College London, London, UK

H. Bertilsdotter Rosqvist and D. Jackson-Perry (eds.), *The Palgrave Handbook of Research Methods and Ethics in Neurodiversity Studies*,
https://doi.org/10.1007/978-3-031-66127-3_24

field (e.g., Bertilsdotter Rosqvist et al., 2023; Botha, 2021; Jones, 2021; Nuwer, 2020; Poulsen et al., 2023). In addition to academic expertise, these researchers bring to their work insights gained through their own lived experience of neurodivergence. They can also bring rich knowledge of neurodivergent experiences beyond their own, gained through deep immersion in the autistic and/or broader neurominority communities. Though openly autistic researchers remain a minority amid the global cohort of scholars researching autism, we have seen some of the more conventional structures in the field of autism research make space for autistic scholars. One such example is the International Society for Autism Research's (INSAR) Autistic Researchers Committee, established in 2020 to advance the interests of autistic researchers within INSAR (Poulsen et al., 2023).

Along with this shift in *who* is producing autism research, change is afoot in *what* and *how* we research. Australia, for example, has seen considerable changes to the national landscape of autism research investment, with a pattern of disproportionate investment in biological research (seen in many Western Countries; OARC, 2019; Pellicano et al., 2013) giving way to more equitable distribution of funding across research areas (den Houting & Pellicano, 2019). Encouragingly, there has been increased investment in research areas aligned with established community priorities, giving precedence to research with tangible, real-world impact as has long been called for by the autistic community and other stakeholder groups (Gatfield et al., 2016; Robinson, 2010; see also Chapters 10, 13, 14, 20, 21, 22, 25, this volume).

More recently, conventional autism science has increasingly drawn inspiration from the experiential expertise of autistic people. The Double Empathy Problem, proposed by autistic scholar Damian Milton (2012), was initially theorised largely on the basis of personal experience and observation (Milton et al., 2022), and has since become a defining work. Arguably, Milton's (2012) landmark paper deployed unconventional scientific methods. It did not put forth novel empirical findings. Rather, it offered an alternative perspective of a well-established phenomenon; casting aside the trappings of pathology to suggest that the autistic experience is, at its core, an aspect of shared human experience. Milton's Double Empathy Problem has now prompted a wide range of empirical investigations seeking to shed light on the mechanics of cross-neurotype social interaction (Crompton et al., 2020; Sasson et al., 2017; Sheppard et al., 2016). In recent years, there has been rapid growth in the body of autism research that is rooted in lived experience, investigating concepts first identified in the autistic lay community (see Chapter 19, this volume). Stronger ties between researchers and the autistic community have facilitated growing understanding of important community-driven phenomena, including autistic burnout (Higgins et al., 2021; Raymaker et al., 2020), masking/camouflaging (Cage & Troxell-Whitman, 2019; Hull et al., 2017), and inertia (Buckle et al., 2021; Rapaport et al., 2023).

Scanning the most-read and most-cited work published by leading autism-specific journals in recent years suggests that readers are embracing these emerging foci. Extremely widely cited papers feature an empirical evaluation of the Double Empathy Problem (Crompton et al., 2020) and an examination of autistic masking (Pearson & Rose, 2021), as well as guidelines for avoiding ableist language (Bottema-Beutel et al., 2021), an intervention aimed at increasing autism acceptance (Jones et al., 2021) and recommendations for promoting inclusive tertiary education (Dwyer et al., 2023). Notably, openly autistic authors are represented in eight of the 10 most-read publications across the field's two leading journals (Mary Ann Liebert, Inc., 2024; SAGE Publications, 2024).

Following these trends, recent years have also seen significant developments initiated by dominant journals in the field. In 2020, *Autism* joined the likes of *The British Medical Journal* in requiring that scholars report on stakeholder engagement (or lack thereof) in their work (Tan et al., 2024b). The journal *Autism in Adulthood*, established in response to calls for greater focus on autistic needs across the lifespan (Roche et al., 2021), aims to represent both scientific and community perspectives, with lay autistic people contributing to both peer review and authorship (c.f. Chapter 19, this volume). When *Autism in Adulthood* recorded its first impact factor in 2023, it outranked all other autism-specific journals, attesting to the shifts evident in the field.

Continuing Challenges

All of this progress is extremely welcome. Yet, early analysis suggests autism research that involves community members in some or all aspects of the research does not always live up to its stated objectives.

As many of this book's authors attest, community engagement can lead to more accessible, acceptable, and rigorous research that better reflects the realities of people's everyday lives and more closely aligns with community priorities (Cornwall & Jewkes, 1995; Lloyd & White, 2011; see also chapters 4, 10, 19, this volume). There are also ethical and moral prerogatives for participatory research approaches (Cascio & Racine, 2018); community engagement can facilitate a degree of self-determination for marginalised groups (including the autistic community), who have often been denied this right. Furthermore, effective participatory research can generate long-term cultural change, through (a) the production of epistemic resources that reflect community knowledge and values, and (b) the bias-reducing effects of relationship-building across disparate stakeholder groups (Cornish et al., 2023).

An umbrella term encompassing diverse methods and methodologies, *participatory research* refers broadly to research produced through collaboration between academic and community stakeholders. Grounded in the understanding that "to address complex problems, scientific expertise alone is not sufficient", participatory research approaches meld academic expertise with community insights to generate knowledge that is "not only of scientific high quality, but also socially robust" (Turnhout et al., 2020, p 16). Commonly conceptualised as a hierarchy or 'ladder' of participation (Arnstein, 1969), the degree of engagement by community stakeholders spans from relatively limited involvement via consultation, through collaborative co-production partnerships, to—at the highest levels—community-led and community-controlled research endeavours (den Houting, 2021; see also Chapters 5, 8, this volume).

Sharing power between academic and non-academic stakeholders is the cornerstone of effective participatory research, and co-production in particular (Israel et al., 2005; Turnhout et al., 2020; see Chapters 5, 6, 7, 8, 12, 14, this volume). Deep-rooted power imbalances are embedded in the systems and structures that govern knowledge production (e.g., Boveda & Annamma, 2023; Ocloo & Matthews, 2016). Equitable research co-production, therefore, necessitates careful navigation of power dynamics and a concerted effort to understand and mitigate inequities between research partners (see Chapters 8, 12, this volume).

Despite heartening increases in community engagement over the past decade, these crucial efforts to disrupt established power structures remain largely absent in autism science. Though Tan et al. (2024a) found 312 published examples of community engagement in autism research, these were drawn from a corpus of more than 40,000 papers. Participatory research, therefore, represents less than 1% of published autism research. Tan et al. (2024a, 2024b) also highlight the relatively limited nature of community engagement across much of this work; community members most commonly acted in consultative roles, with input often limited to research design activities such as the development and/or modification of interview or survey questions (see Chapters 4, 9, 10, 19, 20, 22, this volume). Notably, community engagement was distinctly lacking in basic scientific research. While applied research including lifespan, intervention, and service-focused research was relatively well represented, none of the included studies reported on genetic research.

It is unsurprising, therefore, that many in the autistic and autism communities continue to feel let down by research. Recent findings (Haar et al., 2024) indicate that community members feel the autistic experience continues to be poorly understood and pathologised by non-autistic researchers, and by society more broadly. Autistic people describe feeling an ongoing pressure to conform in the face of barriers posed by a neurotypical society that remains reluctant to accommodate autistic needs (see Chapter 5, this volume). Community members also report feeling ignored by non-autistic researchers, despite having valuable expertise to share—and a keen desire to co-produce research. Haar et al.'s participants call for research "with our voices centred, our experiences validated, and with our control" (p. 15).

Our own work (den Houting et al., 2021, 2022) offers some insight as to why this call remains unanswered. Among the Australian autism research community, we found that academics and community members alike were strongly supportive of community engagement in research, but felt that many factors within the academy posed considerable barriers to meaningful engagement. This concern was shared by autism researchers in the United Kingdom

(Pickard et al, 2022). While these factors, including constraints on funding and time, can impede participatory research efforts, our findings also highlighted the possibility of "bottom up" (Pellicano, 2020) change enacted by individuals (see Chapters 10, 18, this volume). Crucially, few of our academic participants identified navigation of power dynamics as a fundamental element of research co-production over which they could, themselves, effect change. Of those who did acknowledge issues of power, most described firmly entrenched inequities, with academics retaining the bulk of power even in ostensibly co-produced research. These deep inequities were reflected in participants' language, with many academics blurring the distinctions between different kinds of community collaboration and engagement. Concerningly, some academics explicitly expressed resistance and doubt regarding power sharing, perceiving it as impractical and (in some cases) inappropriate (see also Pickard et al., 2020).

We believe that further progress in participatory autism research requires us all to challenge existing power dynamics. Although structural obstacles are real, the academic role remains inherently imbued with considerable social power, especially in research contexts. Contrastingly, marginalised communities—frequently the focus of participatory research—contend with commonplace oppression, injustice, and disempowerment. In autism research, efforts to confront prevailing structures of power are particularly imperative in light of the ways that research has contributed historically to the disempowerment of autistic people.

Next Steps

Throughout this book, epistemic injustice (Fricker, 2007) is an aptly ubiquitous theme. Through neuronormative narratives of deficit and incapacity, the field of autism research has unequivocally characterised autistic people as uncredible epistemic agents, systematically and wrongfully undermining autistic capacity to contribute knowledge (Botha, 2021; Catala et al., 2021). With many in the field now seeking to embrace a more participatory autism science, the enduring impacts of this epistemic injustice have significant implications.

To achieve equitable co-production, stakeholders must work to establish a partnership culture in which a diversity of knowledge and skills are valued (see Chapters 5, 7, this volume). In particular, the cultural intelligence (Ghanbarpour et al., 2020) of community stakeholders should be recognised and understood as a valuable form of expertise. In autism research, though, aspects of the prevailing culture have for too long devalued autistic expertise and disempowered those who hold it. If this prevailing culture is perpetuated within a co-production partnership, community stakeholders' insights may continue to be dismissed, or utilised to serve academic—not community—interests (see Chapters 2, 22, this volume). As a result, project processes and outputs—now legitimised as products of "co-production"—may wrongfully reinforce conventional knowledge and structures of power (Turnhout et al.,

2020). How, then, can we challenge these long-established inequities and foster egalitarian research contexts? What, in other words, are the crucial next steps?

Capitalising on the current enthusiasm for community engagement in autism research begins with capability building. It is essential that academic and community stakeholders alike develop the expertise and skills needed to conduct high-quality participatory research. To achieve equitable research co-production, it is not sufficient to merely implement the practicalities of a participatory research process; a deep commitment to the principles and values underpinning co-production is required. In recent years, a range of resources have been published, providing guidance and insights specific to participatory research conducted with the autistic and autism communities (e.g., den Houting, 2021; Nicolaidis et al., 2019; Pellicano et al., 2020). Several training programmes have been developed (Autism CRC, 2024; The University of Edinburgh, 2017), each aiming to upskill autistic people and/or autism researchers in research co-production. Beyond the autism research field, there exist many useful guides and toolkits designed to inform participatory and co-produced research (e.g., Cornish et al., 2017; Farr et al., 2020; Strnadova et al., 2020).

Autism research is, of course, neither the first nor only field to grapple with academic-community power inequities and to attempt to build the requisite capabilities in response. Valuable lessons can be learned by turning to expertise from fields with long-established traditions of community engagement in research. First Nations research, for example, demonstrates how communities can assert sovereignty over the research conducted within their communities, resulting not only in community-led research, but community-controlled research policies, practices, and methodologies (Drawson et al., 2017). As First Nations communities have enhanced research capacity, many within them have also developed a range of Indigenous research methods and methodologies based in cultural values and ways of knowing, thereby grounding research in community strengths. Autism research can learn from these examples.

Capability building alone will not be enough, however. Establishing a new norm of power sharing requires a concerted effort over time. Effective co-production partnerships are developed across a long-term process of relationship-building (see chapters 5, 6, this volume). These processes will, ideally, be undertaken independently of any specific research project, with the partnership established *before* a research project is planned and maintained *after* the project concludes, according to the mantra developed originally in community organising practice that "relationship *precedes* action".

Establishing and maintaining trusting relationships between disparate stakeholders of this sort, therefore, requires deliberate attention, with long-term partnership processes planned as thoughtfully as the research itself (Jagosh et al., 2015). Equitable partnership processes can be developed through deliberative communication among stakeholders; that is, creating space to discuss the partnership itself (Wallerstein et al., 2019). This may include, for example, collaboratively establishing explicit partnership practices and values such as methods of communication and decision-making; stakeholder roles and responsibilities; and partnership purpose and goals. At the same time, it is important to recognise that these partnerships will not always be smooth or straightforward. Incorporating a diversity of perspectives can, after all, draw implicit tensions and disagreements to the surface (see Chapters 5, 6, this volume). As such, a key element of building sustainable partnerships will involve learning to cope with disagreements and generating respectful exchange in the face of them. In time, it should become clear that tensions within relationships can enable new ideas to emerge and new challenges to prevailing orthodoxies to be taken seriously.

This last theme further reminds us that efforts both to build capability and to establish long-term partnerships must be paired with explicit efforts to grapple with the power differentials described above. Researchers themselves can initiate this process by identifying their own potential roles in the complex power dynamics present in the research process (see Chapters 11, 9, this volume). Just as the chapters in this volume outline, strategies of participatory research are far more likely to be successfully embedded when it becomes commonplace for all stakeholders to engage in difficult conversations about past and present injustices, in addition to future possibilities. Practical tools are already available to enable this. Techniques like "power mapping", once again borrowed from community organising, enable partners to examine the distribution of power within their partnership; reflect on the contextual factors that shape these power structures; and identify strategies to facilitate more equitable redistribution of power (Arleskog et al., 2021; Roper et al., 2018; Wallerstein et al., 2019). Some academic researchers might wish to go further still and use their power to actively advocate against inequitable research systems and structures on behalf of other members of the partnership, thereby siding with the non-academic partners instead of the conventional positions of the academy, or, as Wallerstein puts it, "taking an oppositional stand against their own privilege and toward sharing power with the community" (Wallerstein et al., p. 29S).

Taken together, these three tasks—building capability, sustaining partnerships, and challenging power differentials—can enable each of us to make a positive contribution to the broader shift towards participatory research approaches. And, as this volume demonstrates, that should be to the benefit of everyone.

References

Arleskog, C., Vackerberg, N., & Andersson, A.-C. (2021). Balancing power in co-production: Introducing a reflection model. *Humanities and Social Sciences Communications*, *8*(108). https://doi.org/10.1057/s41599-021-00790-1

Arnstein, S. R. (1969). A ladder of citizen participation. *Journal of the American Institute of Planners, 35*(4), 216–224. https://doi.org/10.1080/01944366908977225

Autism CRC (2024). *Sylvia Rodger Academy Research Program.* https://www.autismcrc.com.au/sylvia-rodger-academy/research-program

Bertilsdotter Rosqvist, H., Botha, M., Hens, K., O'Donoghue, S., Pearson, A., & Stenning, A. (2023). Cutting our own keys: New possibilities of neurodivergent storying in research. *Autism, 27*(5), 1235–1244. https://doi.org/10.1177/13623613221132107

Botha, M. (2021). Academic, activist, or advocate? Angry, entangled, and emerging: A critical reflection on autism knowledge production. *Frontiers in Psychology, 12.* https://doi.org/10.3389/fpsyg.2021.727542

Bottema-Beutel, K. (2023). We must improve the low standards underlying "evidence-based practice." *Autism, 27*(2), 269–274. https://doi.org/10.1177/13623613221146441

Boveda, M., & Annamma, S. A. (2023). Beyond making a statement: An intersectional framing of the power and possibilities of positioning. *Educational Researcher, 52*(5), 306–314. https://doi.org/10.3102/0013189X231167149

Buckle, K. L., Leadbitter, K., Poliakoff, E., & Gowen, E. (2021). "No way out except from external intervention": First-hand accounts of autistic inertia. *Frontiers in Psychology, 12.* https://doi.org/10.3389/fpsyg.2021.631596

Cage, E., & Troxell-Whitman, Z. (2019). Understanding the reasons, contexts and costs of camouflaging for autistic adults. *Journal of Autism and Developmental Disorders, 49*(5), 1899–1911. https://doi.org/10.1007/s10803-018-03878-x

Cascio, M. A., & Racine, E. (2018). Person-oriented research ethics: Integrating relational and everyday ethics in research. *Accountability in Research, 25*, 170–197.

Catala, A., Faucher, L., & Poirier, P. (2021). Autism, epistemic injustice, and epistemic disablement: A relational account of epistemic agency. *Synthese, 199*, 9013–9039. https://doi.org/10.1007/s11229-021-03192-7

Cornish, H., Fransman, J., & Newman, K. (2017). *Rethinking research partnerships: Discussion guide and toolkit.* https://www.christianaid.org.uk/resources/our-work/rethinking-research-partnerships-discussion-guide-and-toolkit

Cornish, F., Breton, N., Moreno-Tabarez, U., et al. (2023). Participatory action research. *Nature Reviews Methods Primers, 3*, 34. https://doi.org/10.1038/s43586-023-00214-1

Cornwall, A., & Kewkes, R. (1995). What is participatory research? *Social Science and Medicine, 41*, 1667–1676. https://doi.org/10.1016/0277-9536(95)00127-S

Crompton, C. J., Ropar, D., Evans-Williams, C. V., Flynn, E. G., & Fletcher-Watson, S. (2020). Autistic peer-to-peer information transfer is highly effective. *Autism, 24*(7), 1704–1712. https://doi.org/10.1177/1362361320919286

den Houting, J., Higgins, J., Isaacs, K., Mahony, J., & Pellicano, E. (2021). 'I'm not just a guinea pig': Academic and community perceptions of participatory autism research. *Autism, 25*(1), 148–163. https://doi.org/10.1177/1362361320951696

den Houting, J., & Pellicano, E. (2019). A portfolio analysis of autism research funding in Australia, 2008–2017. *Journal of Autism and Developmental Disorders, 49*(11), 4400–4408. https://doi.org/10.1007/s10803-019-04155-1

Drawson, A. S., Toombs, E., & Mushquash, C. J. (2017). Indigenous research methods: A systematic review. *The International Indigenous Policy Journal, 8*(2). https://doi.org/10.18584/iipj.2017.8.2.5

Dwyer, P., Mineo, E., Mifsud, K., Lindholm, C., Gurba, A., & Waisman, T. C. (2023). Building neurodiversity-inclusive postsecondary campuses: Recommendations for leaders in higher education. *Autism in Adulthood, 5*(1), 1–14. https://doi.org/10.1089/aut.2021.0042

Farr, M., Davies, R., Davies, P., Bagnall, D., Brangan, E., & Andrews, H. (2020). *A map of resources for co-producing research in health and social care.* https://arc-w.nihr.ac.uk/publications/a-map-of-resources-for-co-producing-research-in-health-and-social-care/

Fricker, M. (2007). *Epistemic injustice: Power and the ethics of knowing.* Oxford University Press. https://doi.org/10.1093/acprof:oso/9780198237907.001.0001

Gatfield, O., Mangan, C., Haar, T., Kinniburgh, A., & Rodger, S. (2016). *2016 Research Priorities Survey Report.* https://www.autismcrc.com.au/sites/default/files/inline-files/Research%20priorities%20survey%20-%20Final%20report.pdf

Ghanbarpour, S., Noguez Mercado, A. P., & Palatai, A. (2020). A language justice framework for culturally responsive and equitable evaluation. *New Directions for Evaluation, 166*, 37–47.

Haar, T., Brownlow, C., Hall, G., Heyworth, M., Lawson, W., Poulsen, R., Reinisch, T., & Pellicano, E. (2024). "We have so much to offer": Community members' perspectives on autism research. *Autism*, in press.

Higgins, J. M., Arnold, S. R. C., Weise, J., Pellicano, E., & Trollor, J. N. (2021). Defining autistic burnout through experts by lived experience: Grounded Delphi method investigating #AutisticBurnout. *Autism, 25*(8), 2356–2369. https://doi.org/10.1177/13623613211019858

den Houting, J. (2021). *Participatory and inclusive autism research practice guides.* Autism CRC.https://www.autismcrc.com.au/best-practice/sites/default/files/resources/Participatory_and_Inclusive_Autism_Research_Practice_Guides.pdf

den Houting, J., Higgins, J., Isaacs, K., Mahony, J., & Pellicano, E. (2022). From ivory tower to inclusion: Stakeholders' experiences of community engagement in Australian autism research. *Frontiers in Psychology, 13.* https://doi.org/10.3389/fpsyg.2022.876990

Hull, L., Petrides, K. V., Allison, C., Smith, P., Baron-Cohen, S., Lai, M.-C., & Mandy, W. (2017). "Putting on my best normal": Social camouflaging in adults with autism spectrum conditions. *Journal of Autism and Developmental Disorders, 47*(8), 2519–2534. https://doi.org/10.1007/s10803-017-3166-5

International Society for Autism Research (INSAR). *Autistic researchers committee (ARC).* https://www.autism-insar.org/page/InsarARC

Jagosh, J., Bush, P. L., Salsberg, J., et al. (2015). A realist evaluation of community-based participatory research: Partnership synergy, trust building and related ripple effects. *BMC Public Health, 15*, 725. https://doi.org/10.1186/s12889-015-1949-1

Jivraj, J., Sacrey, L. A., Newton, A., Nicholas, D., & Zwaigenbaum, L. (2014). Assessing the influence of researcher-partner involvement on the process and outcomes of participatory research in autism spectrum disorder and neurodevelopmental disorders: A scoping review. *Autism, 18*(7), 782–793. https://doi.org/10.1177/1362361314539858

Jones, D. R., DeBrabander, K. M., & Sasson, N. J. (2021). Effects of autism acceptance training on explicit and implicit biases toward autism. *Autism, 25*(5), 1246–1261. https://doi.org/10.1177/1362361320984896

Jones, S. C. (2021). Let's talk about autistic autism researchers. *Autism in Adulthood, 3*(3), 206–208. https://doi.org/10.1089/aut.2021.29012.scj

Mary Ann Liebert, Inc. (2024). *Autism in Adulthood: Most read articles.* https://www.liebertpub.com/action/showMostReadArticles?journalCode=aut

Lloyd, K., & White, J. (2011). Democratizing clinical research. *Nature, 474*(7351), 277–278. https://doi.org/10.1038/474277a

Milton, D. E. M. (2012). On the ontological status of autism: The 'double empathy problem.' *Disability and Society, 27*(6), 883–887. https://doi.org/10.1080/09687599.2012.710008

Milton, D., Gurbuz, E., & López, B. (2022). The 'double empathy problem': Ten years on. *Autism, 26*(8), 1901–1903. https://doi.org/10.1177/13623613221129123

Nicolaidis, C., Raymaker, D., McDonald, K., Dern, S., Ashkenazy, E., Boisclair, C., Robertson, S., & Baggs, A. (2011). Collaboration strategies in nontraditional community-based participatory research partnerships: Lessons from an academic-community partnership with autistic self-advocates. *Progress in Community Health Partnerships: Research, Education, and Action, 5*(2), 143–150. https://doi.org/10.1353/cpr.2011.0022

Nicolaidis, C., Raymaker, D., McDonald, K., Dern, S., Boisclair, W. C., Ashkenazy, E., & Baggs, A. (2013). Comparison of healthcare experiences in autistic and non-autistic adults: A cross-sectional online survey facilitated by an academic-community partnership. *Journal of General Internal Medicine, 28*(6), 761–769. https://doi.org/10.1007/s11606-012-2262-7

Nuwer, R. (2020). *Meet the autistic scientists redefining autism research.* Spectrum News. https://www.spectrumnews.org/features/deep-dive/meet-the-autistic-scientists-redefining-autism-research/

Ocloo, J., & Matthews, R. (2016). From tokenism to empowerment: Progressing patient and public involvement in healthcare improvement. *BMJ Quality and Safety, 25*(8), 626–632. https://doi.org/10.1136/bmjqs-2015-004839

Office of Autism Research Coordination. (2019). *2016 IACC Autism Spectrum Disorder Research Portfolio Analysis Report.* https://iacc.hhs.gov/portfolio-analysis/2016/index.shtml

Pearson, A., & Rose, K. (2021). A conceptual analysis of autistic masking: Understanding the narrative of stigma and the illusion of choice. *Autism in Adulthood, 3*(1), 52–60. https://doi.org/10.1089/aut.2020.0043

Pellicano, E. (2020). Commentary: Broadening the research remit of participatory methods in autism science—A commentary on Happé and Frith (2020). *Journal of Child Psychology and Psychiatry, 61*(3), 233–235. https://doi.org/10.1111/jcpp.13212

Pickard, H., Pellicano, E., den Houting, J., & Crane, L. (2022). Participatory autism research: Early career and established researchers' views and experiences. *Autism, 26*(1), 75-87. https://doi.org/10.1177/13623613211019594

Poulsen, R., Brownlow, C., Lawson, W., & Pellicano, E. (2022). Meaningful research for autistic people? Ask autistics! *Autism, 26*(1), 3–5. https://doi.org/10.1177/13623613211064421

Rapaport, H., Clapham, H., Adams, J., Lawson, W., Porayska-Pomsta, K., & Pellicano, E. (2023). 'I live in extremes': A qualitative investigation of Autistic adults' experiences of inertial rest and motion. *Autism, 28*(5), 1305-1315. https://doi.org/10.1177/13623613231198916

Raymaker, D. M., Teo, A. R., Steckler, N. A., Lentz, B., Scharer, M., Delos Santos, A., Kapp, S. K., Hunter, M., Joyce, A., & Nicolaidis, C. (2020). "Having all of your internal resources exhausted beyond measure and being left with no clean-up crew": Defining autistic burnout. *Autism in Adulthood, 2*(2), 132–143. https://doi.org/10.1089/aut.2019.0079

Robertson, S. M. (2010). Neurodiversity, quality of life, and autistic adults: Shifting research and professional focuses onto real-life challenges. *Disability Studies Quarterly, 30*(1). https://doi.org/10.18061/dsq.v30i1.1069

Roche, L., Adams, D., & Clark, M. (2021). Research priorities of the autism community: A systematic review of key stakeholder perspectives. *Autism, 25*(2), 336–348. https://doi.org/10.1177/1362361320967790

Roper, C., Grey, F., & Cadogan, E. (2018). *Co-production: Putting principles into practice in mental health contexts.* https://healthsciences.unimelb.edu.au/departments/nursing/about-us/centre-for-psychiatric-nursing/resources

SAGE Publications. (2024). *Autism: Most read.* https://journals.sagepub.com/home/aut

Sasson, N. J., Faso, D. J., Nugent, J., Lovell, S., Kennedy, D. P., & Grossman, R. B. (2017). Neurotypical peers are less willing to interact with those with autism based on thin slice judgments. *Scientific Reports, 7*, 40700. https://doi.org/10.1038/srep40700

Sheppard, E., Pillai, D., Wong, G. T., Ropar, D., & Mitchell, P. (2016). How easy is it to read the minds of people with autism spectrum disorder? *Journal of Autism and Developmental Disorders, 46*(4), 1247–1254. https://doi.org/10.1007/s10803-015-2662-8

Strnadova, I., Dowse, L., & Watfern, C. (2020). *Doing research inclusively: Guidelines for co-producing research with people with disability.* https://www.disabilityinnovation.unsw.edu.au/inclusive-research/guidelines

Tan, D. W., Crane, L., Haar, T., Heyworth, M., Poulsen, R., & Pellicano, E. (2024b). Reporting community involvement in autism research: Findings from the journal, *Autism.* OSF Preprints. https://doi.org/10.31219/osf.io/g89w4

Tan, D. W., den Houting, J., Banerjee, M., D'Aloya, L. P., Jayawardene, L., Rabuka, M., Winspear-Schillings, H., Haar, T., & Pellicano, E. (2024a). A comprehensive systematic review of participatory autism research. Manuscript in preparation.

The University of Edinburgh (2017). *Learning about research: A course for autistic people.* https://dart.ed.ac.uk/research/learning-about-research/

Turnhout, E., Metze, T., Wyborn, C., Klenk, N., & Louder, E. (2020). The politics of co-production: Participation, power, and transformation. *Current Opinion in Environmental Sustainability, 42*, 15–21. https://doi.org/10.1016/j.cosust.2019.11.009

Wallerstein, N., Muhammad, M., Sanchez-Youngman, S., Rodriguez Espinosa, P., Avila, M., Baker, E. A., Barnett, S., Belone, L., Golub, M., Lucero, J., Mahdi, I., Noyes, E., Nguyen, T., Roubideaux, Y., Sigo, R., & Duran, B. (2019). Power dynamics in community-based participatory research: A multiple-case study analysis of partnering contexts, histories, and practices. *Health Education and Behavior, 46*(1_suppl), 19S–32S. https://doi.org/10.1177/1090198119852998

CHAPTER 25

Neurodivergent Futures

Dora M. Raymaker and Christina Nicolaidis

The Weight of History and the Push of the Present

In 2010, we (Dora and Christina, this chapter's co-authors) took part in a critical autism studies workshop at the University of Ottawa. The workshop brought together academics who were engaged in critical theory and had a personal interest in autism, but whose scholarship largely centered in fields other than autism research (e.g., classics, political science). As, essentially, a group of outsiders to autism research, the workshop created a unique lens on how the epistemologies of autistic people might interact with scholarship without the theoretical and philosophical assumptions of autism research at that time, both in its "normal science" (i.e., status quo pathology framing with an eradication and/or normalization goal) and in the rising push back to that science by the marginalized voices of the autistic and disability rights communities. Despite our activist connections to the neurodiversity movement, we, too, felt like outsiders to the world of autism research, with our scholarship resting in action research and public health, and our research focusing on improving healthcare access and quality for autistic adults.

At the time of the workshop, four years after co-founding the Academic Autism Spectrum Partnership in Research and Education (AASPIRE), and a year into our first externally funded autism-related study, we were still struggling to convince the scientific community that it was possible to do

D. M. Raymaker (✉) · C. Nicolaidis
School of Social Work, Portland State University, Portland, Oregon, USA
e-mail: draymake@pdx.edu

H. Bertilsdotter Rosqvist and D. Jackson-Perry (eds.), *The Palgrave Handbook of Research Methods and Ethics in Neurodiversity Studies*,
https://doi.org/10.1007/978-3-031-66127-3_25

community-based participatory research with autistic people. We were even struggling to convince people, who largely saw autism as a childhood affliction, that autistic adults existed, let alone had developed autistic culture and community. The paper we presented (ultimately published as Chapter 7 "Participatory Research with Autistic Communities: Shifting the System" in *Worlds of Autism* (Raymaker, 2013)) focused on our theory of how participatory research can shift paradigms by centering and empowering neurodivergent epistemologies in knowledge production. Would those dynamics play out in practice though? We thought they might, based on how participatory research generated similar paradigmatic shifts for other oppressed communities. But at the time it was a grounded vision—an untested, though plausible, theory.

There has been a tremendous infusion of neurodiversity frameworks and paradigms into research since that 2010 conference asked attendees whether there even existed a critical autism studies. An analysis of that shift—including the impact of participatory and inclusive methods as we theorized—could easily occupy its own textbook. A loose "follow up" chapter to "Shifting the System" exists in "Shifting the System: AASPIRE and the Loom of Science and Activism" in the book *Autistic Activism and the Neurodiversity Movement* (Raymaker, 2020)—a book that likely would not exist—at least not from an academic publisher—had the shift not occurred. The chapters in this book also present ample evidence of the shift. For example, discourse about neurodivergent scientists requires that the world acknowledges the existence of autistic adults, and discourse about participatory research requires the existence of more than just one participatory research group. We are continuously delighted and amazed by how quickly paradigms in academics have shifted away from pathology framing and toward neurodiversity framing in the past near-twenty years.

However, the weight of historical injustices and the push of present oppressions are inevitably a part of what we carry into the future. Our present progress in participatory methods, inclusive research, cultural recognition, and connecting community priorities to researchers and research agendas does not erase decades of marginalization and outright harm and abuse, especially for those of us who came of age in a strongly eugenics-oriented and medical model culture. The present pushes us toward equity and justice while the past encumbers us not just with the inertia of the status quo within science and society, but with its own individual and collective trauma. These weights are as evident as progress in the substance of this book (e.g., in discourses around epistemic in/justice or discussion of tokenism in research, see Chapters 5, 9, 10, this volume).

But futures aren't forged by the weight of the past and the push of the present alone. Futures have their own pull, and no future, no matter how imminent, is fully in our control. We can, however, consciously work toward the kinds of futures we would like through our dreams and visions, and how we respond with action to opportunities and threats along the way.

Approach

The futures field has a saying that "the future is plural"; indeed we have been careful to pluralize the word throughout. Until something happens, it is not fixed; although the more the distance closes between present and future the smaller the number of possible outcomes.

Futures thinking asks six basic questions: (1) What will the future be like? (2) What future do we fear? (3) What are we assuming about the future ? (4) What are some alternatives to the future we fear? (5) What futures do we prefer? And (6) How might we get to those preferred futures? (Inayatullah, 2008). These questions entwine and influence each other. As such we have approached our own futures thinking holistically.

Frameworks

In his seminal 1962 book *The Structure of Scientific Revolutions*, Thomas Kuhn observed that science seems to go through patterns of stability and disruption. The status quo of "normal science"[1]—which defines the underlying paradigms, theories, and norms for an area of inquiry—is a usually lengthy period of stable growth. Then, something revolutionary perturbs the science—a new discovery, the nullification of a key assumption—and it changes everything. After a period of instability from the disruption, the science again settles and what was once revolutionary becomes the new "normal science" (Kuhn, 2012). An example of this from the natural sciences is when the discovery of radioactivity changed the science of geological dating.

This chapter explores the neurodiversity paradigm as a disruption to the normal science of neurodivergence. We have structured it using a futures triangle framework to describe the nature of the disruption, its impact on the present state of the science, and the types of futures it could pull us into as it continues to develop. Specifically, the futures triangle considers how the past weighs on any forward motion, the present provides us with a view of trends and motivators—an indication of where we may be pushing toward—and the future pulls us forward into its potential both terrible and hopeful. A holistic examination of these forces results in a vision of plausible futures at the center (Inayatullah, 2008).

Aims

Initially, we started this project with the aim of a broad exploration of neurodivergent futures. What might they look like? We intended to use a futures cone framework (Voros, 2017) to help us imagine a range of futures from likely to preposterous, and then identify pathways to desirable futures and wrong turns

[1] When "normal science" appears in quotes in this article, we intend this special meaning as defined by Kuhn.

to undesirable ones. The conversation, however, quickly turned both existential and dark. For example, did these futures include biological malleability, such as pharmaceuticals to induce any arbitrary neurology? And, if so, would that enable people to freely experience and celebrate different ways of being in the joyous sense of Walker's neuroqueer (Walker, 2021), or would it erase neurodivergent identities or even neurodivergence itself? How much do our neurological experiences contribute to our personalities? And how likely are such futures to lead to horrific exploitation?

At the end of an increasingly disturbing (and wildly associative—this is both a power of our combined neurodivergence and, at times, a massive distraction) two hours, we realized we had only imagined a small number of increasingly catastrophic futures in which evil masterminds nefariously manipulated people's neurodivergent traits to create dystopias. We needed more structure.

However, finding more structure for that original aim proved elusive. Neurodivergence is a fundamental way of experiencing embodied existence; it is an epistemology and thus applies to everything that people can experience. Aiming to describe "neurodivergent futures" is like aiming to describe "futures." Our aim was vastly underspecified.

So, we narrowed our aim to futures in neurodiversity research (i.e., research about topics related to neurodiversity or neurodivergence). Specifically, (1) If we could imagine a best possible future for neurodiversity research, a new "normal science" that operates from within a neurodiversity paradigm, what would that science look like? (2) What could facilitate or jeopardize that vision? How do we seize current opportunities to get there and what should we, as a field, guard against as we become "normal science?".

Positionality and Philosophy

Dora (Autistic) and Christina (ADHD) are the founding co-directors of AASPIRE and interdisciplinary/transdisciplinary scientists and educators who have worked together since 2006. Much of our collaborative and individual work explicitly aims to infuse neurodiversity principles into the "normal science" of autism and public health research to generate a paradigm shift both in science and in society more broadly. Dora is a systems scientist and disability and queer rights activist with theoretical, applied, and grassroots interests in the dynamics of social change. Christina is an internal medicine physician and community-based participatory researcher who has partnered with multiple marginalized communities, including those defined by race, ethnicity, disability, chronic illness, or trauma history, to conduct action research to improve health or well-being. We are epistemological insiders and disciplinary outsiders with respect to autism and neurodiversity research.

As interdisciplinary and transdisciplinary scientists whose flavor of neurodivergence favors nonlinear thinking and a resistance of absolutes, we are philosophical and methodological omnivores. We strongly advocate for appropriately matching paradigm and inquiry over the elevation of any one philosophy of science or scientific methodology. Our philosophical positionality for this paper uses aspects of constructivism with a critical realist ontology (Rees et al., 2020), critical theory (Paradis et al., 2020), and transformative paradigms (Mertens, 2007) as we engage with qualitative futures thinking methods. Specifically, we acknowledge the co-creation of reality through the legitimacy of multiple epistemologies interacting with the generative structures of reality, in conversation with power, and with a goal and praxis of continuous movement toward greater social justice.

Methods

As our research questions are largely descriptive (what would desirable futures for neurodiversity research look like and what could get us there?) with intent to develop ideas for strategy, we selected two "Tools for describing what the future might be like" from *The Futures Toolkit: Tools for futures thinking and foresight across UK Government* (Government Office for Science, 2017).

The first tool is a two-part visioning exercise with questions designed to stimulate futures thinking. Part 1 asks participants to imagine that they successfully created a desired future, and to talk about their accomplishments. Questions include, "What have we achieved?" "Who are our stakeholders? How have they benefited from what we've done?" and "What are the challenges we face now?" Part 2 asks participants to form consensus about the current reality relative to the vision. Questions include, "How close are we to our vision?" "Which changes are in our control? Which aren't?" and "Who will be the winners and losers in this change? How do we bring people with us?".

The second tool is a SWOT Analysis to identify relevant Strengths, Weaknesses, Opportunities, and Threats moving from the present into desired futures. A SWOT Analysis asks participants to fill a grid with one field for each of the four items. Strengths and weaknesses are internal factors to realization; opportunities and threats are external factors to realization. While SWOT Analysis is a tool commonly used in business and policy planning, in a futures thinking context it can be helpful in strategic planning about the future. For the purposes of our SWOT Analysis, we defined internal forces as the people who are actively working toward our vision of desirable futures for neurodiversity science. We defined external forces as people who are working in neurodiversity research or consuming neurodiversity research broadly including other researchers, clinicians, policymakers, and communities.

We held three two-hour meetings to complete our futures exercises. The first meeting centered our futures cone exercise with the broader aim of "neurodivergent futures," which led us to narrow our scope (see Aims above). The second meeting, re-focused on the futures of neurodiversity research, centered the visioning exercise. In the third meeting, we conducted our SWOT Analysis.

At the start of the visioning meeting, we talked about where we came from and where we are. That discussion comprises the narrative in "The Weight of History and the Push of the Present" section above. We then used that conversation as input to our visioning exercise. We imagined the first part of the visioning as having succeeded at bringing the neurodiversity-framed neurodiversity research we've been working toward since 2006 from its current state of disruption into equilibrium as the new "normal science." We then used that imagining of the new normal science as input the second part of the visioning exercise. We took notes on a white board and recorded our conversation. After the meeting, Dora organized the visioning data into coherent concepts and then together we further organized the concepts into (1) a coherent and explicit futures vision for neurodiversity research and (2) an initial set of recommendations for how to get there.

In our third meeting, we used the futures vision as input to the SWOT Analysis and identified items for each quadrant of the SWOT table. After the meeting, we organized the bullets in our SWOT table into themes and synthesized the individual items into a short narratives for each theme. Lastly, we synthesized the narratives and the recommendations from the visioning into a final set of actionable recommendations that include both how we can leverage our existing strengths and opportunities, and ways we need to guard against internal and external risks to futures that, while well-intended, could undermine a long-term movement toward justice.

These findings gave us the third point of the triangle, the pull of the future.

The Pull of the Future

In our futures vision for neurodiversity research...

Neurodivergence is normalized. Political activism cools as neurodivergence normalizes. As discrimination and stigma fade, the full range of neurodivergent experience emerges—both the affirming and the destructive, the idiosyncratic and the stereotypical. As the power/knowledge dynamic (Foucault, 1980) begins to reinforce neurodivergence as a legitimate way of being, people stop dismissing neurodivergent epistemologies in neurodiversity research as either irreparably biased or untouchably sacred, reducing their risk of promoting of bad or harmful science.

Complexity is celebrated. With the normalization of neurodivergence and its subsequent de-politicization, neurodivergent experience no longer needs to be over-simplified to counter harmful social narratives or to gain legitimacy. Both researchers and broader society appreciate neurodivergent people with the same range of nuance as they do neurotypical people; outgroup homogeneity bias dissolves.

Intersectionality is assumed. Research and intervention design accounts for and incorporates the synergetic interactions between neurodivergent and other identities (race, gender, religion, etc.). Scientific inquiry and praxis include intersectional epistemologies within the neurodivergent experience.

Ableism is socially and scientifically unacceptable. Therapy supports function and quality of life, not identity change, normalization, or masking. Research domains once closed due to a high risk of eugenics open to new opportunities; researchers make advances in precision medicine. The mental and physical well-being of neurodivergent populations increases. Health disparities related to neurodivergence are largely eliminated.

Neurodivergent voice is always included. This inclusion is authentic, equitable, considered, and deliberate. Inclusion happens anywhere knowledge production happens, including on research teams, in education, in research policy, on institutional review boards, in academic journals and other dissemination venues, and among research participants. The systems that scaffold knowledge production shift their processes and environments to support neurodivergent presence. Strategies exist to include people with all types and sub-types of neurodivergence, all functional needs, and all types of capabilities.

Neurodivergent cultures are centered, respected, and integrated. Neurodivergent cultures shape how systems and environments shift to include neurodivergent populations in neurodiversity research. The workday includes time for silliness, joy, and stimming. Projects have longer timelines to run on neurodivergent time. Multiple modalities of communication enrich scientific collaborations. A culture of accessibility supports everyone without anyone needing to expend emotional or physical labor.

Epistemology is multiple, simultaneous, and fluid. Science celebrates all ways of knowing. Like neurodivergent minds, researchers may move fluidly between different epistemologies, acknowledging their simultaneous, nonlinear, and associated truths through many philosophical lenses—or allowing the multiplicity to exist while focusing on a single point within the larger universe through a single philosophical lens.

In reconciliation of that futures vision with the present…

Positioning the future. Understanding how close we are to the future gives us an idea of the distance we need to cover. There is most certainly a tipping point between the current more neurotypical-led "normal science" and the revolutionary more neurodiversity-fueled science. But where are we relative to that tipping point? And after we tip, will we equilibrate with neurodiversity as the new “normal science” (in Kuhn’s sense of it becoming the accepted status quo for the field) or continue to oscillate between today’s normal science and neurodiversity until something new disrupts the dynamic? Without a rigorous study of the dynamics, it’s hard to tell where we are now, how close to any tipping point into a neurodivergent normal science. However, we are sure that we are much closer than we were when we first met in 2006, and that this present moment is fertile with both opportunities and dangers. Could we tip too slowly and be crushed by the status quo? Could we tip too quickly and make mistakes that cost us trust with the community?

Controlling the future. Humans cannot control highly complex systems like knowledge production. Learning to relax the urge to control is an early and difficult lesson for anyone interested in the science of social change. However, such systems can be, as Donella Meadows says, "danced with" (Meadows, 2002). Knowledge production is a system designed to shift paradigms. The process is typically slow and incremental, but “advancing the state of the science” can be a fast leap too. In our opinion, what we are witnessing right now in neurodiversity research is one of Kuhn’s scientific revolutions. In 2010’s critical autism studies workshop, we theorized that by infusing autistic voice into the machine of knowledge production we could shift the system toward neurodiversity paradigms. In 2023, we seen substantial evidence that shift in practice. Science is how we’ve been dancing our vision future-ward, and we still strongly feel it has powerful leverage.

Leading the future. One of the ten Disability Justice Framework principles is “Leadership of the most impacted” (Berne, 2018). This means that those who have been most marginalized, most left out, should be the ones to lead since they have the strongest epistemological grounding in the issues, having lived them. In order to reach our vision of the future, it is essential that those whose voices have been most silenced, most left out, be at the center of change. This, too, has been happening steadily over the past few decades, helping to push us (near? to? over?) around the tipping point. We still have further to go in elevating the most marginalized of neurodivergent voices.

In assessing the facilitators and risks to the futures vision given our current positioning…

The results of our SWOT Analysis organized under six themes. The first five ran across all quadrants: (1) Accumulation of Knowledge; (2) Workforce Development; (3) Philosophical Considerations; (4) Infrastructure and Resources; (5) Neurodivergence and Inclusion. The sixth theme (6) Social Pressures was only present in the external Opportunities and Threats, as it involved external contextual forces.

Accumulation of Knowledge. While participatory autism researchers have nearly 20 years of experience and insight working toward the vision with rigorous and impactful research, we can undermine that strength by incompletely or inauthentically implementing what we know, or by reinventing participatory approaches instead of building on existing knowledge. Externally, new domains in related science such as implementation science, as well as growing interest and advances in participatory research and neurodiversity as a paradigm (e.g., (Sonuga-Barke, 2023)), present opportunities. However, 20 years isn’t that long, and we can all be threatened by the sheer force of inertia.

Workforce Development. The growing number of neurodivergent researchers is a tremendous strength, as well as the growing credibility and capacity of researchers of all neurotypes working toward the vision. However, the largest number of neurodivergent researchers is still in the early part of their training and careers, and visionary leaders and senior researchers risk exhaustion and burnout, particularly as a very small number supporting a large number of trainees. The opportunities presented by the growing visibility and respect for researchers who hold the vision externally are threatened by smallness of the field and even smaller numbers working within the field. Additionally, dominant clinical theories of neurodivergent conditions rest on a foundation of ableism which permeates education, training, policy, and professional development at all levels.

Philosophical Considerations. We have strength in the interdisciplinary and transdisciplinary nature of our work and our willingness to tackle complex problems; this gives us access to multiple philosophies of science, including those with subjective, multiple, or interactive epistemologies—and the wide range of qualitative and mixed-methods methodologies available to them. An increasing acceptance of such philosophies and methodologies is also a part of broader opportunities. We need to take care, however, not to develop dogmatic adherence to narrow conceptualizations of neurodiversity or any one paradigm or methodology, or to reject any simply because it was used by someone to marginalize neurodivergence in the past. We also need to be aware of weaknesses in our ability to balance creation and destruction, or to accept inevitable imperfection—particularly in the early, experimental stages of building a vision when perfection is impossible, if only because we are actively still learning. If our internal weakness is going too far in rejecting the philosophies of "normal science," the external threat is in the inertia of normal science itself, especially with the power/knowledge construct resisting change at the center.

Infrastructure and Resources. We are seeing increasing research infrastructure to support our vision (e.g., *Autism in Adulthood* journal, INSAR Autistic Researchers Committee, Facebook group ARRA, AIR-P's neurodiversity-imagined network on health and well-being, textbooks like this), as well as maturing advocacy infrastructure (e.g., ASAN, AWNBN, AANE), and maturing community-campus collaborations (e.g., AASPIRE, PARC, Autism

CRC).[2] As we grow however, we need to protect against the potential for infighting over limited resources and/or innovative ideas instead of generating a culture of sharing. Externally, there are some opportunities in an increased interest in collaborative research by funders, including U.S. federal agencies such as NIDILRR, NIH, and PCORI.[3] However, this remains an area of substantial external threat in that, first, current educational and research systems are not structured to support participatory research, or to support neurodivergent researchers. Second, there remains very limited funding for the types of research the vision most prioritizes.

Neurodivergence and Inclusion. Our neurodivergence is, in itself, one of our greatest strengths. Our neurodivergent creativity, and our culture of creativity, accessibility, playfulness, and justice, is a powerful force for achieving its own infusion into neurodiversity research. We are also strengthened by the growing expansion of the who is included in the neurodiversity movement and consideration for people with greater support needs and intersections of marginalization. Despite that, we currently have uneven development of the vision across those same needs and intersections, and they are insufficiently included. We cannot fully realize the vision until we address these weaknesses. Externally, advances in disability rights, disability justice, and other social justice movements, as well as increased visibility, present opportunity. However, misunderstandings and myths of what neurodiversity means, both in ways that justify opposing it and as a superficial "credentialing" of research that does not use a neurodiversity approach, present significant threats. Additionally, the compulsion to simplify and classify people into binaries or mutually exclusive categories (e.g., high/low functioning, profound autism, visual/verbal thinkers, by diagnostic label) is a substantial threat to inclusion and equity.

Social Pressures. Our current external context provides a number of opportunities. The neurodivergent children diagnosed in the 1990s and early 2000s are now becoming adults, increasing pressure on research to find solutions for the disparities they are experiencing. The neurodiversity paradigm has also permeated the public consciousness through social and other media, particularly for younger generations. However, there are just as many threats. In

[2] Autism in Adulthood https://home.liebertpub.com/publications/autism-in-adulthood/646; INSAR – International Society of Autism Research https://www.autism-insar.org/page/InsarARC; ARRA - Autistic Researchers Researching Autism FaceBook group; AIR-P – Autism Intervention Research Network on Physical Health https://airpnetwork.ucla.edu; ASAN – Autistic Self Advocacy Network https://autisticadvocacy.org; AWN – Autistic Women and Non-Binary Network https://awnnetwork.org; AANE – Association for Autism and Neurodiversity https://aane.org; AASPIRE – Academic Autism Spectrum Partnership in Research and Education https://aaspire.org; PARC – Participatory Autism Research Collaborative https://participatoryautismresearch.wordpress.com; CRC - Cooperative Research Centre https://www.autismcrc.com.au

[3] NIDILRR – National Institute on Disability, Independent Living, and Rehabilitation Research; NIH – National Institutes of Health; PCORI – Patient Centered Outcomes Research Institute.

particular, there has been an alarming increase in ableist policy and political will, as well as an increasing push back to disability rights. We are living in a highly disruptive time globally, which makes its own opportunities and threats, and our vision may have to quickly adapt.

In taking action to reach the futures we would like to have...

Given our reconciliation of the futures vision with the present and our SWOT Analysis, we developed the following recommendations for reaching the vision we would like to have.

- Continue growing the participation, number, and leadership capacity of neurodivergent scientists and activists in existing systems of research and knowledge production.
- Become conversant with and build on the literature for conducting participatory neurodiversity research that has grown since the 2000s, as well as the substantial literature on participatory and emancipatory science since the 1940s.
- Leverage the creative and cultural strengths of neurodivergent scientists and activists to envision new systems for conducting neurodiversity research and producing knowledge. Locate resources to scaffold and implement them, especially by supporting infrastructures created and maintained by the neurodiversity community.
- Support scholarship by neurodivergent and other allied scientists that challenges "normal science," both in theory and in application.
- Support policy initiatives that allocate resources toward neurodiversity research, and advocate for neurodiversity framing in broader settings, including education, clinical practice, popular media, community living, employment, and politics.
- Fearlessly explore interdisciplinary and transdisciplinary spaces, collaborations, and approaches to address the complexity of neurodiversity research.
- Be patient, deliberate, and open to change and adaptation in confronting the inertia of the status quo; strive for "best for now" over "perfect forever."
- Develop strategies to consistently and authentically include people with the most marginalized functional or support needs, or who exist at the most marginalized intersections of identity, in all aspects of neurodiversity research.
- Center neurodivergent epistemologies in all neurodiversity research.
- Recognize that understanding the lived experience of those who oppose neurodiversity paradigms is essential to reconciling past, present, and future and develop strategies for including contradictory opinions and unpopular epistemologies in neurodiversity research.

- Resist the urge to have "winners and losers" in the paradigm shift. Take care not to demonize specific philosophies of science and their methodologies (e.g., positivism and quantitative methods). Take care not to marginalize individuals within the neurodivergent population who do not resonate with the neurodiversity paradigm. Neurodiversity accepts and celebrates all ways of knowing; this is how we bring everyone with us into the kinder, more epistemologically curious, and just future we would love to have.

The Plausible Desired Futures at the Center

In the center of the futures triangle, plausible futures emerge suspended on the tension created by the weight of the past, the push of the present, and the pull of the future. In the specific futures triangle we have envisioned, we focused on futures for neurodiversity research that we felt were desirable to achieve.

The center of our triangle describes an authentically, fully inclusive science, led by the people who best understand neurodivergence through lived experience and balanced by multiple perspectives. It is a fluid and associative science, embracing many ways of knowing simultaneously, or pausing to deeply appreciate the facets of just one without having to extinguish the others. It emerges from neurodivergent culture and is a part of neurodivergent culture. It provides insight into the human experience and strategies for problem solving that would be impossible by neurotypical thinking alone. It legitimizes all neurodivergent epistemologies and disrupts the systems of oppression generated by their marginalization. The world shifts toward justice and greater well-being.

Are these futures plausible? Given the current state of the world and all that the neurodiversity community has achieved in the past two decades, we feel they are, even if they may take time. The current state of the world is, however, fragile, and the pull of the present could easily shift in unpredictable ways to emerge a different center. We hope that whatever centers emerge they include parallel play, gloriously geeky monologues, and all the best stim toys.

References

Berne, P., Levins Morales, A., Langstaff, D., & Invalid, S. (2018). Ten principles of disability justice. *Women's Studies Quarterly, 46*(1–2), 227–230.

Foucault, M. (1980). *Power/knowledge: Selected interviews and other writings, 1972–1977*. Vintage. http://www.amazon.com/Power-Knowledge-Selected-Interviews-1972-1977/dp/039473954X

Government Office for Science. (2017). The futures toolkit: Tools for futures thinking and foresight across UK government. In (1 ed.). https://www.gov.uk/government/publications/futures-toolkit-for-policy-makers-and-analysts: Government Office for Science.

Inayatullah, S. (2008). Six pillars: Futures thinking for transforming. *Foresight*, *10*(1), 4–21. https://doi.org/10.1108/14636680810855991

Kuhn, T. S. (2012). *The structure of scientific revolutions* (Fourth Edition ed.). University of Chicago Press.

Meadows, D. (2002). Dancing with systems. *The Systems Thinker*, *13*(2).

Mertens, D. M. (2007). Transformative paradigm: Mixed methods and social justice. *Journal of Mixed Methods Research, 1*(3), 212–225. https://doi.org/10.1177/1558689807302811

Paradis, E., Nimmon, L., Wondimagegn, D., & Whitehead, C. R. (2020). Critical theory: Broadening our thinking to explore the structural factors at play in health professions education. *Academic Medicine, 95*(6), 842–845. https://doi.org/10.1097/ACM.0000000000003108

Raymaker, D. M. (2020). Shifting the system: AASPIRE and the loom of science and activism. In S. Kapp (Ed.), *Autistic activism and the neurodiversity movement* (pp. 133–144). Palgrave Macmillan.

Raymaker, D. M., & Nicolaidis, C. (2013). Participatory research with autistic communities: Shifting the system. In J. Davidson, & orsini, m. (ed.), *worlds of autism.* University of Minnesota Press.

Rees, C. E., Crampton, P. E. S., & Monrouxe, L. V. (2020). Re-visioning academic medicine through a constructionist lens. *Academic Medicine, 95*(6), 846–850. https://doi.org/10.1097/ACM.0000000000003109

Sonuga-Barke, E. J. S. (2023). Paradigm 'flipping' to reinvigorate translational science: Outlining a neurodevelopmental science framework from a 'neurodiversity' perspective. *Journal of Child Psychology and Psychiatry, 64*(10), 1405–1408. https://doi.org/10.1111/jcpp.13886

Voros, J. (2017, 29 November). The futures cone, use and history. *The Voroscope: Exploring the Totality of Human Knowledge.*

Walker, N. (2021, 29 November). Neuroqueer: An introduction. *Neurocosmopolitanism: Dr. Nick Walker's notes on neurodiversity, autism, and self-liberation.* https://neurocosmopolitanism.com/neuroqueer-an-introduction/

CHAPTER 26

Research Methods and Ethics in Neurodiversity Studies: Looking Back, Moving Forward

Hanna Bertilsdotter Rosqvist and David Jackson-Perry

Introduction

Editing this volume has been very much a journey of discovery. From the beginning, we knew that we wanted to hear from researchers at various stages of their academic careers and to include both academic and non-academic and variously neurodivergent contributors. This inevitably meant that we were accepting, seeking even, an element of unknowability as to what sort of volume we would end up with, what sorts of lessons we would be learning, and so what sorts of reflections—and thoughts moving forward—would find their ways into this final chapter.

We noted in the introduction that an important barrier to doing neurodiversity research differently has been a relative paucity of detailed descriptions of—and reflections on—collaborative, participatory, or other creative methods in the field: hence this volume. In very different ways, all the chapters in this collection respond to that observation. In this final chapter, we have therefore chosen to keep theoretical musings to a minimum and to stay as close as possible to our ambition of producing an accessible handbook for those looking to use alternative methods in neurodiversity research. What follows is organised around four principal methodological themes that have emerged

H. Bertilsdotter Rosqvist (✉)
School of Social Sciences, Södertörn University, Huddinge, Sweden
e-mail: hanna.bertilsdotterrosqvist@sh.se

D. Jackson-Perry
Queen's University, Belfast, UK

H. Bertilsdotter Rosqvist and D. Jackson-Perry (eds.), *The Palgrave Handbook of Research Methods and Ethics in Neurodiversity Studies*,
https://doi.org/10.1007/978-3-031-66127-3_26

from preceding chapters: 'Saying hello': Befriending Communities from the Start; Navigating Collaborative and Participatory Processes; Facilitating Participation; and 'Saying Goodbye': Final Steps. Each section ends with a brief, hands-on toolbox, drawing on lessons learnt through the contributions of individual chapters. These toolboxes are of course far from definitive, and we encourage the reader to play with, adapt or reject them as they feel fit for their research context and their own functioning/access needs. We conclude with a consideration of the potential and reach of 'unknowing' as a foundational tool for research methods and ethics in Neurodiversity Studies.

'Saying Hello': Befriending Communities from the Start

Befriending Neurodivergent Communities

Several chapters in this volume note the importance of shaping research projects through early community engagement and consultation (Chapters 4,9, 10, 19, 20, this volume; see also Milton & Green, 2024). We refer to this as *befriending communities*. This is an important step even if the project involves neurodivergent researchers or advisors, which may not necessarily be enough to give a broader community view of, for example, research priorities or accessible methods. While this befriending may take various forms, depending on researchers' resources and preferences, the underlying principle is to create a more or less active, and more or less formalised, *early* engagement with communities.

Different approaches may bring researchers a feel for the community and its research priorities, such as familiarising themselves with published community priority literature and informal conversations on social media. Those with more time could carry out formal investigation into community research priorities and accessible methods, such as the online survey Jackson-Perry (Chapter 20, this volume) employed, and as is increasingly being done elsewhere (see, for examples, Davies et al., 2024; Dey et al., 2024). However, various low-hanging fruits exist for those with more limited resources. Thom-Jones and Lowe (chapter 10, this volume) suggest that for students or others with limited resources, simply becoming familiar with the literature describing community priorities or preferred methods (see, for example, Emerson et al., 2023; Chown et al., 2023; Cage et al., 2024), is a good low-cost entry point. Another low-cost/high-benefit approach is to seek out—and cite—neurodivergent theorists regardless of the form of dissemination, for example looking to blogs and forums, which can be referred to as community grey sources. This constitutes what Hillary Zisk (Chapter 19, this volume) refers to as an "ethical engagement with the community," which is about "substantively engaging with work by neurodivergent people at all stages of research." Beyond the advantages of bringing a sense of proximity with community ideas and preoccupations, this also provides the opportunity to reconsider meanings of data and theory, legitimising neurodivergent contributors from beyond academia, who can thus be understood as "sources of theory" and "not only as sources of data" (Zisk, Chapter 19, this volume).

If time and resources permit, more in-depth engagement and collaboration with relevant stakeholders may be invaluable. Jackson-Perry, for example, invested heavily in getting to know autistic communities, initially online and one-to-one, and then through participating in autistic-led spaces that welcome researchers from all neurotypes (Chapter 20, this volume). There exist, for example, autistic-led spaces, such as the Participatory Autism Research Collective (PARC, 2020), which welcome all researchers to their conferences and events. This process of engagement and collaboration is pertinent regardless of one's neurotype. Whether the community being researched is that of your own or another neurotype, engaging with broader communities, to the extent possible within the constraints of your project, will enrich both your research and your researcher journey. While some forms of befriending might be relatively high cost in terms of resources, we have noted here some simple and low-cost options and consider that the benefits of all these processes far outweigh the costs.

Toolbox for befriending neurodivergent communities

- *Get involved*: as much as possible, as early as possible within the constraints of your project, engage with neurodivergent communities. This will be invaluable to your research journey.
- *Tailor it*: if you do not have the time and resources to do this actively (e.g., joining PARC, attending neurodivergent-led conferences, etc.), then do it theoretically, through familiarising yourself with published community priority literature.
- *Cite community resources*: search for neurodivergent blogs and other grey sources (see, for example, Zisk, 2023 for an annotated list in the field of autism). In this case, try to contact neurodivergent creators to ask if they are happy for their work to be used in research, and if so in what way they prefer credit and citation. Allow sufficient time for creators to think about this question and respond ahead of any deadlines.
- *Proactively seek community input*: if you have the resources, and to be as pertinent as possible to your specific project, carry out your own investigation through an online survey, for example.

Befriending, Navigating, and Resisting Academic Norms

Befriending communities can also be seen in terms of another relationship: befriending—in this case meaning becoming familiar with, navigating, and resisting—a largely neuronormative academia as a neurodivergent researcher. Sometimes this starts with a deconstruction of academic normativity,` what

Raymaker and Nicolaidis (Chapter 25, this volume) following Kuhn, refer to as "normal science," and how it may clash with researchers' own neurodivergence (see Chapters 5, 17, 18, this volume). Working to change existing systems and paradigms is time and energy consuming. It requires becoming what Ahmed (2023) calls a "killjoy," being willing and able to be perceived as the problem, when in fact you are simply pointing out that a problem exists (see Chapter 5, this volume). Goetz and Adams (Chapter 17) note the "toll" of "trying to live up to the pressure of being the voice in those rooms, advocating for those who are not." The work of changing academic codes, be they concerned with ethics applications or disciplinary codes that pathologise neurodivergence, is all the more complex given that these codes may generally be invisible to neurotypical academics. However, resolving and navigating this clash offers the potential to make academia work better for all researchers, regardless of neurotype. Remember, this should not be left to neurodivergent academics or students to do: making academia a more inclusive and user-friendly environment is everyone's responsibility.

Research ethics applications are one example of this, including formal expectations of filling out research ethics applications, often with little or no guidance (see Chapter 18, this volume). Researchers may be expected to complete applications in unfamiliar or unwelcome language. There may, for example, be an expectation that neurodivergence be described through deficit (when presenting risks and benefits of research in relation to ethics committee assumptions of neurodivergence) rather than describing neurodivergent experiences in line with one's intended research project (Limburg, 2022; c.f. Chapter 10, this volume). Commonly, (neurotypical) ethics committee members, based on their assumptions of neurodivergence, may expect clarifying answers to questions such as "Does your research include anyone who is intellectually, mentally or physically impaired?" or whether research includes "People who may be vulnerable or unable to give fully informed consent." Jones and Lowe (Chapter 10, this volume) use a proactive approach to preemptively address the concerns raised by ethics committees. They suggest creating a proforma statement outlining their position on the ethical conduct of autism research using national or international research ethics guidelines. Adopting a similarly pragmatic approach, Munday (Chapter 18, this volume) deconstructs the UK ethical application process. The author suggests recommendations to improve ethical assessment processes, including making them more accessible to all students and researchers by creating comprehensible guidelines, offering individualised support, and providing access to examples of past ethics applications and forms. These two chapters not only suggest solutions to problems, but also illustrate what Hume (Chapter 13, this volume) refers to as "cripping methodology" or "a neurodiversity-affirming methodology." What all these chapters suggest is to be pragmatic and proactive; where you find a problem, suggest a solution to mitigate the problem for future researchers. This befriending, cripping, or making academia more neurodiversity-affirming is perhaps a pillar of what Raymaker and Nicolaidis (chapter 25, this volume) refer to as neurodiversity science, which they suggest may "infuse neurodiversity principles into the 'normal science' of autism and public health research to generate a paradigm shift both in science and in society more broadly." This is a pleasing reversal of "epistemic infection," whereby neurodivergent individuals might become "infected" by dominant discourses positioning them and their experience as deficitary (Bertilsdotter Rosqvist & Jackson-Perry, 2021). Here, the tables are turned, and neurodivergent researchers are resisting and influencing (neurotypical) modes of functioning rather than submitting to them. We think of befriending academia, or "taking up space," as Munday (Chapter 18, this volume) puts it, as a way of troubling neurotypical "business as usual" from within academia (c.f. Bertilsdotter Rosqvist et al., 2019, p. 1082).

Toolbox for 'befriending academia'

- *Take up space:* to the extent that you feel it is possible for you, be willing to take up space. This might mean asking for support, asking for examples (in the case of ethics applications), or even pointing out to supervisors and colleagues ways in which existing systems and codes might operate to disable you or others. This is the job of all researchers, not just neurodivergent ones.
- *Prioritise time management:* all parts of your research will probably take more time than you think (see Chapters 5, 8 20, this volume). Familiarise yourself with ethics processes as early as possible, leaving you time to take up space without also having to worry about running out of time.
- *Look after yourself:* sometimes we feel able to be the killjoy, sometimes it just takes too much energy and can even be traumatising. Do what you can when you can and feel no guilt when you can't.

Navigating Collaborative and Participatory Processes

We think of collaboration and participatory methods (for example, having an advisory group, or a community advisor) as having much in common, often involving—as they frequently do in this volume—working across neurotypes and academic/non-academic actors. Both therefore require actively navigating "varying and variable needs" (Chapter 8, this volume), and an awareness—and, where possible, correction—of potential power inequalities (see Chapters 4,7,8,9,12,17, this volume). Co-creation, collaboration, and supportive groups of researchers may loosen and unsettle dominant and traditional notions of research processes and expertise, pointing to new directions for research. They may also be a challenge when working within academic institutions and with external funding. In these cases, funders or institutional constraints (such as those for a master's or PhD thesis) largely decide the rules of the game, stressing research outcome deliveries, "fast science" (Alleva, 2006), and limiting the extent to which outside advisors or collaborators are permitted to intervene (c.f. Bertilsdotter Rosqvist et al, 2019, Chapters 5,6,8,20, this volume).

Several chapters focus on relationships between academic researchers and community researchers, or academic researcher/community member dynamics conditioned by academic institutions (see chapters 5, 6, 8, this volume). Academic researchers may be paid as part of their position, or have built-in benefits associated with this type of work (such as funding their own positions), which may not be the case for non-academic community members (see Chapters 6, 19, this volume). Sometimes, as in the case of Grant and co-authors (Chapter 4, this volume), it is possible to include non-academic community members in the development of a funding application. We feel strongly that payment of non-academic community members is necessary in research processes, given that making funding applications and carrying out and writing up research are expected tasks of researchers (along with peer-reviewing and editing and contributing to volumes such as this). However, it can be complex to arrange funding for non-academic community members at this stage, as budgets are not yet assigned. It will be useful when completing funding proposals to lean on published guidance stating the importance of offering "payment at a level that is commensurate with the nature and demands of the activity and is fair when compared to other members of the research team, to acknowledge the value placed on public involvement" (NIHR, 2023).

Collaborative challenges are not limited to tensions implicit in academic researcher/community member dynamics. The question of varying and variable needs, for example, amongst researchers with both different *and* similar neurological positionings is recurrent in chapters here. The co-creation process described by Hultman and co-authors (Chapter 8 see also chaptEr 12, this volume) required navigating between varied and variable needs, particularly around notions of time and energy. Informed by Kafer's (2013) notion of crip time, they stress the importance of "'a reorientation to time' that: requires reimagining our notions of what can and should happen in time or recognizing how expectations of 'how long things take' are based on very particular minds and bodies" (for a similar discussion see Chapters 5,6,7, this volume).

Hume (Chapter 13, this volume) draws on the disability justice movement to address the question of differing needs through the notion of "access intimacy" which, for Mingus (2017, para 21):

> recognizes and understands the relational and human quality of access, while simultaneously deepening the relationships involved. It moves the work of access out of the realm of only logistics and into the realm of relationships.

Recognising the relational nature of navigating needs has, says Hume (Chapter 13, this volume), "the power to bridge the emotional distance between two people, transforming both of them rather than merely adding a ramp one of them can climb while the other takes the stairs." A prerequisite for access intimacy, then, is relationship building for all members of a group to feel at ease and enable the others to express themselves authentically. In the duoethnography of Goetz and Adams (Chapter 17, this volume), the authors explore and reflect upon their positions as differently neurodivergent researchers and their resultantly different experiences of academia. At the same time, their duoethnography forms a caring and supportive space for their reflections and co-writing, in which they explore and mirror each other's experiences in a friendly and curious way. We see this type of process as holding powerful "epistemic healing" (Khan & Naguib, 2019) potential for neurodivergent researchers faced with an often ableist and hostile research environment, however unconscious.

However, as Pearson et al. (Chapter 7, this volume) note, "being nice" to each other is not sufficient to bring about change of practice: rather, as Hultman and Hultman (chapter12, this volume) found, it requires hard, ongoing work throughout the research process, as to accommodate varying and variable needs over a sustained period. Désormeaux-Moreau and Courcy (chapter 9, this volume) refer to this as "navigating *together.*" The authors complicate binary assumptions of sharing or not the same neurological positioning, rather stressing the importance of dialogue and reflection to be able to meet at the crossroads of different experiences. They refer to this as "cross-fertilizing knowledge" valuing the diversity and plurality of knowledge and standpoints and developing a joint political engagement, which can also be seen as a form of access intimacy. Désormeaux-Moreau and Courcy also stress the importance of rebalancing power relations in the collective, where researchers from a dominating neurotype need to consciously practice "self-disempowerment," giving up space and power in order to make space for researchers of a minority neurotype. Something similar may be useful when navigating inequalities in power informing the research collaboration more generally, for example due to age, different disciplinary backgrounds, and different levels of previous academic experience (see Chapters 7, 12, this volume). Given scholarship that increasingly notes interest-based motivation as a powerful resource for both autistic and ADHD folk (Dodson, 2022; Hallowell & Ratey, 2021; Murray, 2019; Murray et al., 2005), what Jones and Lowe (Chapter 10, this volume) refer to as engaging in one's "passion projects," it is perhaps unsurprising that high levels of interest in both research topics and different parts of research processes can overcome or mitigate access difficulties (see Chapters 7, 12, 17, this volume). Indeed, some empirical research suggests that high levels of interest in a given subject may largely bypass both sensory and social challenges (Jackson-Perry, 2023, p.223). Where collaborators have high levels of interest in either a topic or a part of the research process, this plays to strength. However when or if interest wanes, it becomes something of a barrier to the whole project (see

Chapter 12, this volume). As a way to counter the challenges of interest-based research, Hultman and Hultman (Chapter 12, this volume) stress the importance of a pragmatic collective approach in research: "utilizing the group members' collective strengths to create individualized working processes within a collective writing process."

Finally, while participatory research has become increasingly prominentin autism research, some contributors to this volume have questioned how meaningful much community engagement may be. Several chapters here critique the extent to which participation or community involvement is necessarily meaningful, as opposed to tokenistic (see Chapters 5, 9, 10, 17, this volume). Jones and Lowe (Chapter 10, this volume), for example, note that:

> Autistic academics are all too familiar with the proposition of an informal zoom meeting to discuss autistic perspectives about a study that is already underway, being later positioned as a participatory approach or autistic community consultation.

This is not specific to neurodiversity research, being apparent in other fields, such as HIV research (Chapter 22, this volume). This is perhaps the cost of success and brings us back to the risk of "neurodiversity lite" discussed in the introductory chapter (Chapter 1, this volume): when participatory and collaborative approaches become a funding requirement, for example, researchers unfamiliar or unengaged with principles and processes of those approaches may be tempted to do what is necessary to tick the boxes. This is perhaps one of the risks Orsini (2022, p.5) refers to when he says that:

> bringing autistic voices or perspectives to the table may reproduce what many have decried in the past: participation or engagement for the sake of crafting a thin veneer of legitimacy.

While Jackson-Perry (Chapter 22, this volume) is a proponent of participatory research, he is also hesitant about expecting it:

> to fix all our problems, to move from deficit to difference, from 'questions important to researchers' to 'questions important to autistic communities,' from individualist medical narratives to more social leaning models.

The author questions how meaningful participation can be in a research environment still largely dominated by deficit readings of, here, autism, whereby "autism's essence," as Yergeau (2013, p.8) reminds us:

> has been clinically defined as a disorder that prevents individuals from exercising free will and precludes them from accessing self-knowledge and knowledge of human others.

If we agree that this is the case, that neurodivergent folk are to varying extents considered to be unreliable witnesses of their own experience, then the impact of the inclusion of their voices in research processes will necessarily remain limited and tokenistic. To change this requires a reconsideration of everything we have been taught to believe about neurodivergence, a point to which we return below.

Collaborative and participatory toolbox

- *Take small steps:* each project has constraints that need to be considered: we are mindful that collaborative and participatory processes can be hard—if enriching—work, particularly for a novice researcher operating within a resource-poor framework. Jackson-Perry (Chapter 20, this volume) recalls an informal conversation with Damian Milton in 2017. His advice, given the constraints of a PhD, was something like this: even if you only do one small thing that is participatory during your study, and you do it well, this is already more than most people are doing and is better than nothing.
- *Avoid collaboration and participation as an add-on:* to avoid tokenism—build collaboration/participation into your project as early as possible, preferably at the planning stage, including considering pertinent research questions, etc.
- *Compensate when possible, make noise when not:* look into funding mechanisms to appropriately and equitably compensate neurodivergent collaborators for whom research is not part of their jobs. Where this is not possible, lean on existing guidelines to make some noise about this, with your project manager, institution, or ethics committees: you might make a change for those coming after you.
- *Be patient with each other and see challenges as a learning process:* different group members may have varying and variable needs. It is important to discuss these before starting up the research process since it affects planning and time management.
- *Practice access intimacy within your research group:* the time you spend building relationships is time well spent.
- *Time is of the essence:* all parts of the research process will take longer than you had expected—expect this! Further, within the project, different actors will have differing approaches to time and energy management—take this into account, discuss it explicitly, and work with it.
- *Play to interest:* as neurodivergent folk, we often have a complex relationship to interest. Whether concerning the actual subject of the research project, or our interest in particular parts of the research process, our engagement is likely to be significantly impacted by our interest levels. Access intimacy will help make this explicit early on in the collaboration—work to the individual strengths of members rather than a normative idea of what each contributor should be doing.
- *Rethink failure:* as this volume has shown, scholars informed by neurodiversity approaches are often trying out new things. This requires

risk-taking, and not everything will work the first time—if ever! Be willing to embrace failure as a learning process.

Facilitating Participation

Similar tensions as those noted above can be found when it comes to facilitating access for those who have volunteered to be participants in your research project. Access intimacy, an acknowledgement that researcher and participant are inter-dependent, that they create a relationship that needs looking after, is also central here (see Chapter 13, this volume). Indeed, in ethnography, as Taneja-Johansson (Chapter 11, this volume) reminds us, "access to lives and places has been shown to be determined by the time invested in developing trust." The author cites Tavory (2019, p. 727), who describes ethnography as a discipline "made of relationships." We can think of this relationship as starting before a meeting with participants (in whatever form that might take) has even taken place. Indeed, even before the research project has formally started, a broader community relationship can be worked on, through participating in neurodivergent spaces (providing this is allowed: see 'Befriending communities' above and Chapter 20, this volume). This is all the more important given as the distrust that may exist "from autistic persons towards researchers" (Woods and Waltz, 2019, p. 2) and the broader "history of distrusting researchers" amongst "minority communities" (Raymaker & Nicolaidis, 2013, p. 169). This early process of building community relationships will help steer you away from unconscious stereotyping and enrichen your reflections as you move through the entire research process. Further, investing time in these early stages of befriending will be of help when you seek participants for your research, as there is a chance that people will be familiar with your name and approach.

When navigating participation itself, a first step is to identify and adapt the level and type of participation that are appropriate and possible given the expectations and limits of a specific research project. To participate in research can be an opportunity for participants to have a voice, something which in itself can be valuable for people who may not be used to being heard, being listened to (see Chapter 11, this volume). For this to be as emancipatory and meaningful as possible, what can be described as an "*ethics of choice*" may be a valuable tool (see Chapter 4, this volume). An ethics of choice includes choices in terms of individually adapted ways of participation, for which Humes' (Chapter 13, this volume) discussion of access intimacy will also be valuable, including the suggestion to be proactive with proposing accommodation for everyone rather than leaving the burden of requesting accessibility on the participants' shoulders: the onus for providing accessible and comfortable forms of participation is the researchers' responsibility.

Indeed, in inclusive research literature, accessible research opportunities and "maximizing participation" (Harrington et al., 2014; see also Chapters 5,

14, 15, this volume) are commonly referred to when it comes to enabling participant's participation through offering participants a flexible palette of data collection possibilities adapted to the participant's sensory modalities and ways of communication. This can be summed up by the imperative of Jones and Lowe (Chapter 10, this volume) to "Minimise barriers, maximise flexibility" as well as to "Accept autistic participants' personal preferences to engage." Chapman (Chapter 21, this volume) suggests email interviewing, a form of asynchronous communication between a researcher and participant where text-based information is repeatedly exchanged without strict temporal and spatial limitations, as an accessible research method enabling participation of participants who may better express themselves in writing. Participation through written communication may give neurodivergent people time to think, provide structure and predictability, and reduce both feelings of anxiety and negative sensory experiences. However, when choosing tools for data collection, the most important thing is not to assume that a single method will work for all people sharing a neurotype. Rather, as Jones (2022) has suggested in the context of autism research, researchers should: "start from scratch and develop measures that accurately assess the topic of interest, developing appropriate tools in collaboration with autistic people" (Jones, 2022, p 108). Informed by Jones' thinking, Chapman stresses the importance of the role of the researcher in making sure that each interview is tailored to the individual, or "mirroring" the participants, ensuring that the researcher is not responding generically, but to their distinct experiences and in relation to their way of expressing themselves. However, accessibility and access intimacy are also a matter of relationality, of interdependence between participant and researcher; as Hume (Chapter 13, this volume) puts it, "We navigated our, at times divergent, access needs together." Following several chapters (Chapters 11, 13, 21, this volume), this can be summed up as the responsibility of the researcher to build a trusting relationship and provide a space where participants feel at ease and able to share their experiences in the most appropriate way for them.

However, as the challenges of collaboration above illustrated, this is not a task without challenges when the participant's accessibility needs might clash with the accessibility needs of the researcher. For example, in some of the encounters with participants, Jackson-Perry (Chapter 20, this volume) notes that, "in our interaction, what was accessible for Susan was inaccessible for me...access for one person is 'access denied' for another" (c.f. also Chapters 7, 13, this volume). An ethics of choice also therefore requires taking into consideration the energy costs and benefits for participants and weighing those up with what is possible for the researcher, from both the perspective of their own access capabilities and the constraints of the individual project.

Facilitating participation toolbox

- *Invest in your relationship with participants and participant communities:* here, as with collaboration, engage in access intimacy. Think about building both individual and community relationships, before you engage in fieldwork or even plan your project. This will prove invaluable both to inform your substantive and methodological approach and when you seek participants. Remember too that for many people participating in research is a profound and novel experience: recognise and value this.
- *Proactively propose access possibilities to all participants:* do not wait for participants to ask for adaptations—provide a variety of choices and leave yourself open to other possibilities suggested by participants.
- *Think of participant payback and costs/benefits:* seek methods that are accessible, comfortable, and enjoyable for the participants whilst generating the information required to inform the overall research project. Ensure flexibility and choice in timing, location, structure, and modality of interviews.
- *Provide as much clarity as possible:* give participants clear guidance on how the method works in practice, and expectations for them as a participant to help manage any anxieties possible when engaging in research.
- (Here too) *accept and embrace failure:* navigating individual access needs is a process. You won't always get it right. Don't bet yourself up about this. See it as a sign that you are trying something new…and try something new!
- *Crip it up!* Work with crip time in interviews. Follow the participant's flow in conversations. Leave room for circumambulatory conversations rather than pushing for immediate, direct answers. This will enable you to elicit rich storied experiences of participants and avoid gathering only basic demographic information. It will also leave you freer to be surprised, creating space for your participants to tell you something about their experience beyond echoing what you already know, one risk if your questions are too directive.

'Saying Goodbye': Final Steps

She gave me a hug when she left. That was the end. (Taneja-Johansson, Chapter 11, this volume).

As researchers, we spend a great deal of time reading pertinent literature, working on a theoretical framework, and planning out the recruitment of participants and the gathering and analysis of data. As we start our research projects, these are immediate and sometimes urgent requirements for funders, ethics committees, and supervisors (for students), and we give them corresponding levels of attention. At this stage, our levels of energy and motivation are high. However, two or three years down the line, sometimes longer, there is another step that is often completely ignored. This step is not the subject of an institutional requirement and comes when we may be exhausted and at the end of our research tether: saying goodbye to participants. This is not helped by the fact that guidance on this is remarkably sparse in existing literature (Delamont & Smith, 2023), perhaps for the very reason that it is often not on institutional radars and, not being a formal requirement, the researcher concentrates their energy elsewhere. You may have spent considerable time and energy setting up your research tent, you have befriended the community and your participants, but at a certain point your research time and energy runs out and you need—perhaps hastily—to grab your things and leave the field.

And yet, while we have previously discussed an ethics of choice, which is of course as pertinent for this section as for the previous one, we must also consider the "*ethics of care*" in research (Ellis, 2016). Your participants have given their time, their energy, and often their intimacy to your research project. If you have engaged in access intimacy, if you acknowledge and experience the often-intense inter-personal relationship between yourself and your participants, even those you may never have met in person, saying goodbye with care is as important for the researcher as for participants. But "How does one withdraw from the relationships one has so carefully shaped together?", asks Taneja-Johansson (Chapter 11, this volume).

This is perhaps all the truer for long-term studies. Taneja-Johansson, who had numerous in-person meetings with participants over a three-year period, notes that for some participants she "had become a 'friend' and 'confidant,' a person with whom they openly shared what they were struggling with now." In turn, she found herself "looking forward to these regular meetings, engaged in what was happening in their lives." The author wondered "whether a clean break signaling the end was really the most caring way of ending the study for the participants" and decided on a "gradual withdrawal" approach. This involved increasingly mentioning her role as a researcher during meetings and gradually spacing meetings at longer intervals, "timing this to periods when participants were transitioning into a different life phase."

Time management difficulties were an ongoing concern for Jackson-Perry (Chapter20, this volume) throughout the participatory process of his PhD. His data collection and analysis took two years rather than the nine months he had expected, making him extremely uncomfortable about expecting the Research Advisory Group to stay engaged for such a long period. Finally, he sent a final email to advisors thanking them, stating that he would no longer be contacting them but inviting them to contact him if they had any comments, and directing them to the research website for updates. Similarly, he followed up with participants, also leaving the possibility for future communications and saying that the website would contain links to any publications based on the research and give results in an accessible fashion. In this way, just as his website had functioned as a space where potential participants could get information about him and his study and approach *prior* to participating, so it potentially prolonged the relationship beyond the duration of the study. This said, his feeling at the time was one of defeat: he had not anticipated the importance or the difficulty of this stage, not planned for it, and not built in the question of saying goodbye when he was saying hello.

Saying goodbye toolbox

- *Build 'saying goodbye' into your initial planning:* when the moment comes, you will likely be up against an institutional, energetic, and motivational deadline—not the best time to be taking care of this stage of the research process! Again, take into account that all parts of your process will likely take longer than expected.
- *Create a research website:* this can be a valuable way of presenting yourself to participants and the broader community and may function as a point of maintaining contact when the process is over.
- *Lay out the process:* when communicating what will be expected of participants include how and when the process will end as well as the usual 'you will be asked to respond to an interview that should last about one hour.'
- *Consider gradual withdrawal:* if your project involves multiple meetings, think about how to leave the field gradually with the same care you gave to entering it.

(In)Conclusion

> The transition from a paradigm in crisis to a new one from which a new tradition of normal science can emerge is far from a cumulative process, one achieved by an articulation or extension of the old paradigm. Rather it is a reconstruction of the field from new fundamentals, a reconstruction that changes some of the field's most elementary theoretical generalizations as well as many of its paradigm methods and applications. During the transition period, there will be a large but never complete overlap between the problems that can be solved by the old and by the new paradigm. But there will also be a decisive difference in the modes of solution. When the transition is complete, the profession will have changed its view of the field, its methods, and its goals (Kuhn, 1997, p 84-85).

Informed by Kuhn's (1997) thinking, Pellicano and den Houting (2022, p.381) describe current autism research as in a transition period; an ongoing shift from "normal science" to "neurodiversity in autism science" (see also Sonuga-Barke, 2023, for a similar discussion in ADHD research). Also following Kuhn (1997), Raymaker and Nicolaidis (Chapter 25, this volume) envision a future in which "Epistemology is multiple, simultaneous, and fluid. Science celebrates all ways of knowing." As we come to an end of working with this volume, we think of our current time concerning research on all neurodivergencies as one of transition; marked by instability, of disruption. Our hope is that this volume and Neurodiversity Studies more broadly contribute both to the disruption and to the settling process; to a new tradition of normal, neurodiversity science in line with the imagined future of Raymaker and Nicolaidis.

Much has been written here and elsewhere in neurodiversity informed approaches about different forms of knowledge and different types of knowers; of scientific, professional, or experience-based knowledge. These are important debates, doing their invaluable job of disrupting academic business as usual. However, as we have gone through our own process of editing this book, unknowing has come to the fore for us: it is with unknowing that we have decided to conclude this volume. Following Jackson-Perry (Chapter 22, this volume), unknowing of the type we are thinking "is not a lack of knowledge, but rather an intentional, creative, curious, and playful act rejecting a priori premises (notably of deficit) with the aim of coming to new understandings rather than confirming and contributing to existing assumptions." Jackson-Perry leans on queer theorist Halberstam's (2011) work on failure, as well as notions from Ignorance Studies and Epistemic Injustice, to propose unknowing as a way of approaching research, a process of unlearning and rereading, of going back to the sources with a new eye and reinterpreting findings while choosing to remain strategically ignorant of what is assumed to be known about autism.

However, as we consider themes that have emerged in this volume, we have come to think that unknowing is a useful framework beyond individual research processes. There have been, for example, long-standing and ongoing calls for—and recognition of the importance of—the inclusion of neurodivergent voices in neurodiversity research. This is largely framed as seeking ways "to support including neurodivergent perspectives in knowledge production" (Hens & van Goidsenhoven, 2023, p.10), and several chapters in this volume engage, rightly so, with that question (see Chapters 5, 8, 14, 15, this volume). Some critical autism studies scholars push back on this idea, with Elmadagli (2023, p.75), for example, suggesting that rather than seeking inclusion we should consider "limiting the influence of non-autistic scholars within CAS" (see Chapter 2, this volume, for discussion). It seems that the notion of inclusion—as meaning 'assumed to be neurotypical academics including neurodivergent knowers' in their research processes—has come to be so fixed that it is difficult to go beyond either maintaining or rejecting the notion outright. And, as Orsini (2022, p. 5) points out, the current logic of inclusion risks promoting "participation or engagement for the sake of crafting a thin veneer of legitimacy."

We do not believe that neurodivergent folk have a monopoly in understanding all aspects of neurodivergence that they hold a magic connection to the truth (or even that such a thing exists). Indeed, Shildrick (2009, p. 9) reminds us of the importance of "non-disabled people" working in the field of disability studies, of their being well-positioned "to interrogate their own culture and psycho-social location as non-disabled." Further, as Orsini (2022,

p. 11) notes, unknowing "is not a matter of lining up 'for' or 'against' expertise, per se." As Brown (Chapter 23, this volume) points out, unknowing may at times take us to uncomfortable places. The author suggests that we "may have to unlearn much of what the individual 'experts' have told us, even if some of those 'experts' are ADHDers themselves." However, for meaningful change to be brought about in fields so dominated by pathology and deficit, it may be useful in this time of disruption to re-imagine inclusion as conditional inclusion *of* neurotypical researchers rather than *by* them (which, as Struyf et al. (Chapter 5, this volume) point out, brings us to the question of the "'thinness' of neurotypical-led conditionality"), and the parallel elevation of neurodivergent voices. Here, then, we would like to redefine inclusion, unknow its long-accepted sense, and re-read it: we can then conceive of inclusion within Neurodiversity Studies as meaning the inclusion of neurotypical scholars within neurodiversity research.

We can also think of troubling and resisting neurotypical academic norms from the perspective of unknowing. Several chapters (see ex Chapters 5, 16, 17, 18, this volume) have engaged (although not necessarily in these terms) with the complexities of navigating (neurotypical) academic norms. What they describe could be cast as a process of unknowing: first, learn to read and understand existing norms, and then, re-imagine those norms in ways that are a better fit with neurodivergent ways of being. Finally, communicate these re-imaginings to academic hierarchies—help them, in effect, to unknow what they had assumed to be fixed, immutable processes to see them as they could be rather than what they are. The winners, as those authors point out, are not just neurodivergent, but all students—this then becomes a form of institutional unknowing. Troubling neurotypicality is of course in synergy with a diverse range of fellow killjoys and troublemakers. For example, Goodley proposes the "common goal of disturbing those normative homelands that all of us are forced to populate" (Goodley, 2017, p. 194), an invitation to join forces, to share tools and frameworks. Neuro-queering, which "uncovers, upsets, and unsettles power structures in normative spaces" (Yergeau, 2013, p. 205), is also an ally here. Unknowing seems to us to hold powerful potential as we become aware of—and correct for—the risk of the "neurotypical gaze" (McDermott, 2022) in research processes.

Some methods described in this volume seem to intrinsically lend themselves to unknowing. Pearson and co-authors (Chapter 7, this volume) suggest that "the experience of collective knowledge production can lead us to unknow what we had 'known': we started to question our own biases, grow our knowledge and build bridges between different standpoints." This underlines the importance given to working across neurotypes, but also, importantly, across disciplines that has been a hallmark of both Neurodiversity Studies and Critical Autism Studies (and for the future directions of Critical ADHD Studies): "disciplinary silos," as Gross and McGoey (2023, p.3) point out, "can compound ignorance." This is not the creative ignorance of unknowing, but that ignorance which "confirms what is already known according to approved methods of knowing" within disciplinary silos (Halberstam, 2011, p. 6).

Phenomenology too offers much unknowing potential. Autistic feminist Joanne Limburg (2022, p.3) seeks to "try and find a language for all kinds of moments that make up human existence," considering "the accumulation of moments, however trivial they were in themselves, that can determine the course of a life." Finding a language for all kinds of moments also implies changing the questions asked (c,f. also Jones, 2022). Limburg suggests a move from "'How might we recognize you in our [neurotypical world]?' (In what way are you [neurodivergent]? How did they diagnose you? You don't seem [neurodivergent] at all.)" into "How do you experience your world?" (Limburg, 2022, p 63). Shaughnessy et al. (Chapter 14, this volume) and Redmore (Chapter 15, this volume) align themselves with the field of phenomenological inquiries into neurodivergence (see Murray et al., 2023). They suggest both a systematic phenomenological approach to neurodivergence and a flexible approach, allowing for a range of ways for researchers to explore neurodivergence from the perspective of "inner experiences" rather than the "expressive signs" proposed and reproduced through research drawing on traditional diagnostic criteria.

We feel, then, that unknowing has the potential to be an integral part of a palette of foundational methodological approaches when working within Neurodiversity Studies. This necessarily implies a level of comfort with discomfort: trying to do new things comes with the risks of failure—indeed, in this volume we have encouraged contributors to describe their failures as well as their successes. But here too, we could think about unknowing: unknowing the very notion of failure. Autistic people, for example, have long been described through deficit, through failure to understand, to empathise, to fit into widely accepted system-level norms: the autistic person as "the disordered and damaged other" (Milton & Moon., 2012, p. 5). However, we could also turn this around, to think not of the failure of autistic (or otherwise neurodivergent) person, but of neurotypical, dominant norms as representing 'imperfect systems' that require challenging (Jackson-Perry, 2020). Halberstam (2011, p. 2) reminds us that:

> under certain circumstances failing, losing, forgetting, unmaking, undoing, unbecoming, not knowing may in fact offer more creative, more cooperative, more surprising ways of being in the world.

Finding surprising ways of being in the (research) world is one way of working towards the paradigm shift with which we started this section. While we have only sketched out, in this volume, some of the potential of unknowing for the field of Neurodiversity studies, we encourage you to set out to fail, to forget, to unlearn and relearn, as we move together towards a more epistemologically and ethically robust body of neurodiversity research.

References

Ahmed, S. (2023). The feminist killjoy handbook: The radical potential of getting in the way. Seal Press.

Alleva, L. (2006). Taking time to savour the rewards of slow science. *Nature, 443*, 271.

Bertilsdotter Rosqvist, H., & Jackson-Perry, D. (2021). Not doing it properly? (Re) producing and resisting knowledge through narratives of autistic sexualities. *Sexuality and Disability, 39*(2), 327–344.

Bertilsdotter Rosqvist, H., Kourti, M., Jackson-Perry, D., Brownlow, C., Fletcher, K., Bendelman, D., & O'Dell, L. (2019). Doing it differently: Emancipatory autism studies within a neurodiverse academic space. *Disability and Society, 34*(7–8), 1082–1101.

Cage, E., Crompton, C. J., Dantas, S., Strachan, K., Birch, R., Robinson, M., Morgan-Appel, S., MacKenzie-Nash, C., Gallagher, A., & Botha, M. (2024). What are the autism research priorities of autistic adults in Scotland? *Autism*, 13623613231222656.

Chown, N., Beardon, L., Murphy, S. L., Suckle, E., & Baker-Rogers, J. (2023). Autism community research priorities: The potential of future research to benefit autistics. *Canadian Journal of Educational and Social Studies, 3*(2), 15–32.

Davies, J., Romualdez, A. M., Malyan, D., Heasman, B., Livesey, A., Walker, A., Pellicano, E., & Remington, A. (2024). Autistic adults' priorities for future autism employment research: Perspectives from the United Kingdom. *Autism in Adulthood, 6*(1), 72–85.

Delamont, S., & Smith, R. J. (2023). Leaving the field: An editors' introduction. In Leaving the field (pp. 1–30). Manchester University Press.

Dey, I., Chakrabarty, S., Nandi, R., Shekhar, R., Singhi, S., Nayar, S., Ram, J. R., Mukerji, S., & Chakrabarti, B. (2024). Autism community priorities in diverse low-resource settings: A country-wide scoping exercise in India. *Autism, 28*(1), 187–198.

Dodson, W. (2022). Secrets of your ADHD brain. ADDitude. https://www.additudemag.com/secrets-of-the-adhd-brain/ Downloaded April 27, 2024

Ellis, C. (2016). Compassionate research: Interviewing and storytelling from a relational ethics of care. In The Routledge international handbook on narrative and life history (pp. 441–455). Routledge.

Emerson, L. M., Monk, R., van der Meer, L., Sutherland, D., & McLay, L. (2023). Autistic co-led community priorities for future autism research in Aotearoa New Zealand. Autism in Adulthood.

Goodley, D. (2017). Disability Studies: An Interdisciplinary Introduction. 2nd edn. London: SAGE Publications Ltd.

Halberstam, J. 2011. The queer art of failure. Duke University Press.

Hallowell, E. M., Ratey, J. J. (2021). ADHD 2.0: New science and essential strategies for thriving with distraction-from childhood through adulthood. Ballantine Books.

Halperin, D. (1997). *Saint foucault: Towards a gay hagiography*. Oxford University Press.

Harrington, C., Foster, M., Rodger, S., & Ashburner, J. (2014). Engaging young people with autism spectrum disorder in research interviews. *British Journal of Learning Disabilities, 42*, 153–161. https://doi.org/10.1111/bld.12037

Hens, K., & Van Goidsenhoven, L. (2023) Developmental diversity: Putting the development back into research about developmental conditions. *Frontiers in Psychiatry, 13*, 2994.

Jackson-Perry, D. (2020). The autistic art of failure? Unknowing imperfect systems of sexuality and gender. *Scandinavian Journal of Disability Research, 22*(1).

Jackson-Perry, D. (2023). Exploring autistic accounts of sexuality, intimacy, and authenticity (Doctoral dissertation, Queen's University Belfast). Available at: https://pure.qub.ac.uk/en/studentTheses/exploring-autistic-accounts-of-sexuality-intimacy-and-authenticit

Jones, S. C. (2022). Measuring the wrong thing the right way? Time to rethink autism research tools. *Autism in Adulthood: Challenges and Management, 4*(2), 104–109. https://doi.org/10.1089/aut.2021.0050

Khan, F. R., & Naguib, R. (2019). Epistemic healing: A critical ethical response to epistemic violence in business ethics. *Journal of Business Ethics, 156*, 89–104.

Limburg, J. (2022). Letters to my weird sisters. On autism, feminism and motherhood. Atlantic Books.

McDermott, C. (2022). Theorising the neurotypical gaze: Autistic love and relationships in The Bridge (Bron/Broen 2011–2018). *Medical Humanities, 48*(1), 51–62.

Milton, D., & Moon, L. (2012). The normalisation agenda and the psycho-emotional disablement of autistic people. Autonomy, the Critical Journal of Interdisciplinary Autism Studies, 1(1); Mingus, M. (2011). Access intimacy: The missing link. Leaving Evidence. https://leavingevidence.wordpress.com/2011/05/05/access-intimacy-the-missing-link/

Milton, D. E., & Green, J. (2024). Theorising autism. *Autism, 28*(4), 795–797.

Mingus, M. (2017). Access intimacy, interdependence and disability justice. Leaving Evidence. https://leavingevidence.wordpress.com/2017/04/12/access-intimacy-interdependence-and-disability-justice/

Murray, D., Lesser, M., & Lawson, W. (2005). Attention, monotropism and the diagnostic criteria for autism.*Autism, 9*(2), 139–156.

Murray, D., Milton, D., Green, J., & Bervoets, J. (2023). The human spectrum: A phenomenological enquiry within neurodiversity. *Psychopathology, 56*(3), 220–230.

Murray, F. (2019). Me and monotropism. *Psychologist, 32*, 44–48.

NIHR. (2023). Payment guidance for researchers and professionals. Available at https://www.nihr.ac.uk/documents/payment-guidance-for-researchers-and-professionals/27392

Orsini, M. (2022). Who needs to (un)know? On the generative possibilities of ignorance for autistic futures. *International Journal of Qualitative Studies in Education*, 1–18.

PARC. (2020). The participatory autism research collective. The Participatory Autism Research Collective.

Pellicano, E., & den Houting, J. (2022). Annual research review: Shifting from 'normal science' to neurodiversity in autism science. *J Child Psychol Psychiatr, 63*, 381–396.

Poulsen, R., Dwyer, P., Gassner, D., Heyworth, M., & Williams, Z. J. (2023). The INSAR community collaborator request: Using community–academic partnerships to enhance outcomes of participatory autism research. *Autism Research, 16*(11), 2071–2076.

Raymaker, D., & Nicolaidis, C. (2013). Participatory research with autistic communities: Shifting the system. In Worlds of autism: Across the spectrum of neurological difference (pp. 169–188). University of Minnesota Press.

Shildrick, M. (2009). Dangerous Discourses of Disability, Subjectivity and Sexuality. New York: Palgrave Macmillan.

Sonuga-Barke, E. J. (2023). Paradigm 'flipping' to reinvigorate translational science: Outlining a neurodevelopmental science framework from a 'neurodiversity' perspective. *Journal of Child Psychology and Psychiatry*, *64*(10), 1405–1408.

Tavory, I. (2019). Beyond the calculus of power and position: Relationships and theorizing in ethnography. *Sociological Methods and Research*, *48*(4), 727–738.

Woods, R., & Waltz, M. (2019). The strength of autistic expertise and its implications for autism knowledge production: A response to Damian Milton. *Autonomy, the Critical Journal of Interdisciplinary Autism Studies*, *1*(6).

Yergeau, M. (2013). Clinically significant disturbance: on theorists who theorize theory of mind. *Disability Studies Quarterly*, *33*(4).

Zisk, A. H. (2023). Critical autism studies beyond academia: An annotated list. *Ought: The Journal of Autistic Culture*, *5*(1), Article 6.

Index

H. Bertilsdotter Rosqvist and D. Jackson-Perry (eds.), *The Palgrave Handbook of Research Methods and Ethics in Neurodiversity Studies*, https://doi.org/10.1007/978-3-031-66127-3

R

S

T

U

V

W

The manufacturer's authorised representative in the EU is Springer Nature Customer Service Centre GmbH, Europaplatz 3, 69115 Heidelberg, Germany. If you have any concerns regarding our products, please contact ProductSafety@springernature.com

Printed and bound by CPI Group (UK) Ltd, Croydon, CR0 4YY
01/10/2026
02225903-0020